The Royal Entomological Society
Book of British Insects

The Royal Entomological Society
Book of British Insects

I suppose you are an entomologist? – I said with a note of interrogation.

Not quite so ambitious as that, sir. I should like to put my eyes on the individual entitled to that name! A *society* may call itself an Entomological Society, but the man who arrogates such a broad title as that to himself, in the present state of science, is a pretender, sir, a dilettante, an imposter! No man can be truly called an entomologist, sir; the subject is too vast for any single human intelligence to grasp.

Oliver Wendell Holmes, 1882, *The Poet at the Breakfast Table*

The Royal Entomological Society Book of British Insects

Peter C. Barnard

A John Wiley & Sons, Ltd., Publication

This edition first published 2011
© 2011 by Royal Entomological Society

Blackwell Publishing was acquired by John Wiley & Sons in February 2007. Blackwell's publishing program has been merged with Wiley's global Scientific, Technical and Medical business to form Wiley-Blackwell.

Registered office:
John Wiley & Sons, Ltd, The Atrium, Southern Gate, Chichester, West Sussex PO19 8SQ, UK

Editorial offices: 9600 Garsington Road, Oxford OX4 2DQ, UK
The Atrium, Southern Gate, Chichester, West Sussex PO19 8SQ, UK
111 River Street, Hoboken, NJ 07030-5774, USA

For details of our global editorial offices, for customer services and for information about how to apply for permission to reuse the copyright material in this book please see our website at www.wiley.com/wiley-blackwell.

The right of the author to be identified as the author of this work has been asserted in accordance with the UK Copyright, Designs and Patents Act 1988.

Library of Congress Cataloging-in-Publication Data
Barnard, Peter C. (Peter Charles), 1949–
  The Royal Entomological Society book of British insects / Peter C. Barnard.
    p. cm.
  Includes bibliographical references and index.
  ISBN 978-1-4443-3256-8 (hardcover : alk. paper)
  1. Insects—British Isles—Identification.   I. Royal Entomological Society of London.   II. Title.   III. Title: Book of British insects.
  QL482.G8B37 2011
  595.70941—dc23

                                                                                            2011019098

A catalogue record for this book is available from the British Library.

This book is published in the following electronic formats: ePDF 9781444344950; Wiley Online Library 9781444344981; ePub 9781444344967; Mobi 9781444344974

Set in 9.5/12 pt Palatino by Toppan Best-set Premedia Limited
Printed and bound by CPI Group (UK) Ltd, Croydon, CR0 4YY

C9781444332568_270724

Contents

CONTENTS

Preface

In 2007 the Royal Entomological Society (RES) took the momentous decision to move from central London to a more rural environment near St Albans in Hertfordshire, the first time that the Society had left London since its foundation in 1833. The move was seen as controversial by some, although the new headquarters are close to the London Orbital M25 motorway and only just outside the Greater London conurbation. Apart from the financial relief of no longer having to maintain an impressive but aged building in South Kensington, the sale of 41 Queen's Gate realized a substantial capital sum that enabled the RES to press ahead with several new projects, such as the appointment of the first full-time entomologist-in-residence.

This book is one of the tangible products funded by the move out of London and it represents part of the re-positioning of the RES as the central hub of information on British insects, as well as maintaining its role as the premier society for professional entomologists. The book is designed as a key reference work for professional entomologists as well as being a readable and attractively illustrated account for the interested student of insects. It helps to bridge the gap between the popular but incomplete guides to the more conspicuous groups of British insects and the highly specialized works that currently can cover only a fraction of the entire fauna. Of these detailed works the most prestigious are, of course, the Society's own long-standing series *Handbooks for the Identification of British Insects*.

Inevitably such a book highlights the contrast between popular groups such as Lepidoptera and Coleoptera, relatively well-served by identification literature, and the lesser studied groups such as many of the Hymenoptera. However, the aim is not to bemoan the lack of information on these less popular orders of insects, but to encourage potential authors to fill some gaps. This call to arms is as much aimed at the keen amateur entomologist as it is at the professional, since few of the latter are encouraged or even allowed to prepare taxonomic works as part of their duties. In these sadly unenlightened times the perpetual need for basic taxonomic information is overshadowed by the political insistence that good science can be defined by its ability to attract external funding, which may reflect the inability of some influential managers to make informed and independent decisions about the real value of science. With luck, the overwhelming need for better taxonomic information, coupled with the development of systems to coordinate it, will cause this short-sighted approach to be overturned. It will be gratifying if this book helps to restore the rightful place of systematic entomology at the centre of insect studies in this country.

Peter C. Barnard
Royal Entomological Society
May 2011

Foreword

One could proclaim the study of insects to be the study of biodiversity and be at least half right. The more than 1 million insect species described to date account for fully half of all kinds of plants and animals, but no more than 25% of the estimated total number of living insect species (Foottit & Adler, 2009; Chapman, 2009; Grimaldi & Engel, 2005). No full understanding of either evolutionary history or the function of terrestrial or freshwater ecosystems is possible without detailed knowledge of insects, yet our ignorance of insect diversity, and our inadequate description and classification of their kinds, prohibit rapid progress. So how are we to proceed?

In the long term, we need to accelerate fundamental taxonomic and natural history investigations and make relentless progress toward a complete exploration of insect diversity. This will involve investments in museums, a cyberinfrastructure specifically engineered to meet the unique needs of taxonomists, and the inspiration and education of the next generation of insect specialists. We need to ensure also that insects are integral to research into all aspects of basic and applied biology, from agriculture to behaviour, developmental biology and ecosystem science. Our immediate need is for natural laboratories where insects are more approachable than they are on the global stage. What might such natural laboratories look like? First, they would have insect faunas with enough species and diversity to be broadly representative, but not so many species as to be unmanageably large. Second, their fauna would be relatively well known, their species reasonably well described and mirrored in comprehensive and well-curated museum collections. Third, they would have a vibrant community of amateur and professional entomologists. Fourth, they would have visionary organizations to encourage and enable insect studies. And finally, they would have authoritative and comprehensive books that provide an entrée into insect biodiversity for the public and scientists alike. No place on earth meets these requirements as fully as the British Isles.

The concentration of expertise and great institutions, from natural history museums to the Royal Entomological Society (RES), combined with a diverse yet finite assemblage of species make the British insect fauna accessible and approachable on a level unmatched elsewhere. The growth of knowledge of insects is so fast-paced that general summaries of the fauna are soon outdated and become impediments themselves; the number of species recorded for Britain has increased threefold since the time of Stephens (1846). While popular accounts of a few charismatic taxa such as dragonflies and butterflies exist as well as superb treatments of particular groups in journals and the outstanding RES *Handbook* series, there is a need for an up-to-date introduction to the insect fauna of the UK as a whole. This volume fills that gap exceedingly well.

The need for such comprehensive overviews is not new. In 1828, Kirby and Spence were convinced that it was the absence of such an adequately complete and affordable introductory volume on the insect fauna of Britain that was the primary impediment to the advancement of entomology. Their influential book was an answer to this obstruction and was the result of their resolve 'to do what was in their power to remove it, and to introduce their countrymen to a mine of pleasure, new, boundless, and inexhaustible' (Kirby & Spence, 1828: ix).

Thanks to the vision of the RES and Dr Peter Barnard, this need is once again met for a new generation of entomologists. The knowledge and experience required to adequately summarize what we know of 24,000 species representing 558 families are great, and few would be up to the challenge. I can imagine no better choice than Dr Barnard whose research and curatorial work at the Natural History Museum and dedicated service to the RES span more than 30 years and have prepared him for this singularly important challenge.

Insects remain a largely untapped mine of pleasures for those who take the time to identify, observe and enjoy them. The discoveries yet to be made through a more thorough study of entomology remain similarly new, boundless and inexhaustible. Circumstances threatening biodiversity and the environment make the importance and timeliness of this volume and increased attention to research on insects obvious. Of equal importance is increasing understanding and appreciation of insects among the general public. There is no surer or more effective means to that end than making the insects around us more easily accessible. Thanks to Dr Barnard and the RES, the British fauna is poised to play its unique role as a natural laboratory of biodiversity exploration.

Quentin D. Wheeler, PhD, FRES, FLS
Vice President and Dean
Arizona State University

REFERENCES

CHAPMAN, A. 2009. *Numbers of living species in Australia and the World* (2nd edn.). Australian Biological Resources Study, Canberra.

FOOTTIT, R.G. & ADLER, P.H. (eds.). 2009. *Insect biodiversity: science and society*. Blackwell Publishing, Oxford.

GRIMALDI, D.A. & ENGEL, M. 2004. *Evolution of the insects*. Cambridge University Press, Cambridge.

KIRBY, W. & SPENCE, W. 1828. *An introduction to entomology or elements of the natural history of insects*. Vol. I. Longman, Rees, Orme, Brown, and Green, London.

STEPHENS, J.F. 1846. *Illustrations of British entomology*. Supplement. Henry G. Bohn, London.

Acknowledgements

My initial thanks must go to the Council of the Royal Entomological Society and the Registrar, Bill Blakemore, for entrusting me with the task of writing this book on their behalf. Val McAtear, the RES librarian, has helped me on numerous occasions with bibliographic queries, and Mike Claridge has provided a generous supply of useful advice.

Much of the value of this book lies in the range of insect photos, and I take great pleasure in thanking the four photographers for their unique contributions:

Roger Key is a well-known entomologist, formerly of English Nature/Natural England, who now lives in North Yorkshire. As a freelance lecturer, ecologist and field entomologist he specializes in inspiring young people to appreciate wildlife; he is also an experienced broadcaster and his photos of all groups of insects have been published widely in many books and magazines.

Robin Williams, from Wedmore in Somerset, is an authority on the Hymenoptera, but his long experience as a naturalist and photographer is evident in the wide range of insect groups and behaviour portrayed in his pictures. He is one of the few people who can recognize the numerous species associated with oak galls, as well as other families of the 'parasitic' wasps.

Colin Rew lives near Huddersfield in West Yorkshire; he qualified in Electrical Engineering and then worked in industrial insurance, where his photographic skills were in frequent use. Following retirement in 2008 he took a course in macro-photography and became hooked on photographing the insects in his garden; the stunning results of his new-found interest are here for all to see, including on the front cover.

Stuart Crofts is a lifelong fisherman from South Yorkshire who has represented England in fly-fishing competitions; he has a keen interest in fresh-water insects and has written many articles on fly-fishing, entomology and river management, all illustrated by his own photos.

I am very grateful to Peter Chandler for sending me a copy of his *Dipterist's Handbook* before it was published so that I could make use of the invaluable information therein.

Quentin Wheeler's tireless championing of insect taxonomy has always been a source of inspiration, and I thank him for his stirring Foreword; his clear understanding of the needs and potential of British entomology should ideally help to shape the research priorities in this country.

My wife Jane has been very patient with my entomological obsession, though if she hadn't kept the kitchen so clean it might have been easier to find some domestic psocids to photograph. I should also thank my greyhound Reene for providing a supply of fleas (now safely exterminated).

Several staff at Wiley-Blackwell have made the planning and production of this book a much smoother process than it might otherwise have been, and I particularly thank Ward Cooper, Kelvin Matthews, Kathy Palmer and Delia Sandford for their advice and support. In addition, freelance copy-editor Lewis Derrick and freelance project manager Nik Prowse have made invaluable improvements with their editing of the manuscript, saving me from some potentially embarrassing slips. I take responsibility for any remaining errors that the assiduous reader will undoubtedly find.

Many modern authors owe a debt of gratitude to the automatic spell-checkers in their word-processing software, but not in a book with so many scientific names. I quickly ran out of patience when my computer kept trying to change Adelidae to Adelaide, Gerridae to Gertrude, Phoridae to porridge and Sesiidae to seaside.

1 General introduction

THE SYSTEMATIC APPROACH

Systematic studies form the basis of all scientific work, and this is no coincidence or artificial contrivance. Classifying natural objects is an essential instinct in human beings that enables us to make sense of the world, and the origins of this ability are not hard to trace. For an off-beat and thought-provoking look at this subject see Yoon (2009); as Quentin Wheeler says in a review of that book, taxonomy is the 'oldest profession practiced by people with their clothes on', a reference to the belief that Adam was given the task of naming all the creatures (*Genesis* 2: 19–20).

Superficial characters like colour, shape and even some behaviours are often quickly seen to be of little use in predicting the unknown, which is the value of systematics, even at a domestic level. From careful observations about a few animals or plants we can make generalizations about other species that we have not yet encountered, and this ability to predict the unknown is what makes taxonomy a science rather than simply a technical procedure. Recognizing that plants related to nightshades are likely to be poisonous is clearly valuable, whereas wrongly classifying bats as a kind of bird might entail a long and fruitless wait for them to lay eggs!

Classifying organisms can, of course, be done for different purposes: there is a well-known cartoon by Charles Keene in *Punch* magazine (1869, vol. 56, p. 96) in which a railway porter is informing a lady of the charge for carrying a tortoise on a train; he tells her that, 'Cats is "dogs" and rabbits is "dogs" and so's parrots, but this 'ere "Tortis" is an insect . . .'.

The terms classification, taxonomy and systematics are often used as if synonymous, but there are some important differences. A classification is basically a way of groupings organisms in a logical way: although it claims to be based on similarities it often ends up emphasizing differences in order to define the boundaries of groups, and no assumptions about processes or underlying causes of any perceived patterns are made. Such systems are often phenetic in nature, based on morphological differences to divide a large taxon into manageable pieces. Systematics uses the principle that similarities are based on shared, derived characters; it therefore assumes an evolutionary, phylogenetic cause, and importantly it makes predictions about the characteristics of groups, whether morphological, behavioural, ecological or physiological. Taxonomy has traditionally been regarded as all about naming things, though it is linked to a classification or a systematic study, because any system of naming has to be based on principles of relationships, whether real or artificial. This narrow definition has led to the perception that taxonomy is a somewhat mechanical process, which simply produces names and groupings on behalf of the real scientists who are carrying out systematic analysis. It is more useful to see taxonomy as the overarching activity of studying phylogenetic relationships, postulating testable hypotheses about distributions of characters, and hence producing useful classifications with meaningful systems of nomenclature that reflect the underlying relationships.

Thus, the need to recognize true relationships between organisms is essential to our perception of the world and to our ability to exist in some kind of

The Royal Entomological Society Book of British Insects, First Edition. Peter C. Barnard.
© 2011 Royal Entomological Society. Published 2011 by Blackwell Publishing Ltd.

harmony with our environment. The importance of taxonomic research is frequently cited, which makes it all the more surprising that support for taxonomy is increasingly hard to find. Even national museums, the custodians of the most important natural history collections in the world, seem to be constantly downgrading the emphasis on taxonomy in their organizational research priorities (Wheeler, 2008), though this is likely to be rooted in the fundamental misunderstanding of the value and true aims of taxonomy, as stated above. This issue is discussed further by Secord (1996).

THE BRITISH INSECTS

The terms 'British' and 'Britain' are deliberately used rather loosely to include Ireland as part of the British Isles, because many published 'British' lists also include Irish species, even if they contain species found in Ireland but not in the UK. However, there are many special Irish lists (see, for example, http://www.habitas.org.uk/invertebrateireland/index.html). With well over a million species of insects known in the world, and many more to be discovered, our fauna of around 24,000 species looks rather insignificant. However, insect diversity is generally much greater in the tropics than in temperate regions (Foottit & Adler, 2009) and the British insect fauna is in the range to be expected at our latitude.

Table 1.1 shows the approximate numbers of species, families and orders of insects known in the world, Europe and the British isles. World figures are based on Foottit and Adler (2009) and Resh and Cardé (2009), European figures are from the Fauna Europaea website, and British figures from the current book. Of course, these numbers can only reflect the species described to date, and are not necessarily indicative of the numbers of species that actually occur. On a world scale, the five largest groups of insects are: (i) Coleoptera; (ii) Lepidoptera; (iii) Diptera; (iv) Hymenoptera; and (v) Hemiptera.

In the British Isles, the order of largest groups is rather different, being: (i) Hymenoptera and Diptera (roughly equal); (iii) Coleoptera; (iv) Lepidoptera; and (v) Hemiptera. The differences may well reflect the fact that the diversity of groups like Coleoptera and Lepidoptera is higher in the tropics than in temperate regions. For those who like impressive statistics, the ten largest families of insects in the British Isles are Hymenoptera: Ichneumonidae (2100 spp.), Braconidae (1045); Coleoptera: Staphylinidae (1000); Diptera: Cecidomyiidae (652); Hemiptera: Aphididae (630); Diptera: Chironomidae

(608); Hymenoptera: Pteromalidae (600), Eulophidae (500); Coleoptera: Curculionidae (475); Diptera: Mycetophilidae (471). Most of these figures are, of course, approximate and it is unlikely that family limits are interpreted in the same way in the different insect orders.

The decision to cover the British insects at family level is not just for the convenience of dealing with 558 families rather than over 6000 genera! In many groups, families have a reality outside scientific taxonomy; grasshoppers, pondskaters, ladybirds, hawk-moths, mosquitoes and ants are just a few examples of well-known groups of insects that correspond to family level taxa. There are, of course, many little-known families that have no common names.

Inevitably the coverage of each group varies because some are much better known than others, but every family is included, not just the common or conspicuous ones. This at least highlights those groups – often quite large and diverse families – that still need far more work done on them.

All the genera occurring in the British Isles are listed under each family, alphabetically within subfamilies where these are commonly used in the larger groups. Tribes are not mentioned because they are used inconsistently in different insect orders, and are often the subject of considerable disagreement between authors. All generic names are listed in the index, along with every higher taxon of insects mentioned in the text, as well as widely accepted vernacular names of groups at family level and above.

The numbers of species listed here are in many cases an approximation; this is because classifications change, species are added or synonymized, and any checklist is out of date almost as soon as it is published. Where the number is preceded by '$c.$' this indicates even less certainty about the total, particularly noticeable in groups like the parasitic Hymenoptera.

Totals normally include UK and Ireland, but not the Channel Islands, though species found only in the Channel Islands are occasionally mentioned where they are of particular interest or importance. The corresponding numbers of taxa in Europe and the rest of the world are given, so that the significance of the fauna of the British Isles can be placed in context. For typical groups such as Lepidoptera our fauna can seem insignificant in terms of world species numbers, but for more cosmopolitan groups like Mallophaga the apparent proportion of British species is larger in comparison with other parts of the world. The reasons for these apparent differ-

Table 1.1 Numbers of described insect species and families in the world, Europe and British Isles (see main text for sources).

Group	World species	World families	European species	European families	British species	British families
Collembola	8,000	29	2,000	23	250	19
Diplura	1,000	10	300	5	11	1
Protura	750	10	175	4	15	3
Archaeognatha	500	2	200	2	7	1
Zygentoma	400	5	60	2	2	1
Ephemeroptera	3,100	40	330	18	51	10
Odonata	5,600	33	130	11	49	9
Dermaptera	2,000	8	80	5	7	4
Dictyoptera	8,500	20	200	10	11	5
Embioptera	450	8	12	2	0	0
Grylloblattaria	30	1	0	0	0	0
Mantophasmatodea	20	1	0	0	0	0
Orthoptera	24,000	29	1,000	15	33	10
Phasmida	3,000	2	15	3	4	2
Plecoptera	3,000	16	400	7	34	7
Zoraptera	35	1	0	0	0	0
Hemiptera	100,000	104	8,000	94	1,830	63
Phthiraptera	5,000	24	800	17	540	17
Psocoptera	5,000	17	200	25	100	19
Thysanoptera	6,000	9	570	6	179	3
Coleoptera	350,000	175	12,500	144	4,000	112
Diptera	152,000	160	19,000	132	7,000	103
Hymenoptera	150,000	90	16,000	66	7,000	57
Lepidoptera	160,000	120	8,500	87	2,570	72
Mecoptera	600	9	23	3	4	2
Megaloptera	330	2	10	1	3	1
Neuroptera	6,000	17	290	12	69	6
Raphidioptera	225	2	75	2	4	1
Siphonaptera	2,600	15	260	8	62	7
Strepsiptera	600	10	30	7	10	4
Trichoptera	13,000	46	1,100	24	198	19
Totals	**1,011,740**	**1,015**	**72,260**	**735**	**24,043**	**558**

ences are discussed in each chapter. Subspecies are not included in the species totals, though they are mentioned in the text where they are significant.

All taxa are in alphabetical order at every level, which may look different to some conventional lists where 'similar' groups are put together, even though not formally linked. The principle adopted here is that classifications are consulted mainly by non-specialists, and they cannot know where to draw the boundaries unless formal higher groups

are created. Thus a traditional list of Trichoptera families might begin:

1 Rhyacophilidae;
2 Glossosomatidae;
3 Hydroptilidae;
4 Philopotamidae;
5 Ecnomidae.

To the specialist this may represent some kind of vague evolutionary sequence but it is not clear whether families 1 and 2 are more closely related to

3 than family 3 is to 4, and so on. Any well-established associations between taxa that are not represented by formal groupings are described in the text; in some cases these are the result of higher groups being split or combined by successive taxonomists. Since all family names and higher taxa are listed in the index they can be traced easily within the book.

There can be differences in interpreting a 'British' species: some recording schemes include any species found, even as an accidental migrant or import, whereas other schemes relegate these to an appendix as 'unestablished' or even omit them altogether. Therefore totals cannot always be compared directly.

In each group of insects outlines of the main biological and morphological features are given. For detailed identification there are references to more specialized literature at the appropriate level. In the larger insect orders this will be at family level, but for the smaller orders one work may cover the entire group. Much of the specialized identification literature was covered in an earlier publication (Barnard, 1999) and these lists have been updated with newer books and websites, but not with papers in scientific journals because the latter are not easily available to the non-specialist. It is very noticeable in the last few years how much more information is available on the internet, though this ease of access brings as many problems as advantages. In general websites are not moderated, so the information they present can be incorrect or inappropriate for certain purposes and the risks of misidentifying specimens from such sites are very real. Inexperienced users are often unaware of the potential for gross errors from using accurate yet inappropriate sites; numerous enquiries sent to the Royal Entomological Society read something like, 'I found this insect in my garden and have discovered a picture that looks like it on an Australian website, so has it been introduced to the UK?' For this reason, websites are mentioned only when they are known to be accurate and useful, though the absence of a site does not necessarily indicate the opposite. There is also the problem of URLs that change over time; some organizations helpfully provide long-term links to their new websites, but not always. Not all sites contain primary data, but they may include valuable metadata, acting as useful portals to other sources of information.

Articles published in scientific journals are becoming more difficult to access as fewer libraries subscribe to the journals. In some cases taxonomic papers are available online, though often needing a paid subscription, but they may also be listed on the recommended websites. This decision to omit the specialist scientific papers will appear biased towards certain insect groups: it will scarcely affect the Lepidoptera references, for example, whereas the hymenopterists will feel neglected because very little of their literature is available in easily accessible books or websites. This is a reflection of how groups like Hymenoptera, especially the 'Parasitica' are seen as the domain of professionals; perhaps this will stimulate the wider dissemination of this information in order to encourage more enthusiastic amateurs. Where identification depends on specialized literature this is clearly noted in the text. Access to the specialist journals is often available through the libraries of societies such as the British Entomological & Natural History Society and Royal Entomological Society. Even with societies that do not maintain libraries such as the Amateur Entomologists' Society, the contact with other specialists is immeasurably valuable to the beginner, so the message is clear: join a society!

Works that can be obtained individually, such as complete issues or supplements of journals are included, as they are frequently advertised by specialist book dealers or are available directly from the publishers. Some of the older and rarer books are of course difficult to obtain, but increasingly they are being digitized and made available online.

Species currently having legal protection in the UK are listed, though the status of these will vary over time; many other species are cited on various lists of conservation status. The exact details of the protection are not listed here, and the latest information should always be checked on the Joint Nature Conservation Committee (JNCC) website (http://www.jncc.gov.uk). Inevitably it tends to be the larger and more conspicuous species that have gained any legal protection, and under current UK and European legislation there are just 55 British species in this category. These comprise ten species of Coleoptera, 38 Lepidoptera (of which 30 are butterflies), three Odonata, three Orthoptera and one Hemiptera. There are a great many more species whose status causes concern; some of these are the subjects of species action plans and more information on these will be found on the appropriate websites listed throughout the book. Perhaps more informative is the List of UK Priority Species, resulting from the UK Biodiversity Action Plan (http://www.ukbap.org.uk). This UKBAP list includes 349 species of insects, comprised as follows:

- Lepidoptera 176 species (of which 152 are moths);
- Coleoptera 78 species;

- Diptera 35 species;
- Hymenoptera 35 species;
- Hemiptera 10 species;
- Orthoptera 4 species;
- Trichoptera 4 species;
- Plecoptera 2 species;
- Ephemeroptera 2 species;
- Odonata 2 species;
- Neuroptera 1 species.

Although this gives better coverage to some of the less conspicuous groups, it will be noted that only eleven of the 24 orders of Insecta are represented. All species on this UKBAP list are noted in the appropriate chapters in the book, under their corresponding family. The authors of scientific names are not given, as these are easily obtained from standard checklists. Species on these various lists of conservation concern have been given common names if they did not already possess one; some of these names have changed over time, and the latest ones are not always correctly quoted on websites. Clearly the scientific name is the definitive one, and should always be used in preference.

CLASSIFICATION OF INSECTS AND THEIR RELATIVES

The exact relationships of the more primitive insects, especially the 'apterygotes' has challenged authors for many decades. This can be seen in the various editions of the well-known *Imms' General Textbook of Entomology*: in the first edition (Imms, 1925) the Thysanura (including Diplura), Protura and Collembola are placed unequivocally in the subclass Apterygota on the grounds that they shared the primitive wingless condition. This was repeated in other early editions, but by the time of the ninth edition (Richards & Davies, 1957) things had changed; the subclass Apterygota was retained but with the admission that it was a 'diverse, perhaps polyphyletic, assemblage' and acknowledging that some authors argued that only the Thysanura seemed closely related to the true insects. In the tenth edition (Richards & Davies, 1977) the discrepancies could no longer be ignored and the authors admitted that the apterygote orders should no longer be grouped together as a single subclass 'as was done in older classifications'. By this time the subdivision of the Thysanura into two distinct monophyletic groups, the Archaeognatha (or Microcoryphia) and Zygentoma, was becoming established though Richards and Davies conservatively retained them as suborders.

Developments continue in this field and the true phylogenetic relationships between the groups of insects are now becoming much clearer; however the story is far from complete. In some cases morphological and molecular data are in conflict, and there is no agreement on an unambiguous higher classification of all the insect orders at the present time. The relationships between the orders of insects are discussed in the appropriate chapters, but any attempt to produce a formal classification at this level generates an awkward hierarchy of names, with some groups of uncertain affinities; therefore a more informal system seems most appropriate at present (Grimaldi & Engel, 2005). These issues are discussed by Cranston and Gullan (2009) and a broad account of the relationships between arthropods and other groups is given by Thorp (2009).

The family classification used in this book is given at the end of this chapter. Several names are deliberately not given any particular rank in order to highlight their informal nature. In some texts, groups such as orthopteroids and hemipteroids are given formal names as the superorders Orthopteroidea and Hemipteroidea, but these should be avoided as they resemble superfamily names, which always take the 'oidea' suffix. Although the International Code of Zoological Nomenclature does not cover the formation of names above the family group level it seems sensible to avoid any possible confusion.

Any rigorous analysis of relationships based on monophyly tends to create new groupings that have no equivalents in the everyday world. At the same time many older convenience groups are lost, a good example being the 'invertebrates'. It is clear that the Entognatha are distinct from the Insecta, but it seems reasonable to retain the common name of 'insects' for all the Hexapoda. Otherwise we have the awkward terminology of 'true insects' for the Class Insecta, and no common name at all for the Class Entognatha or for the Hexapoda, except for the somewhat artificial 'hexapods'.

ORDERS OF INSECTS NOT FOUND IN THE BRITISH ISLES

There are just four insect orders not found in Britain, though one of them occurs in Europe; all are quite small and all are in the Polyneoptera (the Orthopteroid group).

Embioptera (also known as Embiidina): the webspinners are a fairly small group with around 450 known species, found mainly in the tropical regions of all continents; they also extend into some temperate parts of the USA and southern Europe. Around a dozen species are found in southern Europe in the families Embiidae and Oligotomidae. The Embioptera, together with their sister group Zoraptera, seem distantly related to Plecoptera, but also share some characteristics with the Phasmida and Orthoptera. Their most distinctive feature is a large silk-producing gland in each fore tarsus, found in both sexes and in the nymphal stages; the silk is used to build extensive gallery systems in the soil, leaf litter or on tree bark.

Grylloblattaria (also known as Grylloblattodea): a small and obscure order with fewer than 30 species found only in the western USA and Canada, and parts of China, Korea and Japan. Sometimes known as rock crawlers or ice crawlers, they are usually found at high altitudes, and share some characters with the Dictyoptera and Phasmida.

Mantophasmatodea: these small predatory insects are the most recently discovered insect order, recognized only in 2001. Fewer than 20 species are known, all in southern and east Africa. Because they hold the distal ends of their tarsi off the ground when walking they have been dubbed heelwalkers; their phylogenetic relationships are not yet clear though they have strong links with the Grylloblattaria, and some authors place the two groups together in the Notoptera.

Zoraptera: another obscure order with only around 35 known species principally in tropical regions, living in rotting wood and feeding mainly on fungal hyphae. They have no common name, and live in small colonies, superficially resembling termites and Psocoptera, but they are probably the sister group of the Embioptera.

THE ROLE OF THE ROYAL ENTOMOLOGICAL SOCIETY IN BRITISH INSECT TAXONOMY

When the current RES was instituted in 1833, its stated intention was 'the improvement and diffusion of Entomological Science' and at the Jubilee Address in 1883 the then President, J.W. Dunning, asserted that this aim was being successfully executed, a view that is even more true today. The early origins of the Society have been well described before (Neave, 1933). The first Entomological Society of London was actually founded in 1806, with *Transactions* published between 1807 and 1812; although there was little further activity until 1833 there is no evidence that this earlier group was actually disbanded. One can certainly make a case for

the RES really beginning in 1806 (Gardiner, 2002), which would make it the oldest entomological society in the world. At present that record is considered to be held by the Société Entomologique de France, so any further clarification of the early origins of the RES might reverse this position! It has been suggested that the notorious lawsuit brought by J.F. Stephens against J. Rennie in 1832/3 for alleged plagiarism, helped to unite entomologists in a common cause, which led in turn to the formation of the Entomological Society itself (Clark, 2009: 10).

Through the early years it is impossible to say whether the Society collectively influenced the direction of entomological research any more than the individual members did; the membership was relatively small (growing from 127 members in 1834 to 178 in 1848) and their personal interests and efforts effectively set the agenda of the Society's activities. Through the first half of the 19th century the importance of specialists in insect taxonomy was becoming clear; the Society had of course contributed greatly to this advance by publishing papers in its *Transactions* from its founding date but it was decided to compile an authoritative catalogue of British insects. There had been several such attempts made by individual entomologists in previous years, beginning with Forster (1770), who listed just under 1000 species, plus a few related groups of invertebrates. Samouelle's (1819) book also included spiders, mites, crustaceans and myriapods, but his total was around 4000 species. Two important works appeared ten years later; Curtis (1829) mentioned around 15,000 species, but many of these were undescribed or were synonyms. Stephens' (1829) total of around 10,000 species may be a more accurate reflection of the knowledge at the time though he later revised the figure down to nearer 9000 (Stephens, 1846). The British Museum had begun its *Lists of the Specimens of British Animals in the Collection of the British Museum* in 1848, but these were based on the museum's collections only, and the series was never finished. So, in 1851 the Society resolved to produce a work entitled *Insecta Britannica*, using different authors for each major group of insects. In the end the series was published by Lovell Reeve, but only two sections ever appeared, three volumes on the Diptera by Walker (1851–6) and the Lepidoptera Tineina by Stainton (1854); it seems that the Society's strict rules about the arrangement and contents of the works deterred any more authors from contributing (Neave, 1933).

A new catalogue appeared by the Rev. Morris (1865–7), listing 12,454 species and based largely on existing lists, but the complete version of this is a rare book. In 1867 the Entomological Society began a similar project to the one abandoned some years before, but this time with more determination. The *Catalogue of British Insects* was first proposed by the Neuropterist Robert McLachlan. A committee was formed to oversee its production, but again there were arguments about the nomenclature and groups such as Coleoptera were never published; however five volumes did appear (McLachlan & Eaton, 1870; Smith, 1871; Marshall, 1872, 1873; Douglas & Scott, 1876). By the end of this series only around 3400 species had been catalogued, and it was to be many years before a similar project was attempted. In the meantime separate lists of many groups of insects were published by individual authors and inevitably they vary in quality and comprehensiveness, the very difficulties that a coordinated project was designed to avoid.

In 1933, the year that the Society received its 'Royal' status, the Committee on Generic Nomenclature was set up, with the object 'to prepare lists of specific names to be fixed as genotypes of genera of British insects with a view to the suspension where necessary of the law of priority in respect of those generic names' which sounded an ambitious project. Eleven subcommittees were formed to cover all groups of insects, but it was not long before the enormity and complexity of the task were apparent, partly because fixing the type species of genera is clearly a task that requires study of the world fauna, not just the British species. The main committee had to admit in its first report (Committee on Generic Nomenclature, 1933) that their optimism for completing the task in a reasonable time was unrealistic, stating that for some subcommittees 'several years must necessarily elapse before their work is complete'. In the event, eight parts of the proposed lists were published and formed Volume 1, together with the Committee's report, but no more appeared (Hemming, 1934; Cowley, 1935; Killington, 1937; Richards, 1937; Andrewes, 1939; Balfour-Browne, 1940; China, 1943; Tottenham, 1949). One suspects that many readers found the included checklists of insects more useful than the often lengthy discussions on nomenclatural problems.

At the same time as these lists were being prepared, the RES was recognizing the increasing importance of specialized journals to cover taxonomic entomology. Longer revisionary papers had always been published in the *Transactions*, but in 1932 a new journal *Stylops* was begun, named after the Society's logo; it was designed for the rapid publication of short papers and descriptions of new species.

The stalled Generic Nomenclature project was effectively saved by the appearance of the first modern checklist of British insects (Kloet & Hincks, 1945). This was a list begun in the 1930s by the amateur entomologist and Manchester businessman George Sidney Kloet (1904–81) and later aided by Walter Douglas Hincks (1906–61), an authority on Dermaptera with a keen interest in other groups such as Coleoptera, and who eventually became Keeper of Entomology at Manchester Museum. They were well aware of the shortcomings of their work, which listed just over 20,000 species, but its publication stimulated others to bring the critical sections up to scratch. As O.W. Richards perspicaciously put it, 'Naturally, almost as soon as the list was published, it was criticized by the various experts who had always been too busy to produce lists themselves' (Kloet & Hincks, 1964). Although both authors were Fellows of the RES, they had their book published privately, perhaps demonstrating that keen individuals can often be more efficient at finishing projects than committees, however well-meaning the latter may be.

During this same decade, the RES launched its most successful and long-standing series, the *Handbooks for the Identification of British Insects*. The first parts were published in July 1949 and they have appeared at irregular intervals up to the present day, with some of the more popular parts being reprinted or revised. The ambitious intention was to cover all the British orders, and altogether ten volumes were planned, with vol. 1 including the smaller insect orders, vol. 2 the Hemiptera, vol. 3 the Lepidoptera, vols 4 and 5 the Coleoptera, vols 6, 7 and 8 the Hymenoptera, and vols 9 and 10 the Diptera. A projected vol. 11 was set aside for updates to the checklists. Inevitably the numbering system became rather complicated because individual authors varied in the coverage of each handbook. The original intention was no doubt to have simple subsections for each family or Superfamily, so that parts might be numbered vol. 5, part 3, or if necessary vol 5, part 3a. However, difficulties arose when some authors could only cover a subfamily at a time, leading to complicated numbers like vol. 8, part 3(dii); recently this system has been simplified, partly in recognition of the fact that some volumes will never be complete. Because lepidopterists have always published their own books in their preferred ways, vol. 3 has never been used.

The early handbooks all carried lists of books already published by the RES, whether handbooks or separate publications like the Centenary History (Neave, 1933); in fact the *Generic Names of British Insects* project was still being publicized until the early 1960s, with the optimistic statement 'Parts so far published' though no more appeared after 1949. However, the *Handbooks* had not supplanted that series because the early sections did not contain checklists of the groups they covered, an omission that was not put right until the 1970s.

In the early 1960s the decision was made to update the Kloet and Hincks checklist. Hincks had died in 1961, and Kloet was not inclined to carry on with the project alone so all their lists were handed over to the RES, along with the rights to publish all future editions. The task was given to small groups of experts and the work eventually appeared in five volumes spread over 12 years. In recognition of the pioneering work of the original authors the series was still entitled the Kloet and Hincks lists.

In date order these new lists were:

1964, Kloet & Hincks, A check list of British insects, second edition (revised). Part 1: small orders and Hemiptera. *Handbooks for the Identification of British Insects* 11(1): 119 pp.

1972, Kloet & Hincks, A check list of British insects, second edition (revised). Part 2: Lepidoptera. *Handbooks for the Identification of British Insects* 11(2): 153 pp.

1976, Kloet & Hincks, A check list of British insects, second edition (revised). Part 5: Diptera and Siphonaptera. *Handbooks for the Identification of British Insects* 11(5): 139 pp.

1977, Kloet & Hincks, A check list of British insects, second edition (revised). Part 3: Coleoptera and Strepsiptera. *Handbooks for the Identification of British Insects* 11(3): 105 pp.

1978, Kloet & Hincks, A check list of British insects, second edition (revised). Part 4: Hymenoptera. *Handbooks for the Identification of British Insects* 11(4): 159 pp.

A completely new edition of checklists was planned in the 1990s; to date only the Diptera list has been published (Chandler, 1998) and it may be that all future lists in this series will be published online to allow easier updating. Already there are many checklists being published on various websites, and the more reliable of these are listed in the appropriate chapters of this book.

Meanwhile the RES continues to support insect taxonomy in many other ways. The journal *Stylops* became *Proceedings (Series B: Taxonomy)* in 1936 as part of a general re-naming of the Society's journals. In 1971 this was merged with *Proceedings (Series A: General Entomology)* to become the *Journal of Entomology (Series B)*, and the latest journal for taxonomic publications is *Systematic Entomology*, begun in 1976.

For a more general view of the history of entomology in Europe there are several recent books, including d'Aguilar (2006), Clark (2009), Gilbert (1977, 2005, 2007), Salmon (2000) and Salmon and Edwards (2005). The classic work on the earliest entomological history on a world scale is Bodenheimer (1928–9).

GENERAL ENTOMOLOGICAL WORKS

For the identification of British insects there are many useful series, such as the *Naturalists' Handbooks*, *Handbooks for the Identification of British Insects* (published by the Royal Entomological Society) and the *AIDGAP* keys (published by the Field Studies Council); these are listed in their appropriate chapters throughout this book. For identifying insects in particular habitats or on particular plants there are several useful titles in the *Naturalists' Handbooks* series, though these will not always take identifications to species level. They cover topics such as insects on nettles (Davis, 1991), on thistles (Redfern, 1995), on dock plants (Salt & Whittaker, 1998), on cherry trees (Leather & Bland, 1999), on cabbages and oilseed rape (Kirk, 1992), under logs and stones (Wheater & Reed, 1996), and aphid predators (Rotheray, 1989). There are similar titles in the ever-growing *AIDGAP* series. Insects that cause plant galls are particularly well covered by several useful guides at different levels (Redfern & Askew, 1998; Redfern & Shirley, 2002, 2004).

At a more basic level, for those with little previous knowledge of insects, the books by Chinery (1986, 1993, 2009) cannot be bettered; and for a general entomological textbook Gullan and Cranston (2005) is the most comprehensive.

REFERENCES

D'AGUILAR, J. 2006. *Histoire de l'entomologie*. Delachaux & Niestlé, Paris.

ANDREWES, H.E. 1939. The generic names of the British Carabidae, with a check list of British species. Part 6. *The generic names of British insects*. Vol. 1: 151–92. Royal Entomological Society, London.

BALFOUR-BROWNE, W.A.F. 1940. The generic names of the British Hydradephaga, with a check list of British species. Part 7. *The generic names of British insects*. Vol. 1: 193–209. Royal Entomological Society, London.

BARNARD, P.C. (ed.) 1999. *Identifying British insects and arachnids: an annotated bibliography of key works*. Cambridge University Press, Cambridge.

BODENHEIMER, F.S. 1928–9. *Materialen zur Geschichte der Entomologie bis Linné*. W. Junk, Berlin. 2 vols.

CHANDLER, P.J. (ed.) 1998. Checklists of insects of the British Isles (new series). Part 1: Diptera. *Handbooks for the identification of British insects* 12(1): 234 pp.

CHINA, W.E. 1943. The generic names of the British Hemiptera–Heteroptera, with a check list of British species. Part 8. *The generic names of British insects*. Vol. 1: 209–342. Royal Entomological Society, London.

CHINERY, M. 1986. *Collins guide to the insects of Britain and western Europe*. Collins Pocket Guide, HarperCollins, London.

CHINERY, M. 1993. *Insects of Britain and northern Europe* (3rd edn.). Collins Field Guide, HarperCollins, London.

CHINERY, M. 2009. *Collins complete guide to British insects*. HarperCollins, London.

CLARK, J.F.M. 2009. *Bugs and the Victorians*. Yale University Press, New Haven & London.

Committee on Generic Nomenclature. 1933. Recommendations relating to the publication of the Committee's reports. Part 1. *The generic names of British insects*. Vol. 1: 1–6. Royal Entomological Society, London.

COWLEY, J. 1935. The generic names of the British Odonata, with a check list of British species. Part 3. *The generic names of British insects*. Vol. 1: 43–60. Royal Entomological Society, London.

CRANSTON, P.S. & GULLAN, P.J. 2009. Phylogeny of insects. In: RESH, V.H. & CARDÉ, R.T. (eds.) *Encyclopedia of insects* (2nd edn.). Academic Press/Elsevier, San Diego & London, pp 780–93.

CURTIS, J. 1829. *A guide to an arrangement of British insects*. London.

DAVIS, B.N.K. 1991. *Insects on nettles* (2nd edn.). Naturalists' Handbooks no. 1, Richmond Publishing, Slough.

DOUGLAS, J.W. & SCOTT, J. 1876. *A catalogue of British Hemiptera; Heteroptera and Homoptera*. Entomological Society of London.

FOOTTIT, R.G. & ADLER, P.H. (eds.) 2009. *Insect biodiversity: science and society*. Blackwell Publishing, Chichester.

FORSTER, J.R. 1770. *A catalogue of British insects*. Warrington.

GARDINER, B.O.C. 2002. A short account of the Royal Entomological Society and of the progress of entomology in Great Britain (1833–1999). In: PEDERSEN, B. (ed.) *A guide to the archives of the Royal Entomological Society*. Ashgate, Aldershot, pp. 1–30.

GILBERT, P. 1977. *A compendium of the biographical literature on deceased entomologists*. British Museum (Natural History), London.

GILBERT, P. 2005. *The Entomological Club and Verrall Supper: a history (1826–2004)*. The Entomological Club, London.

GILBERT, P. 2007. *A source book for biographical literature on entomologists*. Backhuys Publishers, Leiden.

GRIMALDI, D. & ENGEL, M.S. 2005. *Evolution of the insects*. Cambridge University Press, Cambridge.

GULLAN, P.J. & CRANSTON, P.S. 2005. *The insects: an outline of entomology* (3rd edn.). Blackwell, Oxford.

HEMMING, F. 1934. The generic names of the British Rhopalocera, with a check list of British species. Part 2. *The generic names of British insects*. Vol. 1: 6–40. Royal Entomological Society, London.

IMMS, A.D. 1925. *A general textbook of entomology*. Methuen, London.

KILLINGTON, F.J. 1937. The generic names of the British Neuroptera, with a check list of British species. Part 4. *The generic names of British insects*. Vol. 1: 63–80. Royal Entomological Society, London.

KIRK, W.D.J. 1992. *Insects on cabbages and oilseed rape*. Naturalists' Handbooks no. 18, Richmond, Publishing, Slough.

KLOET, G.S. & HINCKS, W.D. 1945. *A check list of British insects*. Privately published, Stockport.

KLOET, G.S. & HINCKS, W.D. 1964. A check list of British insects (2nd edn., revised). Part 1: small orders and Hemiptera. *Handbooks for the identification of British insects* 11(1): 119 pp.

LEATHER, S.R. & BLAND, K.P. 1999. *Insects on cherry trees*. Naturalists' Handbooks no. 27, Richmond, Publishing, Slough.

MARSHALL, T.A. 1872. *A catalogue of British Hymenoptera; Chrysididae, Ichneumonidae, Braconidae and Evaniidae*. Entomological Society of London.

MARSHALL, T.A. 1873. *A catalogue of British Hymenoptera; Oxyura*. Entomological Society of London.

McLACHLAN, R. & EATON, A.E. 1870. *A catalogue of British Neuroptera*. Entomological Society of London.

MORRIS, F.O. 1865–7. *A catalogue of British insects, in all the orders*. Longmans, Green, Reader & Dyer, London [issued in parts].

NEAVE, S.A. 1933. *The history of the Entomological Society of London, 1833–1933*. Entomological Society, London.

REDFERN, M. 1995. *Insects and thistles* (2nd edn.). Naturalists' Handbooks no. 4, Richmond, Publishing, Slough.

REDFERN, M. & ASKEW, R.R. 1998. *Plant galls* (2nd edn.). Naturalists' Handbooks no. 17, Richmond, Publishing, Slough.

REDFERN, M. & SHIRLEY, P. 2002. *British plant galls: identification of galls on plants and fungi*. Field Studies Council, Preston Montford, OP270.

REDFERN, M. & SHIRLEY, P. 2004. *A guide to plant galls in Britain*. Field Studies Council, Preston Montford, OP91.

RESH, V.H. & CARDÉ, R.T. (eds.) 2009. *Encyclopedia of insects* (2nd edn.). Academic Press/Elsevier, San Diego & London.

RICHARDS, O.W. 1937. The generic names of the British Hymenoptera, Aculeata, with a check list of British species. Part 5. *The generic names of British insects*. Vol. 1: 79–149. Royal Entomological Society, London.

RICHARDS, O.W & DAVIES, R.G. 1957. *Imms' general textbook of entomology* (9th edn.). Methuen, London.

RICHARDS, O.W & DAVIES, R.G. 1977. *Imms' general textbook of entomology* (10th edn.). Chapman & Hall, London.

ROTHERAY, G.E. 1989. *Aphid predators*. Naturalists' Handbooks no. 11, Richmond, Publishing, Slough [reprinted 2003].

SALMON, M.A. 2000. *The Aurelian legacy: British butterflies and their collectors*. Harley Books, Great Horkesley.

SALMON, M.A. & EDWARDS, P.J. 2005. *The Aurelian's fireside companion: an entomological anthology*. Paphia Books, Lymington.

SALT, D.T. & WHITTAKER, J.B. 1998. *Insects on dock plants.* Naturalists' Handbooks no. 26, Richmond, Publishing, Slough.

SAMOUELLE, G. 1819. *A nomenclature of British entomology.* London.

SECORD, J.A. 1996. The crisis of nature. In: JARDINE, N., SECORD, J.A. & SPARY, E.C. (eds.) *Cultures of natural history.* Cambridge University Press, Cambridge, pp. 447–59.

SMITH, F. 1871. *A catalogue of British Hymenoptera; Aculeata.* Entomological Society of London.

STAINTON, H.T. 1854. *Insect Britannica. Lepidoptera: Tineina.* Reeve, London.

STEPHENS, J.F. 1829. *A systematic catalogue of British insects.* London, 2 vols.

STEPHENS, J.F. 1846. Illustrations of British entomology *(Suppl.).* Henry G. Bohn, London.

THORP, J.H. 2009. Arthropoda and related groups. In: RESH, V.H. & CARDÉ, R.T. (eds.) *Encyclopedia of insects* (2nd edn.). Academic Press/Elsevier, San Diego & London, pp 50–6.

TOTTENHAM, C.E. 1949. The generic names of the British Staphylinidae, with a check list of British species. Part 9. *The generic names of British insects.* Vol. 1: 343–466. Royal Entomological Society, London.

WALKER, F. 1851–6. *Insecta Britannica. Diptera.* Reeve, London, 3 vols.

WHEATER, C.P. & READ, H.J. 1996. *Animals under logs and stones.* Naturalists' Handbooks no. 22, Richmond, Publishing, Slough.

WHEELER, Q.D. 2008. Taxonomic shock and awe. In: WHEELER, Q.D. (ed.) *The new taxonomy.* Systematics Association/CRC Press, London/New York, pp 211–26.

YOON, C.K. 2009. *Naming nature: the clash between instinct and science.* Norton, New York.

FAMILY CLASSIFICATION OF THE BRITISH INSECTS

Each group of higher taxa is arranged in alphabetical order

Phylum Arthropoda: Subphylum Hexapoda
Class Entognatha
Order Collembola (springtails)
  Suborder Entomobryomorpha
  Superfamily Entomobryoidea
Families Cyphoderidae, Entomobryidae, Isotomidae
  Superfamily Tomoceroidea
Familes Oncopoduridae, Tomoceridae
  Suborder Neelipleona
Family Neelidae
  Suborder Poduromorpha
  Superfamily Hypogastruroidea
Family Hypogastruridae
  Superfamily Neanuroidea
Families Brachystomellidae, Neanuridae, Odontellidae
  Superfamily Onychiuroidea
Families Onychiuridae, Tullbergiidae, Poduridae
  Suborder Symphypleona
Families Arrhopalitidae, Bourletiellidae, Dicyrtomidae, Katiannidae, Sminthuridae, Sminthurididae
Order Diplura (two-tailed bristletails)
Family Campodeidae
Order Protura (proturans)
  Suborder Eosentomata
Family Eosentomidae
  Suborder Acerentomata
Families Acerentomidae, Protentomidae

Class Insecta 'Apterygota' [= Thysanura]
Order Archaeognatha (bristletails)
Family Machilidae

Order Zygentoma (silverfish and firebrats)
Family Lepismatidae

Insecta Pterygota Palaeoptera
Order Ephemeroptera (mayflies)
  Superfamily Baetoidea
Families Ameletidae, Baetidae, Siphlonuridae
  Superfamily Caenoidea
Family Caenidae
  Superfamily Ephemerelloidea
Family Ephemerellidae
  Superfamily Ephemeroidea
Families Ephemeridae, Potamanthidae
  Superfamily Heptagenioidea
Families Arthropleidae, Heptageniidae
  Superfamily Leptophlebioidea
Family Leptophlebiidae
Order Odonata (damselflies and dragonflies)
  Suborder Anisoptera
  Superfamily Aeshnoidea
Families Aeshnidae, Gomphidae
  Superfamily Cordulegastroidea
Family Cordulegastridae
  Superfamily Libelluloidea
Families Corduliidae, Libellulidae
  Suborder Zygoptera
  Superfamily Calopterygoidea
Family Calopterygidae
  Superfamily Coenagrionoidea
Families Coenagrionidae, Platycnemididae
  Superfamily Lestoidea
Family Lestidae

Insecta Pterygota Neoptera Polyneoptera
Order Dermaptera (earwigs)
  Suborder Forficulina

Families Anisolabididae, Forficulidae, Labiduridae, Spongiphoridae
Order Dictyoptera (cockroaches, termites and mantids)
Suborder Blattodea
Families Blaberidae, Blattellidae, Blattidae
Suborder Isoptera
Family Rhinotermitidae
Suborder Mantodea
Family Mantidae
Order Orthoptera (grasshoppers, crickets and bush-crickets)
Suborder Caelifera
Superfamily Acridoidea
Family Acrididae
Superfamily Tetrigoidea
Family Tetrigidae
Suborder Ensifera
Superfamily Grylloidea
Families Gryllidae, Gryllotalpidae, Mogoplistidae
Superfamily Rhaphidophoroidea
Family Rhaphidophoridae
Superfamily Tettigonioidea
Families Conocephalidae, Meconematidae, Phaneropteridae, Tettigoniidae
Order Phasmida (stick insects)
Families Bacillidae, Phasmatidae
Order Plecoptera (stoneflies)
Superfamily Nemouroidea
Families Capniidae, Leuctridae, Nemouridae, Taeniopterygidae
Superfamily Perloidea
Families Chloroperlidae, Perlidae, Perlodidae

Insecta Pterygota Neoptera Paraneoptera
Order Hemiptera (true bugs)
Suborder Heteroptera
Infraorder Cimicomorpha
Superfamily Cimicoidea
Families Anthocoridae, Cimicidae, Nabidae
Superfamily Miroidea
Families Microphysidae, Miridae
Superfamily Reduvioidea
Family Reduviidae
Superfamily Tingoidea
Family Tingidae
Infraorder Dipsocoromorpha
Superfamily Dipsocoroidea
Families Ceratocombidae , Dipsocoridae
Infraorder Leptopodomorpha
Superfamily Saldoidea
Families Aepophilidae, Saldidae
Infraorder Nepomorpha
Superfamily Corixoidea

Family Corixidae
Superfamily Gerroidea
Families Gerridae, Veliidae
Superfamily Hebroidea
Family Hebridae
Superfamily Hydrometroidea
Family Hydrometridae
Superfamily Mesovelioidea
Family Mesoveliidae
Superfamily Naucoroidea
Families Aphelocheiridae, Naucoridae
Superfamily Nepoidea
Family Nepidae
Superfamily Notonectoidea
Family Notonectidae
Superfamily Pleoidea
Family Pleidae
Infraorder Pentatomomorpha
Superfamily Aradoidea
Family Aradidae
Superfamily Coreoidea
Families Alydidae, Coreidae, Rhopalidae, Stenocephalidae
Superfamily Lygaeoidea
Families Berytidae, Lygaeidae, Pyrrhocoridae
Superfamily Pentatomoidea
Families Acanthosomatidae, Cydnidae, Pentatomidae, Scutelleridae, Thyreocoridae
Superfamily Piesmatoidea
Family Piesmatidae
Suborder Cicadomorpha
Superfamily Cercopoidea
Families Aphrophoridae, Cercopidae
Superfamily Cicadoidea
Family Tibicinidae
Superfamily Membracoidea
Families Cicadellidae, Membracidae, Ulopidae
Suborder Fulgoromorpha
Families Cixiidae, Delphacidae, Issidae, Tettigometridae
Suborder Sternorrhyncha
Superfamily Aleyrodoidea
Family Aleyrodidae
Superfamily Aphidoidea
Family Aphididae
Superfamily Coccoidea
Families Aclerdidae, Asterolecaniidae, Coccidae, Diaspididae, Eriococcidae, Kermesidae, Margarodidae, Ortheziidae, Pseudococcidae
Superfamily Phylloxeroidea
Families Adelgidae, Phylloxeridae
Superfamily Psylloidea
Families Calophyidae, Homotomidae, Psyllidae, Triozidae

Order Phthiraptera (lice)
  Suborder Amblycera
  Superfamily Gyropoidea
  Familiesy Gyropidae, Gliricolidae
  Superfamily Laemobothrioidea
  Family Laemobothriidae
  Superfamily Menopodoidea
  Family Menopodidae
  Superfamily Ricinoidea
  Family Ricinidae
  Superfamily Trimenoponoidea
  Family Trimenoponidae
  Suborder Anoplura
  Superfamily Echinophthirioidea
  Family Echinophthiriidae
  Superfamily Linognathoidea
  Families Enderleinellidae, Hoplopleuridae, Linognathidae, Polyplacidae, Phthiridae
  Superfamily Pediculoidea
  Families Haematopinidae, Pediculidae
  Suborder Ischnocera
  Superfamily Goniodoidea
  Family Goniodidae
  Superfamily Philopteroidea
  Family Philopteridae
  Superfamily Trichodectoidea
  Family Trichodectidae

Order Psocoptera (booklice and barklice)
  Suborder Psocomorpha
  Infraorder Caeciliusetae
  Families Amphipsocidae, Caeciliusidae, Stenopsocidae
  Infraorder Epipsocetae
  Family Epipsocidae
  Infraorder Homilopsocidea
  Families Ectopsocidae, Elipsocidae, Lachesillidae, Mesopsocidae, Peripsocidae, Philotarsidae, Trichopsocidae
  Infraorder Psocetae
  Family Psocidae
  Suborder Troctomorpha
  Infraorder Nanopsocetae
  Families Liposcelididae, Pachytroctidae, Sphaeropsocidae
  Suborder Trogiomorpha
  Infraorder Atropetae
  Families Lepidopsocidae, Psoquillidae, Trogiidae
  Infraorder Psocathropetae
  Family Psyllipsocidae

Order Thysanoptera (thrips)
  Suborder Terebrantia
  Families Aeolothripidae, Thripidae
  Suborder Tubulifera
  Family Phlaeothripidae

Insecta Pterygota Neoptera Endopterygota = Holometabola
Order Coleoptera (beetles)
  Suborder Adephaga
  Superfamily Caraboidea
  Families Carabidae, Dytiscidae, Gyrinidae, Haliplidae, Hygrobiidae, Noteridae
  Suborder Myxophaga
  Superfamily Sphaeriusoidea
  Family Sphaeriusidae
  Suborder Polyphaga
  Infraorder Bostrichiformia
  Superfamily Bostrichoidea
  Families Anobiidae, Bostrichidae, Dermestidae, Lyctidae
  Superfamily Derodontoidea
  Family Derodontidae
  Infraorder Cucujiformia
  Superfamily Chrysomeloidea
  Families Cerambycidae, Chrysomelidae, Megalopodidae, Orsodacnidae
  Superfamily Cleroidea
  Families Cleridae, Dasytidae, Malachiidae, Phloiophilidae, Trogossitidae
  Superfamily Cucujoidea
  Families Alexiidae, Biphyllidae, Bothrideridae, Byturidae, Cerylonidae, Coccinellidae, Corylophidae, Cryptophagidae, Cucijidae, Endomychidae, Erotylidae, Kateretidae, Laemophloeidae, Languriidae, Latridiidae, Monotomidae, Nitidulidae, Phalacridae, Silvanidae, Sphindidae
  Superfamily Curculionoidea
  Families Anthribidae, Apionidae, Attelabidae, Curculionidae, Dryophthoridae, Erirhinidae, Nanophyidae, Nemonychidae, Platypodidae, Raymondionymidae, Rhynchitidae
  Superfamily Lymexyloidea
  Family Lymexylidae
  Superfamily Tenebrionoidea
  Families Aderidae, Anthicidae, Ciidae, Melandryidae, Meloidae, Mordellidae, Mycetophagidae, Mycteridae, Oedemeridae, Pyrochroidae, Pythidae, Ripiphoridae, Salpingidae, Scraptiidae, Tenebrionidae, Tetratomidae, Zopheridae
  Infraorder Elateriformia
  Superfamily Buprestoidea
  Family Buprestidae
  Superfamily Byrrhoidea
  Families Byrrhidae, Dryopidae, Elmidae, Heteroceridae, Limnichidae, Psephenidae, Ptilodactylidae
  Superfamily Dascilloidea
  Family Dascillidae
  Superfamily Elateroidea

Families Cantharidae, Drilidae, Elateridae, Eucne-
midae, Lampyridae, Lycidae, Throscidae
Superfamily Scirtoidea
Families Clambidae, Eucinetidae, Scirtidae
Infraorder Scarabaeiformia
Superfamily Scarabaeoidea
Families Aegialiidae, Aphodiidae, Bolboceratidae,
Cetoniidae, Geotrupidae, Lucanidae, Melolon-
thidae, Rutelidae, Scarabaeidae, Trogidae
Infraorder Staphyliniformia
Superfamily Hydrophiloidea
Families Georissidae, Helophoridae, Histeridae,
Hydrochidae, Hydrophilidae, Spercheidae,
Sphaeritidae
Superfamily Staphylinoidea
Families Hydraenidae, Leiodidae, Ptiliidae, Scyd-
maenidae, Silphidae, Staphylinidae
Order Diptera (true flies)
Suborder Brachycera
Infraorder Asilomorpha
Superfamily Asiloidea
Families Asilidae, Bombyliidae, Scenopinidae,
Therevidae
Superfamily Empidoidea
Families Atelestidae, Dolichopodidae, Empididae,
Hybotidae, Microphoridae
Superfamily Nemestrinoidea
Family Acroceridae
Infraorder Muscomorpha Aschiza
Superfamily Lonchopteroidea
Family Lonchopteridae
Superfamily Platypezoidea
Families Opetiidae, Phoridae, Platypezidae
Superfamily Syrphoidea
Families Pipunculidae, Syrphidae
Infraorder Muscomorpha Schizophora Acalyptratae
Superfamily Carnoidea
Families Braulidae, Canacidae, Carnidae, Chloro-
pidae, Milichiidae, Tethinidae
Superfamily Conopoidea
Family Conopidae
Superfamily Diopsoidea
Families Megamerinidae, Psilidae, Strongyloph-
thalmyiidae, Tanypezidae
Superfamily Ephydroidea
Families Camillidae, Campichoetidae, Diastatidae,
Drosophilidae, Ephydridae
Superfamily Lauxanioidea
Families Chamaemyiidae, Lauxaniidae
Superfamily Nerioidea
Families Micropezidae, Pseudopomyzidae
Superfamily Opomyzoidea
Families Acartophthalmidae, Agromyzidae, Antho-
myzidae, Asteiidae, Aulacigastridae, Clusiidae,

Odiniidae, Opomyzidae, Periscelididae,
Stenomicridae
Superfamily Sciomyzoidea
Families Coelopidae, Dryomyzidae, Phaeomyiidae,
Sciomyzidae, Sepsidae
Superfamily Sphaeroceroidea
Families Chyromyidae, Heleomyzidae, Sphaero-
ceridae
Superfamily Tephritoidea
Families Lonchaeidae, Pallopteridae, Piophilidae,
Platystomatidae, Tephritidae, Ulidiidae
Infraorder Muscomorpha Schizophora Calyptratae
Superfamily Hippoboscoidea
Families Hippoboscidae, Nycteribiidae
Superfamily Muscoidea
Families Anthomyiidae, Fanniidae, Muscidae,
Scathophagidae
Superfamily Oestroidea
Families Calliphoridae, Oestridae, Rhinophoridae,
Sarcophagidae, Tachinidae
Infraorder Tabanomorpha
Superfamily Stratiomyoidea
Families Stratiomyidae, Xylomyidae
Superfamily Tabanoidea
Families Athericidae, Rhagionidae, Spaniidae,
Tabanidae
Infraorder Xylophagomorpha
Superfamily Xylophagoidea
Family Xylophagidae
Suborder Nematocera
Infraorder Bibionomorpha
Superfamily Bibionoidea
Family Bibionidae
Superfamily Sciaroidea
Families Bolitophilidae, Cecidomyiidae, Diado-
cidiidae, Ditomyiidae, Keroplatidae, Myceto-
philidae, Sciaridae
Infraorder Culicomorpha
Superfamily Chironomoidea
Families Ceratopogonidae, Chironomidae, Simu-
liidae, Thaumaleidae
Superfamily Culicoidea
Families Chaoboridae, Culicidae, Dixidae
Infraorder Psychodomorpha
Superfamily Anisopodoidea
Families Anisopodidae, Mycetobiidae
Superfamily Psychodoidea
Family Psychodidae
Superfamily Scatopsoidea
Family Scatopsidae
Superfamily Trichoceroidea
Family Trichoceridae
Infraorder Ptychopteromorpha
Superfamily Ptychopteroidea

Family Ptychopteridae
  Infraorder Tipulomorpha
  Superfamily Tipuloidea
Families Cylindrotomidae, Limoniidae, Pediciidae,
  Tipulidae
Order Hymenoptera (ants, bees and wasps)
  Suborder Apocrita Aculeata
  Superfamily Apoidea
Families Apidae, Crabronidae, Sphecidae
  Superfamily Chrysidoidea
Families Bethylidae, Chrysididae, Dryinidae,
  Embolemidae
  Superfamily Vespoidea
Families Formicidae, Mutillidae, Pompilidae,
  Sapygidae, Scoliidae, Tiphiidae, Vespidae
  Suborder Apocrita Parasitica
  Superfamily Ceraphronoidea
Families Ceraphronidae, Megaspilidae
  Superfamily Chalcidoidea
Families Aphelinidae, Chalcididae, Elasmidae,
  Encyrtidae, Eucharitidae, Eulophidae, Eupel-
  midae, Eurytomidae, Mymaridae, Ormyridae,
  Perilampidae, Pteromalidae, Signiphoridae,
  Tetracampidae, Torymidae, Trichogrammatidae
  Superfamily Cynipoidea
Families Cynipidae, Figitidae, Ibaliidae
  Superfamily Evanioidea
Families Aulacidae, Evaniidae, Gasteruptiidae
  Superfamily Ichneumonoidea
Families Braconidae, Ichneumonidae
  Superfamily Mymarommatoidea
Family Mymarommatidae
  Superfamily Platygastroidea
Families Platygastridae, Scelionidae
  Superfamily Proctotrupoidea
Families Diapriidae, Heloridae, Proctotrupidae
  Superfamily Trigonaloidea
Family Trigonalidae
  Suborder Symphyta
  Superfamily Cephoidea
Family Cephidae
  Superfamily Orussoidea
Family Orussidae
  Superfamily Pamphilioidea
Family Pamphilidae
  Superfamily Siricoidea
Family Siricidae
  Superfamily Tenthredinoidea
Families Argidae, Blasticotomidae, Cimbicidae,
  Diprionidae, Tenthredinidae
  Superfamily Xiphrydioidea
Family Xiphydriidae
  Superfamily Xyeloidea
Family Xyelidae

Order Lepidoptera (moths and butterflies)
  Suborder Glossata
  Superfamily Alucitoidea
Family Alucitidae
  Superfamily Bombycoidea
Families Endromidae, Saturniidae, Sphingidae
  Superfamily Choreutoidea
Family Choreutidae
  Superfamily Cossoidea
Family Cossidae
  Superfamily Drepanoidea
Families Drepanidae, Thyatiridae
  Superfamily Epermenioidea
Family Epermeniidae
  Superfamily Eriocranioidea
Family Eriocraniidae
  Superfamily Gelechioidea
Families Agonoxenidae, Amphisbatidae, Autosti-
  chidae, Batrachedridae, Blastobasidae, Chimaba-
  chidae, Coleophoridae, Cosmopterigidae,
  Depressariidae, Elachistidae, Ethmiidae, Gele-
  chiidae, Momphidae, Oecophoridae, Scythrididae,
  Stathmopodidae
  Superfamily Geometroidea
Family Geometridae
  Superfamily Gracillarioidea
Family Bucculatricidae, Douglasiidae, Gracillariidae,
  Roeslerstammiidae
  Superfamily Hepialoidea
Family Hepialidae
  Superfamily Hesperioidea
Family Hesperiidae
  Superfamily Incurvarioidea
Families Adelidae, Heliozelidae, Incurvariidae,
  Prodoxidae
  Superfamily Lasiocampoidea
Family Lasiocampidae
  Superfamily Nepticuloidea
Family Nepticulidae, Opostegidae
  Superfamily Noctuoidea
Families Arctiidae, Ctenuchidae, Lymantriidae, Noc-
  tuidae, Nolidae, Notodontidae, Thaumetopoeidae
  Superfamily Papilionoidea
Families Lycaenidae, Nymphalidae, Papilionidae,
  Pieridae, Riodinidae
  Superfamily Pterophoroidea
Family Pterophoridae
  Superfamily Pyraloidea
Families Crambidae, Pyralidae
  Superfamily Schreckensteinioidea
Family Schreckensteiniidae
  Superfamily Sesiodea
Family Sesiidae
  Superfamily Tineoidea

Families Psychidae, Tineidae
  Superfamily Tischerioidea
Family Tischeriidae
  Superfamily Tortricoidea
Family Tortricidae
  Superfamily Yponomeutoidea
Families Acrolepiidae, Bedelliidae, Glyphiptery-
  gidae, Heliodinidae, Lyonetiidae, Plutellidae,
  Yponomeutidae, Ypsolophidae
  Superfamily Zygaenoidea
Families Limacodidae, Zygaenidae
  Suborder Zeugloptera
  Superfamily Micropterigoidea
Family Micropterigidae
Order Mecoptera (scorpionflies)
Familie Boreidae, Panorpidae
Order Megaloptera (alderflies)
Family Sialidae
Order Neuroptera (lacewings and antlions)
  Suborder Hemerobiiformia
Families Chrysopidae, Coniopterygidae, Hemero-
  biidae, Osmylidae, Sisyridae
  Suborder Myrmeleontiformia
Family Myrmeleontidae
Order Raphidioptera (snakeflies)
Family Raphidiidae
Order Siphonaptera (fleas)
  Superfamily Ceratophylloidea
Families Ceratophyllidae, Ischnopsyllidae, Lepto-
  psyllidae
  Superfamily Hystrichopsylloidea
Families Ctenophthalmidae, Hystrichopsyllidae
  Superfamily Pulicoidea

Family Pulicidae
  Superfamily Vermipsylloidea
Family Vermipsyllidae
Order Strepsiptera (stylops)
Families Elenchidae, Halictophagidae, Stylopidae,
  Xenidae
Order Trichoptera (caddisflies)
  Suborder Annulipalpia
  Superfamily Hydropsychoidea
Families Ecnomidae, Hydropsychidae, Polycentro-
  podidae, Psychomyiidae
  Superfamily Philopotamoidea
Family Philopotamidae
  Suborder Integripalpia
  Superfamily Leptoceroidea
Families Leptoceridae, Molannidae, Odontoceridae
  Superfamily Limnephiloidea
Families Apataniidae, Brachycentridae, Goeridae,
  Lepidostomatidae, Limnephilidae
  Superfamily Phryganeoidea
Family Phryganeidae
  Superfamily Sericostomatoidea
Families Beraeidae, Sericostomatidae
  Suborder Spicipalpia
  Superfamily Glossosomatoidea
Family Glossosomatidae
  Superfamily Hydroptiloidea
Family Hydroptilidae
  Superfamily Rhyacophiloidea
Family Rhyacophilidae

PART 1 Entognatha

This group has gradually emerged from the confusion of the primitively wingless insects, known previously as the Apterygota, a polyphyletic group that also included the Thysanura (= Archaeognatha + Zygentoma). The latter are now shown to be genuinely related to the winged insects, whereas the Entognatha are a monophyletic group united by the eponymous condition of entognathy in which the mouthparts are recessed in a pouch on the head and can be extruded for feeding. All the rest of the Hexapoda, as well as other Arthropoda such as Symphyla and Diplopoda, have ectognathous mouthparts, not contained in a gnathal pouch.

The relationships between the three groups of Entognatha are still open to debate; the Collembola and Protura share many common features and some authors combine them in a formal subclass Ellipura, leaving the Diplura in their own group.

Members of all three orders are small, with very simple genitalia and indirect fertilization. Their compound eyes are either very reduced or absent, which is probably related to the fact that many are soil-dwelling. Apart from the Collembola they are little known to many people, even professional entomologists.

The Royal Entomological Society Book of British Insects, First Edition. Peter C. Barnard.
© 2011 Royal Entomological Society. Published 2011 by Blackwell Publishing Ltd.

2 Order Collembola: the springtails

c. 250 species in 19 families

Springtails are very small wingless creatures, just a few millimetres long, that live mainly in soil and leaf litter, and are perhaps best known for their forked springing organ, or furca, which enables them to jump considerable distances of up to several centimetres when disturbed, though a few groups have secondarily lost the furca.

Their eyes are reduced, each consisting of at most eight ommatidia, but they may be absent, especially in soil-dwelling species. The antennae are usually short, with just four segments or occasionally five (in *Heteromurus nitidus*) or six (in *Orchesella*), but in a few groups the terminal segments are very long and secondarily subdivided. The entognathous mouthparts consist mainly of elongate mandibles and maxillae, with the labium forming the ventral margin. In some groups the mouthparts become extremely complex in shape, though the reason for this is unknown, and some authors have used these structures taxonomically. A pair of chemosensory structures, the post-antennal organs, is often present on the head and their morphology is important in species identification. The rather simple walking legs may have specialized clavate setae on the tibiotarsus, which are useful for separating some species.

There are two distinct body shapes in the Collembola: the suborders Entomobryomorpha and Poduromorpha have an elongate abdomen in which the six segments are clearly visible, whereas the Neelipleona and Symphypleona have globular bodies in which the segmentation is not apparent (Fig. 2.1). On the first abdominal segment is the ventral tube or collophore, originally thought to be a 'sticky organ' which gives the Collembola their name. It has eversible vesicles, and is now thought to be involved with water regulation and other related functions. The furca, when present, is on the fourth abdominal segment, and is held ready for jumping by a tenaculum, or retinaculum, on the third segment; the detailed structure of the furca is important in identification. Various setae and spines are present over the whole body, many of which are used in identification keys.

As well as being a familiar sight to gardeners in soil, leaf litter and occasionally on living plants, some Collembola are found on the seashore, and several species live on the surface of freshwater, often in large numbers. Despite their small size, the soil-dwelling species may occur in huge populations and can be a significant component of the soil fauna. Some species congregate in huge swarms of up to several million individuals, which can reach pest proportions but, as with so many soil organisms, the details of their ecology and life histories are often very sketchy though they are undoubtedly vital in maintaining soil structure. Most springtails feed on fungi, decaying vegetation, bacteria or algae, though a few species are predatory on smaller organisms such as rotifers and nematodes. Communities of some species (mainly *Hypogastrura viatica* and *Anurida tullbergi*) are important in sewage treatment works where they help remove fungi from filters; they are known colloquially as 'Achorutes' though this should not be confused with the (non-European) genus of the same name.

In general the Collembola need conditions of high humidity in order to thrive, and this is reflected in their usual habitats. However, some species can

The Royal Entomological Society Book of British Insects, First Edition. Peter C. Barnard.
© 2011 Royal Entomological Society. Published 2011 by Blackwell Publishing Ltd.

Fig. 2.1 Symphypleonan springtail next to a ladybird for scale (Photo: Colin Rew)

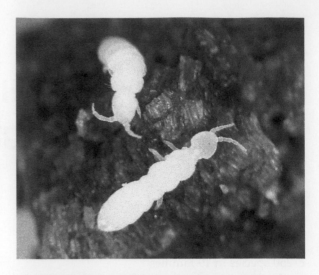

Fig. 2.2 *Folsomia candida* **(Isotomidae) is a pale soil-dwelling species (Photo: Peter Barnard)**

Fig. 2.3 Leaf-litter Collembola are often brightly coloured (Photo: Robin Williams)

survive periods of desiccation by entering a dried apparently lifeless form, from which they quickly recover when moisture is restored. Others routinely go through cycles of inactivity, with a changed external morphology, from which they recover following a moult.

The smaller 'apterygote' groups such as Diplura and Protura, with their small size and few species, have been neglected by most entomologists, but the lack of information on the Collembola, with their large numbers of both species and population sizes, has been an embarrassment for some time. Until recently there were very few keys to the group at any taxonomic level; the numbers of species were quite unknown, with speculations about species complexes, and the whole group was generally considered 'difficult' by most authors. All this changed when Hopkin (2007) brought out a key to the British species, following on from his seminal book on the biology of the group (Hopkin, 1997). Instantly the Collembola became accessible to anyone with a good microscope and some patience, and there is now no excuse for their neglect. There were previously thought to be around 350 species in the British Isles, but this was reduced by Hopkin to nearer 250, partly by a common-sense approach to synonymy and also by showing that many apparent British records could not be confirmed by voucher specimens. The number of British species will undoubtedly change as more work is done, but Hopkin's key has made it much easier for others to study this potentially rewarding group; it includes a synonymic checklist with notes on distribution and abundance.

The Collembola are a surprisingly colourful group, though on a microscopic scale; colour can be useful for identification though it is obviously prone to fading in preservatives such as alcohol. Some

early colour plates were included in Lubbock's (1873) book, and many colour photos have been taken by Hopkin (2007). In general, colour is found in the leaf-litter species, whereas many soil-dwellers are unpigmented (Figs. 2.2 & 2.3). Although some species can be recognized by their characteristic colour patterns or other easily observed characters, even under a hand-lens, most will need microscopic examination of slide-mounted material (Fig. 2.4).

The British total of about 250 species in 19 families compares with around 2000 species in 23 families in Europe, and nearly 8000 species in 29 families worldwide. The worldwide classification and a summary of the biology of the whole group were covered by Hopkin (1997); a brief world overview is provided by Christiansen et al. (2009).

The following higher classification is modified from Hopkin (2007). Some authors have included

Fig. 2.4 Typical microscope preparation of springtail (Photo: Peter Barnard)

the Entomobryomorpha and Poduromorpha in the Arthropleona, but this is a paraphyletic grouping.

HIGHER CLASSIFICATION OF BRITISH COLLEMBOLA

Suborder Entomobryomorpha
  Superfamily Entomobryoidea
    Family Cyphoderidae (1 genus, 2 species)
    Family Entomobryidae (9 genera, 33 species)
    Family Isotomidae (23 genera, 67 species)
  Superfamily Tomoceroidea
    Family Oncopoduridae (1 genus, 1 species)
    Family Tomoceridae (2 genera, 6 species)
Suborder Neelipleona
    Family Neelidae (3 genera, 3 species)
Suborder Poduromorpha
  Superfamily Hypogastruroidea
    Family Hypogastruridae (7 genera, 32 species)
  Superfamily Neanuroidea
    Family Brachystomellidae (1 genus, 1 species)
    Family Neanuridae (10 genera, 23 species)
    Family Odontellidae (2 genera, 2 species)
  Superfamily Onychiuroidea
    Family Onychiuridae (13 genera, 18 species)
    Family Tullbergiidae (6 genera, 19 species)
    Family Poduridae (1 genus, 1 species)
Suborder Symphypleona
    Family Arrhopalitidae (1 genus, 6 species)
    Family Bourletiellidae (3 genera, 9 species)
    Family Dicyrtomidae (3 genera, 4 species)
    Family Katiannidae (1 genus, 7 species)
    Family Sminthuridae (6 genera, 6 species)
    Family Sminthurididae (3 genera, 9 species)

SPECIES OF CONSERVATION CONCERN

No species of Collembola have any legal protection and none is on the UKBAP list.

The Families of British Collembola

SUBORDER ENTOMOBRYOMORPHA

SUPERFAMILY ENTOMOBRYOIDEA

Family Cyphoderidae (1 genus, 2 species)
Cyphoderus albinus is quite common in ants' nests, but *C. bidenticulatus* is rare in Britain.

British genus: *Cyphoderus*

Family Entomobryidae (9 genera, 33 species)
There are several quite common species in genera such as *Entomobrya*, *Lepidocyrtus*, *Pseudosinella* and *Orchesella*. For example, *Entomobrya albocincta* is found under bark and even in the canopy in many woodlands, along with other common species such as *Orchesella cincta*.

British subfamilies and genera:
  Entomobryinae: *Entomobrya*, *Entomobryoides*, *Mesentotoma*, *Sinella*
    Leipidocyrtinae: *Lepidocyrtus*, *Pseudosinella*
    Orchesellinae: *Heteromurus*, *Orchesella*
    Seirinae: *Seira*

Family Isotomidae (23 genera, 67 species)
This is a large family with several common species in *Folsomia*, *Isotoma* and other genera. Like several members of the Entomobryidae, *Anurophorus laricis* is commonly found under tree bark and in the canopy. The identification of species within *Folsomia* is particularly difficult, but the parthenogenetic *Folsomia candida* is a well-known laboratory culture species that is particularly easy to breed (Fig. 2.5).

Fig. 2.5 *Folsomia candida* (Isotomidae) is often cultured in laboratories (Photo: Peter Barnard)

British subfamilies and genera:

Anurophorinae: *Anurophorus, Archisotoma, Ballistura, Cryptopygus, Folsomia, Folsomides, Isotomiella, Isotomodes, Proctostephanus, Proisotoma, Pseudanurophorus, Tetracanthella, Uzelia*

Isotominae: *Agrenia, Axelsonia, Desoria, Halisotoma, Isotoma, Isotomurus, Parisotoma, Pseudisotoma, Vertagopus*

Pachyotominae: *Pachyotoma*

SUPERFAMILY TOMOCEROIDEA

Family Oncopoduridae (1 genus, 1 species)
The only species, *Oncopodura crassicornis*, is an unpigmented, blind soil-dweller that is apparently scarce, but probably overlooked because of its small size of around 0.8 mm.

British genus: *Oncopodura*

Family Tomoceridae (2 genera, 6 species)
Two of the species in *Tomocerus* are common and widespread, but the most distinctive member of this family is Pogonognathellus longicornis (formerly in *Tomocerus*), which rolls its antennae into coils when disturbed; at 6 mm it also the largest species of Collembola in Britain and is often found in tree canopies (Fig. 2.6).

British genera: *Pogonognathellus, Tomocerus*

SUBORDER NEELIPLEONA

Family Neelidae (3 genera, 3 species)
Only *Megalothorax minimus* is common in leaf litter and rotten wood, but is easy to overlook because of its small size of around 0.5 mm.

British genera: *Megalothorax, Neelides, Neelus*

SUBORDER PODUROMORPHA

SUPERFAMILY HYPOGASTRUROIDEA

Family Hypogastruridae (7 genera, 32 species)
A large family with several common species in *Ceratophysella* and *Hypogastrura. Hypogastrura viatica* and *H. purpurescens* are found in rich organic matter, including sewage works as well as leaf litter and compost heaps.

British genera: *Ceratophysella, Choreutinula, Hypogastrura, Mesogastrura, Schaefferia, Willemia, Xenylla*

SUPERFAMILY NEANUROIDEA

Family Brachystomellidae (1 genus, 1 species)
The only UK species, *Brachystomella parvula* is quite common and widespread in soil and has a distinctive blue or violet colour.

British genus: *Brachystomella*

Family Neanuridae (10 genera, 23 species)
There are several common species in *Friesea, Neanura* and *Anurida. Anurida maritima* is a very common, dark blue species found all round Britain's coasts in the intertidal zone (Fig. 2.7). *Anurida tullbergi* is a similar bluish black species, also able to tolerate some salinity, but usually found in sewage filter beds. *Neanura muscorum* is commonly found throughout the year, even under snow.

British subfamilies and genera:
Frieseinae: *Friesea*

Fig. 2.6 *Pogonognathellus longicornis* (Tomoceridae) (Photo: Roger Key)

Fig. 2.7 *Anurida maritima* (Neanuridae) (Photo: Roger Key)

Neanurinae: *Lathriopyga, Monobella, Neanura, Paranura*

Pseudachoreutinae: *Anurida, Anuridella, Micranurida, Pseudachorutella, Pseudachorutes*

Family Odontellidae (2 genera, 2 species)
This family contains only two rare species.

British genera: *Odontella, Xenyllodes*

SUPERFAMILY ONYCHIUROIDEA

Family Onychiuridae (13 genera, 18 species)
A rather diverse family that probably needs further taxonomic refinement; identification to species is particularly problematic, and the exact identities of many of the described species need confirmation. Because of this, the relative scarcity or abundance of individual species need careful reassessment based on accurately identified voucher material As an example, several species of *Protaphorura* have been distinguished by setal patterns, yet individuals are frequently asymmetrical, which would place each side of an animal in different species. Common British species include *Deuteraphorura inermis, Onychiurus ambulans* and *Protaphorura armata*.

British genera: *Allonychiurus, Deharvengiurus, Detriturus, Deuteraphorura, Kalaphorura, Megaphorura, Micraphorura, Oligaphorurus, Onychiurus, Orthonychiurus, Protaphorura, Supraphorura, Thalassaphorura*

Family Tullbergiidae (6 genera, 19 species)
This family contains a few common species, including *Mesaphorura macrochaeta* and *Paratullbergia callipygos*. The species in some genera such as *Mesaphorura* need careful assessment because of previous misidentifications.

British genera: *Mesaphorura, Metaphorura, Neonaphorura, Neotullbergia, Paratullbergia, Stenaphorura*

Family Poduridae (1 genus, 1 species)
The only species, *Podura aquatica*, is popularly known as the common springtail seen on puddles and other freshwater bodies (Fig. 2.8); in fact there are not a great many confirmed records in recent years and many of the sightings, especially on small puddles, probably refer to members of the Hypogastruridae such as *Ceratophysella*.

British genus: *Podura*

SUBORDER SYMPHYPLEONA

Family Arrhopalitidae (1 genus, 6 species)
At least two species of *Arrhopalites* are quite common, especially in caves.

British genus: *Arrhopalites*

Family Bourletiellidae (3 genera, 9 species)
There are a few common species, including the bluish black *Bourletiella hortensis*, which can be a pest species on plant seedlings. Confusingly, *Deuterosminthurus pallipes* can be found in two common colour forms, purple and yellow.

British genera: *Bourletiella, Deuterosminthurus, Heterosminthurus*

Family Dicyrtomidae (3 genera, 4 species)
Two species of *Dicyrtomina* are very common in Britain, with distinctive colour patterns (Fig. 2.9).

British subfamilies and genera:
Dicyrtominae: *Dicyrtoma, Dicyrtomina*
Ptenothricinae: *Ptenothrix*

Fig. 2.8 *Podura aquatica* **(Poduridae) (Photo: Robin Williams)**

Fig. 2.9 Species of *Dicyrtomina* **(Dicyrtomidae) have distinct colour patterns (Photo: Robin Williams)**

Family Katiannidae (1 genus, 7 species);
The status of several species in this family is unclear, but at least two species of *Sminthurinus* are common. There are also some possible accidental imports.

British genus: *Sminthurinus*

Family Sminthuridae (6 genera, 6 species)
Allacma fusca is common in damp habitats and *Sminthurus viridis* is the only other common species in this family. The yellowish green *S. viridis* is a common native of European grasslands and crops such as clover, but when introduced to Australia it became a major pest on 'lucerne grass', or alfalfa (*Medicago sativa*) and is known as the Lucerne flea.

British subfamilies and genera:
Sminthurinae: *Allacma*, *Caprainea*, *Disparrhopalites*, *Lipothrix*, *Sminthurus*
Sphyrothecinae: *Sphyrotheca*

Family Sminthurididae (3 genera, 9 species)
A small family with several common species, especially in the genus *Sminthurides*. The male antennae in this family are modified as claspers to hold the females during mating. As its name suggests, *Sminthurides aquaticus* is often found on the surface of still water and *S. parvulus* is frequently found on damp moss. *Sphaeridia pumilis* is a common soil-dweller.

British genera: *Sminthurides, Sphaeridia, Stenacidia*

REFERENCES

CHRISTIANSEN, K.A., BELLINGER, P. & JANSSENS, F. 2009. Collembola. In: Resh, V.H. & Cardé, R.T. (eds.) *Encyclopedia of insects* (2nd edn.). Academic Press/ Elsevier, San Diego & London, pp 206–10.

HOPKIN, S.P. 1997. *Biology of the springtails (Insecta: Collembola)*. Oxford University Press, Oxford.

HOPKIN, S.P. 2007. *A key to the Collembola (springtails) of Britain and Ireland*. Field Studies Council, Preston Montford (AIDGAP) no. 111: vi + 245 pp.

LUBBOCK, J. 1873. *Monograph of the Collembola and Thysanura*. Ray Society, London.

WEBSITES

http://www.stevehopkin.co.uk
The personal website of the late Steve Hopkin is still accessible. It contains expanded information on distribution, including maps, which supplement the information in his key (Hopkin, 2007).

http://www.roehampton.ac.uk/collembola/index.html
An updated version of the British species is given on Peter Shaw's site.

http://www.collembola.org
The modestly named Checklist of the Collembola of the World, maintained by P.F. Bellinger, K.A. Christiansen and F. Janssens, in fact contains a great deal of other useful information about the group, including many photos.

3

Order Diplura:
the two-tailed or
two-pronged bristletails

11 species in 1 family

Although traditionally included in the 'apterygote' insects, it is now clear that Diplura are hexapodan arthropods within the Class Entognatha, removed from the Insecta along with the Collembola and Protura.

Diplura are small, wingless, unpigmented creatures, living in soil or detritus, often under stones and sometimes under bark. They have no eyes, but can be recognized by the pair of cerci at the end of the abdomen which, in the British species, are long and multisegmented, and easily distinguish the group from other soil-living Hexapoda. Like many soil-dwelling animals they are rarely seen unless specialized collecting techniques are employed; consequently relatively little is known about much of their lifecycles and habits.

Like all the Entognatha, Diplura have indirect fertilization of their eggs. The males deposit stalked spermatophores on the substrate for receptive females to find, and unused spermatophores may be eaten by the males as a kind of recycling. When females collect them, the sperm fertilize the ova internally, and the eggs are then laid in small clusters. In some species the female guards her eggs against predators. Some species seem to live in colonies, but whether there is any real social structure is not yet clear. As in the Collembola moulting continues after sexual maturity is reached but in the Diplura the juvenile stages are epimorphic, i.e. all the main body segments are present in the very earliest stages. Differences between each moult are generally confined to lesser numerical features such as antennal segments and chaetotaxy.

Around 1000 species of Diplura have been described worldwide, belonging to 10 families. There are two main subgroups: the Japygomorpha (or Japygoidea) and Campodeomorpha (or Campodeina), which are easily distinguished by the form of the anal cerci, though there are some small families that do not fit comfortably into this classification. The Japygomorpha have simple unsegmented 'pincers', rather like those of the Dermaptera, whereas the Campodeomorpha have filiform multisegmented cerci. The Japygomorpha include some of the larger predacious species, found predominantly in the tropics and subtropics, and can be over 50 mm long, but the Campodeomorpha are generally smaller (all the British species are between 3 and 5 mm long) and are apparently omnivorous or mainly herbivorous, feeding on living or decaying vegetation. There are other differences between the two suborders, which at one time caused some workers to question the validity of the overall group, but the monophyly of the Diplura is now generally accepted (Koch, 2009).

There are over 300 species in five families in Europe, but the British species are all in the family Campodeidae. Delany (1954) recorded eleven species from Britain, all in the genus *Campodea*; another species was added by Steel (1964), but recent synonymy has apparently put the total back to eleven.

A world overview of the Diplura is given by Koch (2009), and a rather outdated but still useful account and classification is by Paclt (1957).

The Royal Entomological Society Book of British Insects, First Edition. Peter C. Barnard.
© 2011 Royal Entomological Society. Published 2011 by Blackwell Publishing Ltd.

Fig. 3.1 *Campodea staphylinus* (Campodeidae) (Photo: Michel Vuijlsteke, Creative Commons Licence)

SPECIES OF CONSERVATION CONCERN

No species of Diplura have any legal protection and none is on the UKBAP list.

Family Campodeidae (1 genus, 11 species)

Currently 11 species have been recorded from Britain, all in the genus *Campodea*, but there is little information on their status and distribution. In recent years the taxonomy of *Campodea* species has been modified by European workers and it is no longer clear exactly which species occur in Britain, nor how best to identify them. Consequently the group is in urgent need of revision in this country. Delany's (1954) key cannot be relied upon because of these taxonomic changes and a new handbook is needed for this small yet significant group. Species identification is based largely on the chaetotaxy of the body (Fig. 3.1).

British genus: *Campodea*

REFERENCES

DELANY, M.J. 1954. Thysanura and Diplura. *Handbooks for the identification of British insects* 1(2): 7 pp.

KOCH, M. 2009. Diplura. In: Resh, V.H. & Cardé, R.T. (eds.) *Encyclopedia of insects* (2nd edn.). Academic Press/Elsevier, San Diego & London, pp. 281–3.

PACLT, J. 1957. Diplura. *Genera Insectorum* 212: 123 pp.

STEEL, W.O. 1964. Diplura. In: Kloet & Hincks: a check list of British insects. Part 1: small orders and Hemiptera. *Handbooks for the identification of British insects* 11(1): 2.

4 Order Protura: the proturans

c. 15 species in 3 families

In many ways the Protura are one of the strangest groups of arthropods; they are certainly distinctive within the Hexapoda. They are minute, less than 2 mm long, unpigmented, lacking eyes and antennae. Although they can be quite common in soil and leaf-litter, especially in wooded areas, they are rarely seen except by soil specialists and little is known of the biology and life histories of most species. In the few that have been investigated in detail, fungal mycelium seems to form their main food. This lack of knowledge partly results from the fact that until very recently no-one had been able to rear Protura in the laboratory.

Although taxonomically hexapods, the Protura are functionally tetrapods because they normally use only the middle and hind legs for walking; the fore legs have many sensilla and are held above the head or directed forwards, acting as antennae. The head is small and pointed and the entognathous mouthparts are often elongate, which may be an adaptation to feeding on the contents of fungal mycelial cells. As in some myriapods, but uniquely within the Hexapoda, the number of abdominal segments increases during development, a process called anamorphosis: newly hatched prelarvae have eight segments plus a telson, and subsequent instars gain a segment until the adult number of eleven plus a telson is reached. The smaller hind end, or postabdomen, can be telescoped inside the large preabdomen, apparently to aid respiration. Little is known about their reproduction, but males do not produce spermatophores so direct sperm transfer seems likely. Despite all these unique characters, there is evidence that Protura are most closely related to the Diplura.

The group was not discovered until the early years of the 20th century. Being relatively unknown to the layman, they have no common name and usually go by the unimaginative title of proturans. Because of their narrow pointed head, some authors have dubbed them 'coneheads' but this potentially causes confusion with the Orthopteran family Conocephalidae. In German they are appropriately called Beintaster or Beintastler, meaning roughly 'leg-feelers', reflecting the way the front legs are used as antennae.

Although between 12 and 15 species have apparently been recorded from the UK (Tuxen, 1964b) the status of some of these is doubtful and a modern review of the British species is urgently needed. The Protura are represented by around 175 species in Europe in four families and there are about 750 species in 10 families worldwide, making it a relatively small group. Worldwide there are three main suborders, only two of which occur in Europe; the suborder Sinentomata is found only in eastern Asia. Although the monophyly of the Protura is beyond doubt, the relationships between the subgroupings are far from clear.

Inevitably, such small creatures can only be identified using microscopic characters. Most specimens are collected by standard soil-extraction techniques; material is preserved in alcohol and often slide-mounted. There are no recent works covering the British species. The standard world monograph on the group was by Tuxen (1964a), and Nosek (1973) provided keys to the European species, though both works are inevitably out of date. Janetschek's (1970)

The Royal Entomological Society Book of British Insects, First Edition. Peter C. Barnard.
© 2011 Royal Entomological Society. Published 2011 by Blackwell Publishing Ltd.

review is also useful. A recent world overview of the Protura was given by Koch (2009), with a world catalogue by Szeptycki (2007).

Fig. 4.1 *Acerentomon* sp. (Acerentomidae) (Photo: Gregor Znidar, Creative Commons Licence)

HIGHER CLASSIFICATION OF BRITISH PROTURA

Suborder Eosentomata
  Family Eosentomidae (1 genus, 2 species)
Suborder Acerentomata
  Family Acerentomidae (between 2–5 genera, between 6–11 species)
  Family Protentomidae (2 genera, 2 species)

SPECIES OF CONSERVATION CONCERN

No species of Protura have any legal protection and none is on the UKBAP list.

The Families of British Protura

SUBORDER EOSENTOMATA

The Eosentomata are characterized by a number of inconspicuous morphological characters such as the presence of a simple tracheal system for respiration; the most obvious is the retention of toothed mandibles.

Family Eosentomidae (1 genus, 2 species)
The Eosentomidae is the largest family of Protura, with around 340 species worldwide, although the family may not be monophyletic. Only the genus *Eosentomon* is found in the British Isles.

British genus: *Eosentomon*

SUBORDER ACERENTOMATA

The Acerentomata have lost the tracheal system, and the mandibles are reduced to simple elongate stylets.

Family Acerentomidae (between 2–5 genera, between 6–11 species)
Currently it is unclear just how many species in this family are genuinely recorded from the British Isles

and a revision is urgently needed, based on unambiguous voucher material; again the monophyly of this family is in some doubt. A typical species is shown in Fig. 4.1.

British genera: *Acerentomon, Acerentulus* (possibly also *Acerella, Filientomon, Gracilentulus*)

Family Protentomidae (2 genera, 2 species)
As with the other families, the monophyly of the Protentomidae is not certain, and little is known about the two species found in Britain.

British genera: *Protentomon, Proturentomon*

REFERENCES

JANETSCHEK, H. 1970. Ordnung Protura (Beintastler). *Handbuch der Zoologie* 4(2): 72 pp.

KOCH, M. 2009. Protura. In: Resh, V.H. & Cardé, R.T. (eds.) *Encyclopedia of insects* (2nd edn.). Academic Press/ Elsevier, San Diego & London, pp. 855–8.

NOSEK, J. 1973. *The European Protura, their taxonomy, ecology and distribution, with keys for determination.* Muséum d'Histoire Naturelle, Geneva.

SZEPTYCKI, A. 2007. *Catalogue of the world Protura.* Cracow, Poland, 210 pp. [downloadable from http://www.isez. pan.krakow.pl/journals/azc_i/pdf/50B(1)/01.pdf]

TUXEN, S.L. 1964a. *The Protura: a revision of the species of the world, with keys for determination.* Hermann, Paris.

TUXEN, S.L. 1964b. Protura. In: Kloet & Hincks: a check list of British insects. Part 1, small orders and Hemiptera. *Handbooks for the identification of British insects* 11(1): 3.

PART 2 Insecta – 'Apterygota'

The Apterygota, which formerly included the other primitively wingless insects currently placed in the class Entognatha, are now restricted to the two orders Archaeognatha and Zygentoma, which in turn were formerly united as the Thysanura. Despite the superficial similarity of the two groups, it is now clear that they are not closely related, mainly because of fundamental differences in the mouthparts. The Archaeognathous mandibles are monocondylic, having a single articulating point with the head so that the mandible can rotate; Zygentomous mandibles are dicondylic, with two articulating points that restrict the motion to a single plane yet enable the development of a much stronger biting action; this is the type found in all the higher insects, and the Zygentoma may well be the sister group of the Pterygota.

Clearly the 'Apterygota' is not a monophyletic group and is simply retained for convenience in grouping these two orders that superficially resemble each other and have similar life histories.

The Royal Entomological Society Book of British Insects, First Edition. Peter C. Barnard.
© 2011 Royal Entomological Society. Published 2011 by Blackwell Publishing Ltd.

5 Order Archaeognatha or Microcoryphia: the bristletails

7 species in 1 family

The Archaeognatha were formerly grouped with the Zygentoma in a single order, the Thysanura, but it is now clear that the similarities are superficial; the order is characterized by the monocondylic mandibles, having a single attachment to the head. They are active, cylindrical insects up to 18 mm long, with long flagellate antennae, and the abdomen bears a long terminal filament flanked by a pair of cerci, giving a three-tailed appearance. The ectognathous mouthparts are notable for the seven-segmented maxillary palps, which are longer than the legs. Much of the body is covered with flat scales, including the caudal appendages, giving the bristletails a shiny appearance (Fig. 5.1). They have a well-developed pair of eyes, which are conspicuous by being positioned dorsally on the head and touching in the mid-line; there are also three ocelli, the lateral ones being below the main eyes.

Although sometimes found in leaf-litter and detritus, the members of this group often live under or among stones and the genus *Petrobius* is particularly well-known for living on the seashore; they probably feed mainly on algae and lichens.

Like many small primitive Hexapoda the bristletails are preyed upon by a variety of other animals, particularly spiders but they have at least a couple of defence mechanisms that give them some protection. Their scaly bodies make them difficult for a predator to hold, and they have the ability to jump, even in the juvenile stages, by rapidly flicking their abdomen downwards.

Archaeognatha are notable for their elaborate courtship rituals, which are more elaborate than anything seen in the other primitive groups of Hexapoda, and are quite unique amongst the insects. Some males produce silken threads, which they stretch between their parameres and the ground, adding sperm droplets to the thread then guiding the female into a position where she can collect the sperm; in other groups the male places the sperm droplets directly onto the female's ovipositor. The sperm then pass into the female's gonopore, thus fertilization is internal, even though sperm transfer is indirect.

The Archaeognatha is a relatively small group of insects with only two families recognized

Fig. 5.1 The coloured scale patterns on *Dilta* (Machilidae) (Photo: Robin Williams)

The Royal Entomological Society Book of British Insects, First Edition. Peter C. Barnard.
© 2011 Royal Entomological Society. Published 2011 by Blackwell Publishing Ltd.

worldwide, the Machilidae and Meinertellidae, both occurring in Europe, though most Meinertellidae are in the southern hemisphere. There are over 200 species in Europe and around 500 species worldwide. On this scale our fauna of just seven species, all in the family Machilidae, is very small. Most of the British species can still be identified with Delany's (1954) key though there are some name changes that should be noted from the *Fauna Europaea* website. There were also a couple of errors in Delany's key, which do not seem to be widely known: his fig. 4 actually shows *Petrobius brevistylis*, and fig. 11 shows *P. maritimus*, and these changes also need to be made in his couplets 15 and 16. However, a modern revision of this group is needed, as with so many of the 'apterygotes'. For a world overview of Archaeognatha see Sturm (2009); a more detailed account is given by Sturm & Machida (2001).

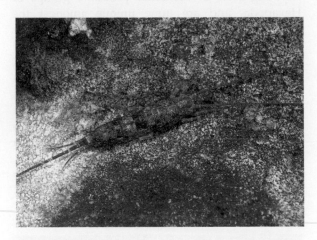

Fig. 5.2 *Petrobius maritimus* **(Machilidae) (Photo: Roger Key)**

HIGHER CLASSIFICATION OF BRITISH ARCHAEOGNATHA

All species are in the family Machilidae.

SPECIES OF CONSERVATION CONCERN

No species of Archaeognatha have any legal protection and none is on the UKBAP list.

Family Machilidae (3 genera, 7 species)
Petrobius maritimus and *P. brevistylis* are both very common on rocky coasts all round Britain (Fig. 5.2) and there are similar, rarer species. Two species of *Dilta* are quite common (see Fig. 5.1) and members of this genus are generally found inland, on heaths and other areas of dense vegetation, although the distribution and status of many species are almost completely unknown.

British genera: *Dilta, Petrobius, Trigoniophthalmus*

REFERENCES

Delany, M.J. 1954. Thysanura and Diplura. *Handbooks for the identification of British insects* 1(2): 7 pp.

Sturm, H. (2009). Archaeognatha (Bristletails). In: Resh, V.H. & Cardé, R.T. (eds.) *Encyclopedia of insects* (2nd edn.). Academic Press/Elsevier, San Diego & London, pp. 48–50.

Sturm, H. & Machida, R. 2001. Archaeognatha. Part 37. *Handbook of zoology*, vol. IV. Walter de Gruyter, Berlin.

6

Order Zygentoma: the silverfish and firebrats

2 species in 1 family

When the Archaeognatha were recognized as a separate order of insects, the remaining group of silverfish, firebrats and their relatives were sometimes still known as the Thysanura, but to avoid confusion with the older broad grouping they are now generally called the Zygentoma. There are many superficial similarities between the two orders, but the dicondylous mandibles clearly place the Zygentoma close to the higher insects.

The general body shape is less cylindrical than in the Archaeognatha, tapering towards the rear, and dorsoventrally flattened. There is a similar covering of scales, and the common name silverfish is derived from this shiny covering and fish-like shape, perhaps coupled with the sinuous body movements made by the fast-running adults. There is a long terminal filament and two lateral cerci, giving a similar three-tailed appearance to the Archaeognatha, but in the Zygentoma the cerci are often held at a wider angle, sometimes even at right-angles to the body. The eyes are reduced (and absent in some non-British species) and there are no ocelli. The long antennae resemble those of the Archaeognatha, but the five-segmented maxillary palps are of normal length. Although silverfish can run fast, they lack the jumping ability of the bristletails, but their slippery scales probably give them similar protection from potential predators. Both British species have a body length of 10–11 mm, excluding the antennae and caudal appendages.

Originally Zygentoma probably had a typical diet of vegetation and plant detritus but several modern species have become adapted to living in human habitations and in some parts of the world they can become pests under suitable conditions. The Common silverfish (*Lepisma saccharina*) can certainly digest cellulose and in houses it finds an abundance of suitable starchy foodstuffs such as cereals, as well as the organic glue and paper of books. Zygentoma are usually more diurnal than Archaeognatha, though they still hide in crevices much of the time, and are also more omnivorous. There is a ritualized courtship involving much antennal contact and repeated movements; eventually the male lays a spermatophore on the substrate which is picked up by the female.

There are around 60 species in two families in Europe, and around 400 species in five families worldwide; half of the world's species are in the family Lepismatidae to which both British species belong. A brief world overview of the Zygentoma is given by Sturm (2009).

HIGHER CLASSIFICATION OF BRITISH ZYGENTOMA

All species are in the family Lepismatidae.

SPECIES OF CONSERVATION CONCERN

No species of Zygentoma have any legal protection and none is on the UKBAP list.

Family Lepismatidae (2 genera, 2 species)
The Silverfish, *Lepisma saccharina*, is a cosmopolitan species that prefers damp habitats. Its flattened

---

The Royal Entomological Society Book of British Insects, First Edition. Peter C. Barnard.
© 2011 Royal Entomological Society. Published 2011 by Blackwell Publishing Ltd.

Fig. 6.1 *Thermobia domestica*, the firebrat
(Lepismatidae) (Photo: Peter Barnard)

body enables it to hide in narrow crevices, which is why it is often common around sinks and baths in houses. The Firebrat, *Thermobia domestica*, is also cosmopolitan (Fig. 6.1) though it is not clear how common or widespread it is in Britain. In warmer regions it lives outside, under rocks, but in more temperate areas such as Britain it is found in warm building such as commercial bakeries and around heating furnaces and boilers. Like *Lepisma* it is an opportunistic feeder, and it is now bred in captivity for feeding to amphibians and reptiles. It has longer antennae than *Lepisma*, and the two species are easily separated with Delany's (1954) key.

British genera: *Lepisma, Thermobia*

REFERENCES

DELANY, M.J. 1954. Thysanura and Diplura. *Handbooks for the identification of British Insects* 1(2): 7 pp.

STURM, H. 2009. Zygentoma. In: Resh, V.H. & Cardé, R.T. (eds.) *Encyclopedia of insects* (2nd edn.). Academic Press/Elsevier, San Diego & London, pp. 1070–2.

PART 3 Palaeoptera

The winged insects, or Pterygota, are undoubtedly a monophyletic group in that wings have developed only once in their evolutionary history; moreover the possession of wings has probably been the single most important factor in the success of insects as a whole. Within this group there are two distinct kinds of wing construction; in the Palaeoptera the wings can only be held upright or outstretched at rest, whereas in the Neoptera the wings can be flexed or folded flat over the body. The difference is simply due to a muscle attached to a small sclerite at the base of the wing. By enabling the twisting and folding of the wings this small muscle has probably enabled insects to evolve the complex wing-cases seen in several groups, as well as other adaptations. It seems clear that the presence of this flexor muscle, along with other related features, means that the Neoptera are a monophyletic group. The situation in the Palaeoptera is less certain, as some characters support the monophyly of the group, while others suggest that it is paraphyletic. In either case, the Palaeoptera are generally considered as the most primitive of the winged insects, comprising the Ephemeroptera and Odonata. The fact that both groups have aquatic juvenile stages is probably the result of convergence, as it is unlikely to be the primitive condition of the insects. The resting positions of these insects, with the wings held upright in the mayflies, held together vertically over the abdomen in the damselflies and held outstretched in the dragonflies, demonstrate their palaeopterous state.

The Royal Entomological Society Book of British Insects, First Edition. Peter C. Barnard.
© 2011 Royal Entomological Society. Published 2011 by Blackwell Publishing Ltd.

7 Order Ephemeroptera: the mayflies or upwing flies

51 species in 10 families

Arguably the most primitive of the winged insects, mayflies are generally recognized by their triangular fore wings held vertically above the body at rest. Adults are a familiar sight by the side of freshwater bodies and they are particularly well known to anglers, who have given names to many of the common species. They are unique among the insects in having a winged, pre-adult stage, known as the subimago (or the 'dun' to anglers), which moults into the adult (or 'spinner') stage.

Adult mayflies have two main functions, to reproduce and to disperse, and morphological diversity between the main families is consequently limited. They all have short antennae and large eyes, especially in the males, and there are no functional mouthparts because neither the subimago nor the adult can feed. The male eyes have a separate dorsal section with enhanced vision for looking upwards during the mating swarm, and this reaches it greatest development in the turbinate eyes of the Baetidae (Fig. 7.1). The large fore wings have a complex network of veins, and the venation is used in the recognition of families and genera; the hind wings are much smaller and are even lost in some families. At the apex of the abdomen are two or three 'tails', formed by a pair of cerci with or without a central caudal filament. The male fore legs are often elongated, and these are used to hold the female during courtship.

Despite the large fore wings, mayflies are generally weak flyers and their dispersal ability is limited to relatively short distances; on occasions they are probably carried further by wind than by active flight. The subimago can fly as well as the final adult stage and its main difference is that the subimaginal wings are translucent, rather than clear, and the body colours are generally duller.

Because the winged stages do not feed, their longevity is necessarily limited. Although many popular texts insist that mayflies live for just a day, the adults of some species do indeed live for only a few hours, whereas other can live for several days. Such a short adult life necessitates a well-coordinated emergence, misleadingly called a 'hatch' by anglers, so that the chances of mating are maximized, and this gives rise to the spectacular swarms seen on some rivers,. Usually it is males that form the mating swarms; the females fly in and are grabbed by the males from underneath. After mating the females drop their eggs into the water.

The greatest part of the lifecycle of Ephemeroptera is spent in the water, and it is in the nymphal stage that the most morphological diversity is seen, as adaptations to different habitats within the freshwater environment. Although they can be found in a wide variety of freshwater types, each species often has a narrow range of tolerance, being particularly sensitive to organic pollution and for this reason the species profiles of Ephemeroptera populations are widely used in the assessment of water quality. As a group they are easily distinguished from most other aquatic insect larvae by the presence of three long caudal filaments and feathery or plate-like gills along the sides of the abdomen. The aquatic stage of the lifecycle can last from a few weeks up to two years, during which time the nymphs can moult up to 50 times in some species. The habitat requirements of a nymph can often be seen at a glance:

The Royal Entomological Society Book of British Insects, First Edition. Peter C. Barnard.
© 2011 Royal Entomological Society. Published 2011 by Blackwell Publishing Ltd.

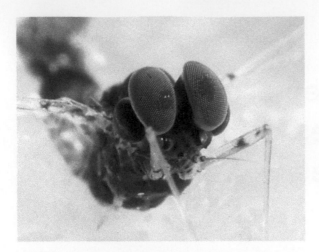

Fig. 7.1 Turbinate eyes in male *Centroptilum* (Baetidae) (Photo: Peter Barnard)

Fig. 7.2 Iron blue, *Baetis* sp.(Baetidae) (Photo: Roger Key)

those that live under stones in fast water have flattened bodies and stout, outstretched legs (many Heptageniidae, see Fig. 7.7); those living in open water or swimming among vegetation have cylindrical bodies and long thin legs (many Baetidae, Fig. 7.3), a form also seen in still-water species; in some small slowly crawling nymphs the second gill forms a cover over all the others (Caenidae); and large burrowing nymphs have strong legs and tusk-like mandibles for digging through the silty substrate and feathery gills kept in constant motion to maintain a current through the burrow (Ephemeridae). Most groups are algal grazers, but a few are carnivorous or even cannibalistic.

Most adult mayflies are diurnal, and are most frequently seen in their mating swarms in the late afternoon, though some are attracted to lights at night. Because of their importance as food for fish, fly-fishermen have a keen interest in this group and they imitate many species with artificial flies. This is why so many species have well-established common names, although there is not always a strict correspondence between the scientifically recognized species and the anglers' names. Some distinct species are 'lumped' under a single anglers' name and some, like *Ephemera danica*, have different names for the different adult stages and even for the two sexes. Many of the common species can be recognized by size and colour patterns in the field, as seen in such books as Goddard (1988, 1991). A useful guide to the British species is by Harker (1989) and the adults can be identified accurately using Elliott and Humpesch (1983), though neither of these books covers the three species recently added to the British list. The nymphs can be identified with the recent key by Elliott and Humpesch (2010). There is a new pictorial guide to the nymphs and adults, including notes on distribution and biology, by Macadam and Bennett (2010). Because of the potential confusion of the name mayfly for the whole group, as well as for the species of *Ephemera*, some authors prefer to call the group upwing (or upwinged) flies, though this is not universally accepted and *Ephemera* may need to be termed the 'true mayfly'.

There are around 330 species of Ephemeroptera in 18 families in Europe, and around 3100 species in 40 families worldwide. A brief world overview of the Ephemeroptera is given by Brittain and Sartori (2009).

HIGHER CLASSIFICATION OF BRITISH EPHEMEROPTERA

Superfamily Baetoidea
    Family Ameletidae (1 genus, 1 species)
    Family Baetidae (4 genera, 14 species)
    Family Siphlonuridae (1 genus, 3 species)
Superfamily Caenoidea
    Family Caenidae (2 genera, 9 species)
Superfamily Ephemerelloidea
    Family Ephemerellidae (2 genera, 2 species)
Superfamily Ephemeroidea
    Family Ephemeridae (1 genus, 3 species)
    Family Potamanthidae (1 genus, 1 species)
Superfamily Heptagenioidea
    Family Arthropleidae (1 genus, 1 species)
    Family Heptageniidae (5 genera, 11 species)
Superfamily Leptophlebioidea
    Family Leptophlebiidae (3 genera, 6 species)

SPECIES OF CONSERVATION CONCERN

No species of British mayflies currently have any legal protection but two are on the UKBAP list: *Nigrobaetis niger* (Baetidae) and *Potamanthus luteus* (Potamanthidae). One species, *Heptagenia longicauda* (Heptageniidae), was removed from the UKBAP list in 2007.

The Families of British Ephemeroptera

SUPERFAMILY BAETOIDEA

Family Ameletidae (1 genus, 1 species)
Our only species, *Ameletus inopinatus*, is confined to upland areas of northern Britain, and requires cool water in which to breed.

British genus: *Ameletus*

Family Baetidae (4 genera, 14 species)
Many species in this family are common and widespread, and therefore have been given common names by fly-fishermen. The male eyes are distinctive in having a separate cylindrical dorsal section, known as turbinate or turret-like, eyes (see Fig. 7.1); this feature is even visible in the nymphal stage. Running water species include several known as Iron blues (see Fig. 7.2) including *Baetis muticus*, Small dark olive (*B. scambus*) and Large dark olive (*B. rhodani*, Fig. 7.3), while the still- or slow-water species include the Pond olive (*Cloeon dipterum*). Note that the genus *Cloeon* is sometimes wrongly spelled '*Chloeon*'. The genus *Baetis* was once split into several smaller genera including *Alainites*, *Labiobaetis* and *Nigrobaetis*, but all the British species are now returned to *Baetis* (Macadam & Bennett, 2010).

Baetis niger (Southern iron blue) is on the UKBAP list.

British genera: *Baetis,   Centroptilum,   Cloeon, Procloeon*

Family Siphlonuridae (1 genus, 3 species)
The three species of *Siphlonurus*, of which *S. lacustris* is the most common, are known as the Summer mayflies.

British genus: *Siphlonurus*

SUPERFAMILY CAENOIDEA

Family Caenidae (2 genera, 9 species)
Apart from *Brachycercus harrisellus*, all the British species belong to the genus *Caenis*, and are known collectively as the Angler's curse. *C. beskidensis* and *C. pseudorivulorum* were recently added to the British list on the basis of nymphal finds; the adults of this genus are not easy to distinguish.

British genera: *Brachycercus, Caenis*

SUPERFAMILY EPHEMERELLOIDEA

Family Ephemerellidae (2 genera, 2 species)
Until recently the Blue-winged olive (often abbreviated to BWO by anglers), *Serratella ignita*, was included in the genus *Ephemerella*; it is a common fast-water species (Fig. 7.4).

British genera: *Ephemerella, Serratella*

Fig. 7.3 Nymph of *Baetis rhodani* (Baetidae) (Photo: Stuart Crofts)

Fig. 7.4 Female Blue-winged olive, *Serratella ignita* (Ephemerellidae) (Photo: Stuart Crofts)

Fig. 7.5 Female mayfly *Ephemera danica* (Ephemeridae) (Photo: Stuart Crofts)

Fig. 7.6 Yellow may dun female, *Heptagenia sulphurea* (Heptageniidae) (Photo: Stuart Crofts)

SUPERFAMILY EPHEMEROIDEA

Family Ephemeridae (1 genus, 3 species)
The three large, well-marked species in *Ephemera* are known as the true mayflies (Fig. 7.5). The commonest species is *E. danica*, which is a common sight on rivers and lakes during the early summer; the subimago is known to anglers as the Green drake, the adult male is simply the Mayfly or Black drake and the adult female is the Grey drake.

British genus: *Ephemera*

Family Potamanthidae (1 genus, 1 species)
Potamanthus luteus (Yellow mayfly) is a rare, bright-yellow species that may now be confined to the River Wye; it is on the UKBAP list.

British genus: *Potamanthus*

SUPERFAMILY HEPTAGENIOIDEA

Family Arthropleidae (1 genus, 1 species)
Our only species, *Arthroplea congener*, is known from just a single, unconfirmed, record, and should almost certainly be removed from the British list.

British genus: *Arthroplea*

Family Heptageniidae (5 genera, 11 species)
A well-known group to fly-fishermen and many of them have common names, though identification to species can be difficult. Common species include the Yellow may dun (*Heptagenia sulphurea*, Fig. 7.6) and the Olive upright (*Rhithrogena semicolorata*). *Electrogena affinis* was recently added to the British list, but *Heptagenia longicauda* may no longer occur here. The nymphs can also be difficult to separate at species level (Fig. 7.7).

Fig. 7.7 Brook dun nymph, *Ecdyonurus* sp. (Heptageniidae) (Photo: Stuart Crofts)

British genera: *Ecdyonurus, Electrogena, Heptagenia, Kageronia, Rhithrogena*

SUPERFAMILY LEPTOPHLEBIOIDEA

Family Leptophlebiidae (3 genera, 6 species)
Another family well known to fly-fishermen, including the Turkey brown (*Paraleptophlebia submarginata*).

British genera: *Habrophlebia, Leptophlebia, Paraleptophlebia*

REFERENCES

BRITTAIN, J.E. & SARTORI, M. 2009. Ephemeroptera (Mayflies). In: Resh, V.H. & Cardé, R.T. (eds.) *Encyclopedia of insects* (2nd edn.). Academic Press/Elsevier, San Diego & London, pp. 328–33.

ELLIOTT, J.M. & HUMPESCH, U.H. 1983. A key to the adults of the British Ephemeroptera. *Scientific Publications of the Freshwater Biological Association* 47: 101 pp.

ELLIOTT, J.M. & HUMPESCH, U.H. 2010. Mayfly larvae (Ephemeroptera) of Britain and Ireland. *Scientific Publications of the Freshwater Biological Association* 66: 152 pp.

GODDARD, J. 1988. *John Goddard's waterside guide*. Unwin Hyman, London.

GODDARD, J. 1991. *Trout flies of Britain and Europe*. A & C Black, London.

HARKER, J. 1989. *Mayflies*. Naturalists' handbooks no. 13. Richmond Publishing, Slough, 56 pp.

MACADAM, C. & BENNETT, C. 2010. *A pictorial guide to British Ephemeroptera*. Field Studies Council, Preston Montford (AIDGAP) OP139.

WEBSITES

http://www.ephemeroptera.pwp.blueyonder.co.uk/
The Ephemeroptera Recording Scheme is the starting point for UK studies on this group; it has links to other related sites around the world.

http://www.insecta.bio.pu.ru/#Ephemeroptera
A world checklist of Ephemeroptera.

8 Order Odonata: the dragonflies and damselflies

49 species in 9 families

Despite being linked with the Ephemeroptera within the Palaeoptera, the Odonata have a very different appearance from the mayflies. The wings are long, narrow and parallel-sided, and the fore and hind wings are the same length. Details of the wing venation are widely used at the higher level classification. There are two distinct suborders of Odonata: the Zygoptera, or damselflies, and the Anisoptera, or dragonflies, although members of the whole order are often known colloquially as dragonflies. A third suborder was formerly recognized, the Anisozygoptera known from just two species in southeast Asia, but these are now considered as Anisoptera.

Adult Odonata are well known to most people; their bright colours, rapid flight and impressive behaviour during territorial defence, courtship and hunting prey always attract attention. In contrast to the Ephemeroptera, dragonflies are very strong and agile fliers and they are fierce and effective aerial predators. They have many morphological features that enhance this predatory role: the large compound eyes give them acute vision (Fig. 8.1); their spiny legs are held forward, basket-like, to catch prey on the wing; and to aid this function the thorax is skewed to bring the leg attachments forward. As a result, the wing attachments are towards the back of the thorax, which enables the wings to keep clear of the legs and the directly attached flight muscles allow the fore and hind wings to beat independently, greatly increasing manoeuvrability in the air (Fig. 8.2). Some adults chase their prey during patrolling flights; others dart out quickly from a favourite perch. These two kinds of flight are also exhibited by the highly territorial males; these either regularly patrol their chosen site for a female to oviposit, or else wait on a perch and fly out quickly to repel competing males.

Some characteristics are concerned with reproduction, which is complex in this group. The first stage is for the male to grab the female's thorax with his legs, and then grasp her behind the head with his anal appendages (see Figs. 8.5 & 8.13). The pair can then continue flying in a straight line, the male in front, in the so-called tandem position. Next the male transfers sperm from the gonopore at the apex of the abdomen to his secondary intromittent organ (often loosely called the penis) on the ventral side of the second abdominal segment. The female then brings the tip of her abdomen round in a circle to pick up the sperm from the male, a position known as the copulatory wheel. The couple may remain united until the female is ready to oviposit, when

Fig. 8.1 Eyes of Southern hawker, *Aeshna cyanea* (Aeshnidae) (Photo: Roger Key)

Fig. 8.2 The independently beating wings of a dragonfly (Photo: Robin Williams)

they either revert to the tandem position (see Fig. 8.12) or else the male remains close by, guarding the female as she lays alone (see Fig. 8.4). This prolonged association obviously increases the male's chances of successfully breeding, and his intromittent organ can even displace sperm that the female has acquired from an earlier mating. The eggs are either laid into plant tissue using a piercing ovipositor, or else are dropped into the water where they often adhere to the outside of aquatic vegetation.

Odonata nymphs are found in a wide range of aquatic habitats, though they are usually restricted to still or slow water. The carnivorous nymphs feed on a wide variety of invertebrates, especially insect larvae, though the larger nymphs can capture small fish. The head has an extensible labium, known as the mask, which can shoot out at high speed to catch almost any prey within range. Most damselfly nymphs have a narrow cylindrical body with three plate-like caudal lamellae at the end of the abdomen; these are gills but can also help with locomotion. The larger dragonfly nymphs often have stouter bodies with five small spines at the abdominal apex; their gills are internal to the rectum and oxygen is extracted from water pumped in and out of the anus. A rapid contraction of the rectal muscles can propel the nymph rapidly by means of a jet of water. Identification of nymphs is not easy and a microscopic examination is needed in some groups. Only the larger nymphs or their exuviae can be identified, using books such as Cham (2007, 2009).

Although up to 50 species have been recorded from the British Isles, only 40 could be considered as resident. Two species (*Coenagrion armatum* and *Oxygastra curtisii*) have become extinct in England during the last half century, one species (*Coenagrion lunulatum*) is known only from Ireland, and many others are regular migrants or vagrants. There is evidence that others have recently become established, so the accepted British total may rise again soon. Every British species has a common name, though some of these were recently contrived for the purpose of species-management plans. A guide to the Irish dragonflies (Nelson & Thompson, 2004) has introduced some entirely different names for some species, based on those appearing in some recent European books, though they have never been widely used in Britain.

There are many useful books on the British species but only a small selection of the very recent ones is included here (Askew, 2004; Brooks, 1997; Corbet & Brooks, 2008; Smallshire & Swash, 2010). As long as the book is up to date, choosing which one to use can be a matter of deciding whether photos or paintings are preferred as illustrations. Because many dragonflies can be recognized in the field, perhaps using binoculars, there is a greater emphasis on watching, rather than catching them although not every species can be identified in this way. This has led to a new approach, very similar to bird-watching, as exemplified by recent books such Dudley, Dudley and Mackay (2007). This increase in dragonfly-watching, enhanced by their complex life-histories and photogenic appearance, has probably made the group second only to butterflies in insect popularity among amateur naturalists.

There are around 130 species in 11 families in Europe, and 5600 species in 33 families worldwide. A brief world overview of the Odonata is given by Tennessen (2009).

HIGHER CLASSIFICATION OF BRITISH ODONATA

Suborder Anisoptera
    Superfamily Aeshnoidea
        Family Aeshnidae (4 genera, 11 species)
        Family Gomphidae (1 genus, 1 species)
    Superfamily Cordulegastroidea
        Family Cordulegastridae (1 genus, 1 species)
    Superfamily Libelluloidea
        Family Corduliidae (3 genera, 4 species)
        Family Libellulidae (5 genera, 14 species)
Suborder Zygoptera
    Superfamily Calopterygoidea
        Family Calopterygidae (1 genus, 2 species)
    Superfamily Coenagrionoidea
        Family Coenagrionidae (6 genera, 13 species)
        Family Platycnemididae (1 genus, 1 species)
    Superfamily Lestoidea
        Family Lestidae (1 genus, 2 species)

SPECIES OF CONSERVATION CONCERN

Two species of Odonata are listed on the Wildlife and Countryside Act 1981: *Aeshna isosceles* (Aeshnidae) and *Coenagrion mercuriale* (Coenagrionidae); both species are also on the UKBAP list. *Oxygastra curtisii* (Corduliidae) is listed on the Bern Convention.

The Families of British Odonata

SUBORDER ANISOPTERA

The dragonflies, in the narrow sense, are distinguished by rather different shapes in the fore and hind wings, the latter having a broader basal section forming an enlarged anal field. The head is about the same width as the thorax, with the eyes usually touching in the mid-line (except the Gomphidae).

SUPERFAMILY AESHNOIDEA

Family Aeshnidae (4 genera, 11 species)
The hawker dragonflies are large and spectacular, well-known for their powerful and almost continuous flight. Their nymphs are found in a wide range of still-water habitats from small ponds to large lakes and canals. None of the British species can be considered as common; *Aeshna juncea* (Common hawker) is perhaps the most widespread, though it is not found everywhere and *A. cyanea* (Southern hawker) is more common in southern England (Fig. 8.3), as is *Anax imperator* (Emperor), another large and striking species (Fig. 8.4). *Hemianax ephippiger* is included on the British list as an increasingly common migrant from north Africa and the Middle East. Other species of *Anax* and *Aeshna* are occasional migrants.

Aeshna isosceles (Norfolk hawker) is listed on the Wildlife and Countryside Act 1981 and is also on the UKBAP list.

British genera: *Aeshna, Anax, Brachytron, Hemianax*

Family Gomphidae (1 genus, 1 species)
Like all the members of this family the Club-tailed dragonfly, *Gomphus vulgatissimus*, is easily distinguished from the other Anisoptera by the widely separated eyes, like those of the Zygoptera. The nymph lives in running water.

British genus: *Gomphus*

SUPERFAMILY CORDULEGASTROIDEA

Family Cordulegastridae (1 genus, 1 species)
This family was originally included within the Aeshnidae. *Cordulegaster* is the only genus in Europe, and the Golden-ringed dragonfly, *Cordulegaster boltonii*, is the sole British species (Fig. 8.5). This distinctive black and yellow species is most common in the north and west of Britain. The nymph is quite different from those of the Aeshnidae in that it breeds in fast-flowing water.

British genus: *Cordulegaster*

Fig. 8.4 Female Emperor dragonfly, *Anax imperator* (Aeshnidae) ovipositing (Photo: Robin Williams)

Fig. 8.3 *Aeshna cyanea* in flight (Aeshnidae) (Photo: Robin Williams)

Fig. 8.5 Mating pair of Golden-ringed dragonfly, *Cordulegaster boltonii* (Cordulegasteridae) (Photo: Roger Key)

Fig. 8.7 Adult Four-spotted chaser, *Libellula quadrimaculata* (Libellulidae) emerging from nymphal skin (Photo: Roger Key)

Fig. 8.6 Libellulid nymph (Photo: Roger Key)

Fig. 8.8 Exuvium of *Libellula quadrimaculata*, with remains of nymphal tracheae (Libellulidae) (Photo: Roger Key)

SUPERFAMILY LIBELLULOIDEA

Family Corduliidae (3 genera, 4 species)
These metallic green Odonata are known as the emerald dragonflies. All are very local and at least two species are considered as threatened. The nymphs of this family all develop in still water such as canals, ponds and bogs.

Oxygastra curtisii (Orange-spotted emerald) is on the British list, but is now considered extinct in this country; it is listed on the Bern Convention.

British genera: *Cordulia, Oxygastra, Somatochlora*

Family Libellulidae (5 genera, 14 species)
This is the largest family of British dragonflies, and also includes many large-sized species and distinctive species (Figs. 8.6, 8.7 & 8.8). The species of *Libellula*, known as chasers, have broad, flattened abdomens, and their nymphs live in still water. The two species of *Orthetrum*, called skimmers, are both rather local in Britain and as their name suggests they fly low over the water; their nymphs develop in either still or slow-flowing water. The status of the species of *Sympetrum*, known as darters

45

Fig. 8.9 Black darter, *Sympetrum danae* (Libellulidae) (Photo: Roger Key)

Fig. 8.10 Female Banded demoiselle, *Calopteryx splendens* (Calopterygidae) (Photo: Roger Key)

(Fig. 8.9) is not always clear. *S. striolatum* (Common darter) is the most widespread resident, but several other species are occasional migrants and some, such as *S. fonscolombei* (Red-veined darter) occasionally breed here. The species in this genus are not easy to separate, especially the females; their nymphs develop in still water. *Pantala flavescens* is included on the British list as a migrant; it is an almost cosmopolitan species known in America as the Globe skimmer, although not commonly seen in Europe.

British genera: *Leucorrhinia, Libellula, Orthetrum, Pantala, Sympetrum*

SUBORDER ZYGOPTERA

The damselflies are generally smaller and more delicate than dragonflies, with a weaker flight. The fore and hind wings are almost identical in shape, and the head is much wider than the thorax with the eyes widely separated.

SUPERFAMILY CALOPTERYGOIDEA

Family Calopterygidae (1 genus, 2 species)
A very distinct group, known as the 'demoiselles', with bright colours on the wings and metallic coloured bodies. The male of *Calopteryx splendens* (Banded demoiselle) has broad blue bands across the centre of each wing, while the females have entirely greenish wings (Fig. 8.10); in *C. virgo* (Beautiful demoiselle) most of the wing is bluish, while the female wings are purplish brown. *C. virgo*

Fig. 8.11 The aptly named Red-eyed damselfly, *Erythromma najas* (Coenagrionidae) (Photo: Roger Key)

nymphs are unusual in preferring fast-running water.

British genus: *Calopteryx*

SUPERFAMILY COENAGRIONOIDEA

Family Coenagrionidae (6 genera, 13 species)
The largest family of damselflies in Britain, containing all the common blue or red species (Figs. 8.11 & 8.12). *Pyrrhosoma nymphula* (Large red damselfly),

Fig. 8.12 Pair of *Erythromma najas* ovipositing (Coenagrionidae) (Photo: Roger Key)

Fig. 8.14 Common blue damselfly, *Enallagma cyathigerum* (Coenagrionidae) (Photo: Peter Barnard)

Fig. 8.13 Mating pair of Azure damselfly, *Coenagrion puella* (Coenagrionidae) (Photo: Roger Key)

Ischnura elegans (Blue-tailed damselfly), *Coenagrion puella* (Azure damselfly, Fig. 8.13) and *Enallagma cyathigerum* (Common blue damselfly, Fig. 8.14) are among the most frequently seen and widespread Odonata in the country. Their nymphs are usually in still or slow water, amongst vegetation. *Coenagrion armatum* has become extinct in Britain; *C. scitulum* (Dainty damselfly) was thought to have suffered the same fate though it seems to be surviving in the Thames estuary, but other species are threatened. On a more positive note, *Erythromma viridulum* (Small red-eyed damselfly) now seems to have become established in this country.

Coenagrion mercuriale (Southern damselfly) is listed on the Wildlife and Countryside Act 1981 and is also on the UKBAP list.

British genera: *Ceriagrion, Coenagrion, Enallagma, Erythromma, Ischnura, Pyrrhosoma*

Fig. 8.15 White-legged damselfly, *Platycnemis pennipes* (Platycnemididae) (Photo: Robin Williams)

Family Platycnemididae (1 genus, 1 species)
The only British species is *Platycnemis pennipes* (White-legged damselfly, Fig. 8.15), a whitish or very pale blue species, whose nymphs live in well-vegetated water bodies.

British genus: *Platycnemis*

SUPERFAMILY LESTOIDEA

Family Lestidae (1 genus, 2 species)
The two British species of *Lestes* are metallic green in colour, giving them the common name of Emerald damselflies and their nymphs prefer still, often

stagnant, water. *L. dryas* (Scarce emerald damselfly) was thought to have become extinct in Britain, but a few surviving colonies are now known. *Sympecma fusca* is an occasional migrant from continental Europe and it was originally thought that *Lestes viridis* (Willow emerald damselfly) and *L. barbarus* (Southern emerald damselfly) had the same status, but these may now be breeding in parts of south and eastern England.

British genus: *Lestes*

REFERENCES

ASKEW, R.R. 2004. *The dragonflies of Europe* (2nd edn.). Harley Books, Colchester.

BROOKS, S.J. 1997. *Field guide to the dragonflies and damselflies of Great Britain and Ireland* (revised edn.). British Wildlife Publishing, Milton on Stour.

CHAM, S. 2007, 2009. *Field guide to the larvae and exuviae of British dragonflies*, 2 vols. British Dragonfly Society, Whittlesey.

CORBET, P.S. & BROOKS, S.J. 2008. *Dragonflies*. HarperCollins, London.

DUDLEY, S., DUDLEY, C. & MACKAY, A. 2007. *Watching British dragonflies*. Subbuteo Natural History Books, Upton Magna.

NELSON, B. & THOMPSON, R. 2004. *The natural history of Ireland's dragonflies*. Ulster Museum, Belfast.

SMALLSHIRE, D. & SWASH, A. 2010. *Britain's dragonflies: a field guide to the damselflies and dragonflies of Britain and Ireland* (2nd edn.). WildGuides, Maidenhead.

TENNESSEN, K.J. 2009. Odonata. In: Resh, V.H. & Cardé, R.T. (eds.) *Encyclopedia of insects* (2nd edn.) Academic Press/Elsevier, San Diego & London, pp. 721–8.

WEBSITES

http://www.dragonflysoc.org.uk
The British Dragonfly Society website is the starting point for information about the British species, latest sightings, recommended books and so on, as well as useful links to other sites worldwide. It also has downloads on creating and managing habitats for dragonflies.

PART 4 Polyneoptera

The higher groups of winged insects, the Neoptera, have the ability to flex their wings so that they can be folded flat over the body. The evolutionary advantage of this was probably to protect the wings from damage while crawling through vegetation or other confined spaces; this led to the development of thicker fore wings as protective covers for the hind wings and abdomen seen in many insect groups and culminating in highly modified elytra in the Coleoptera.

It is relatively easy to recognize members of the Polyneoptera, as they have a very broad, fan-like extension to the hind wings. The largest order is the Orthoptera, and the groups sharing this anal fan were initially known as the Orthopteroidea, but this superorder can no longer be maintained as they share no clear derived characters. There are some better-defined subgroups: for example the Plecoptera, Embioptera and Zoraptera can be united in the Plecopterida; the Orthoptera and Phasmida in the Orthopterida, and the former Blattodea, Mantodea and Isoptera are now united in the single order Dictyoptera. However, this still leaves the Dermaptera with uncertain affinities, so the Polyneoptera is retained as a convenience group until the relationships of all the constituent orders are clarified.

The Royal Entomological Society Book of British Insects, First Edition. Peter C. Barnard.
© 2011 Royal Entomological Society. Published 2011 by Blackwell Publishing Ltd.

9 Order Dermaptera: the earwigs

7 species in 4 families

The terminal, forceps-like cerci make earwigs easily recognizable to most people, and the common *Forficula auricularia* is almost cosmopolitan in distribution, at least in the cooler regions of the world. Although Dermaptera clearly belong in the 'orthopteroid' group, their exact relationships with the Orthoptera and other orders in this group are not clear, although the Dictyoptera may well be the closest.

Earwig antennae are long and simple, the prognathous head is broad with no ocelli, and the thorax and abdomen are generally flattened. The thickened fore wings, or tegmina, are modified as wing-covers, very similar to the elytra of staphylinid Coleoptera, but the hind wings are large, semicircular and membranous (Fig. 9.1). Some species fly readily, others more reluctantly, and the wing-folding mechanism is extremely complex to accommodate the large wings under such small covers (Fig. 9.2). The abdominal cerci are usually straight in females (Fig. 9.3), but more curved and enlarged in males (Fig. 9.4), and they have a multitude of functions. As well as the more obvious functions of defence and prey capture, they also have a role in fighting, courtship and even helping to fold the wings after flight.

Female Dermaptera show parental care of their eggs and newly hatched nymphs, keeping them together and guarding them, at least until after their first moult. Most earwigs are omnivorous, or else are scavengers, and are often regarded as pests when they feed in large numbers on flower buds.

The group is not particularly well studied in Britain, though the existence of the Recording Scheme for the Orthopteroids of the British Isles is helping to improve the situation. Of the seven species currently known in Britain, four are native and three are established aliens or occasional introductions, though the status of some is not clear. All seven species have been given common names, and they are readily identified using Marshall and Haes (1988). There are around 80 species in five families in Europe, and 2000 species in eight families worldwide. A brief world overview of Dermaptera is given by Rankin and Palmer (2009); other useful references are Haes and Harding (1997) and Steinmann (1989).

HIGHER CLASSIFICATION OF BRITISH DERMAPTERA

Suborder Forficulina
    Family Anisolabididae (1 genus, 1 species)
    Family Forficulidae (2 genera, 3 species)
    Family Labiduridae (1 genus, 1 species)
    Family Spongiphoridae (= Labiidae) (2 genera, 2 species)

SPECIES OF CONSERVATION CONCERN

No species have any legal protection and none is on the UKBAP list.

The Royal Entomological Society Book of British Insects, First Edition. Peter C. Barnard.
© 2011 Royal Entomological Society. Published 2011 by Blackwell Publishing Ltd.

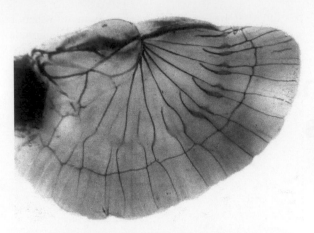

Fig. 9.1 Hind wing of *Forficula*, microscope preparation (Forficulidae) (Photo: Peter Barnard)

Fig. 9.3 Female *Forficula auricularia* (Forficulidae) (Photo: Roger Key)

Fig. 9.2 *Forficula* folding its wings (Forficulidae) (Photo: Colin Rew)

Fig. 9.4 Male *Forficula auricularia* (Forficulidae) (Photo: Colin Rew)

survive in suitable damp and warm habitats under stones.

British genus: *Euboriella*

Family Forficulidae (2 genera, 3 species)
The hind wings of *Forficula auricularia* (Common earwig) are well developed although it flies very rarely (see Figs. 9.1–9.4). In *F. lesnei* (Lesne's earwig) and *Apterygidia media* (Short-winged or Hop-garden earwig) the hind wings are vestigial and the two species are therefore flightless; both are confined to southern England.

British genera: *Apterygidia*, *Forficula*

Family Labiduridae (1 genus, 1 species)
Labidura riparia (Giant or Tawny earwig) is the largest species in Europe, and can be up to 25 mm long or even larger in its native area of southern Europe. It is essentially a coastal species, preferring

The Families of British Dermaptera

Family Anisolabididae (1 genus, 1 species)
The sole British species, *Euboriella annulipes* (Ring-legged earwig) is easily recognized by the complete absence of both fore and hind wings. It is a casual introduction from the Mediterranean region, apparently transported by ship as most records are from sites near ports, and small colonies may be able to

Fig. 9.5 Lesser earwig, *Labia minor* (Spongiphoridae) (Photo: Robin Williams)

sandy habitats; at present it is a rare introduction to Britain but might become established if the climate becomes warmer.

British genus: *Labidura*

Family Spongiphoridae (= Labiidae) (2 genera, 2 species)

Labia minor (Lesser earwig) is apparently quite widespread in Britain, and it flies readily, as well as being attracted to light; it is a very small species at 4–6 mm and it can be easy to mistake for a staphylinid beetle (Fig. 9.5). It lives in warm, moist places like compost heaps or rubbish dumps. *Marava arachidis* is originally a native of the south-east Asian

region that has become cosmopolitan in warmer parts of the world. As an occasional introduction, it has the delightful vernacular name of the Bonehouse earwig, as it became established in the bone stores previously used in glue manufacture.

British genera: *Labia, Marava*

REFERENCES

Haes, E.C.M. & Harding, P.T. 1997. *Atlas of grasshoppers, crickets and allied insects in Britain and Ireland*. ITE Research Publication no. 11. The Stationery Office, London.

Marshall, J.A. & Haes, E.C.M. 1988. *Grasshoppers and allied insects of Great Britain and Ireland*. Harley Books, Colchester.

Rankin, S.M. & Palmer, J.O. 2009. Dermaptera (Earwigs). In: Resh, V.H. & Cardé, R.T. (eds.) *Encyclopedia of insects* (2nd edn.). Academic Press/Elsevier, San Diego & London, pp. 259–61.

Steinmann, H. 1989. *World catalogue of Dermaptera*. Kluwer Academic Publishers, Dordrecht, The Netherlands, 934 pp.

WEBSITES

http://www.orthoptera.org.uk

The starting point for information on the British species is the Orthopteroid recording scheme website.

http://www.earwigs-online.de/

A worldwide site on earwigs, with much information on nomenclature, distributions, etc.

10 Order Dictyoptera: the cockroaches, termites and mantids

11 species in 5 families

The exact relationships of the cockroaches, termites and mantids have been debated for some time, and all three have previously been considered as separate orders. The cockroaches (Blattodea, Blattaria, or even Blattoptera) were recognized as being close to the Mantodea, but now the termites (Isoptera) have been shown to be nested within this clade, called the Dictyoptera, and the three groups can be treated as suborders therein. Termites can thus be regarded as cockroaches that have evolved a highly developed eusocial structure.

Despite this overall grouping into one order, it is convenient to treat each suborder separately, as they have unique characteristics. Cockroaches are almost synonymous with household pests to many people, yet only a small proportion of the world's species deserve this reputation. Most have no impact on humans, and are innocuous omnivores found in a wide range of natural habitats. Nonetheless those few species that have become cosmopolitan pests cause much damage and distress when they invade our conveniently heated homes and commercial premises, though increased hygiene standards can do much to keep them under control. They do not directly transmit diseases, but they can carry a wide range of harmful organisms, which can easily be transmitted to human food. Cockroaches can also cause allergic reactions when present in large numbers; the smell of an established population is both striking and unpleasant. In contrast, our three native cockroaches are small harmless insects, found in a wide variety of habitats including woodland, heathland and the seashore.

Termites would not usually be covered in a book on British insects, but the recent discovery of a single colony necessitates their inclusion. They are essentially a tropical and subtropical group; they all digest cellulose, with or without the aid of intestinal microorganisms, which means that some can be pests on agricultural crops or in timber constructions. Relatively few groups build the famous termite mounds and most build less impressive, though still complex, nests either in rotting wood or in the soil.

Mantids are another easily recognized group, though they can be confused with the family Mantispidae in the Neuroptera (found in continental Europe but not Britain), which have similar raptorial fore legs.

There are nearly 150 species of Blattodea in four families in Europe, and over 4000 species in five families worldwide; 12 species of Isoptera in three families in Europe and over 2500 species in seven families worldwide; 35 species of Mantodea in three families in Europe and nearly 2000 species in eight families worldwide.

HIGHER CLASSIFICATION OF THE BRITISH DICTYOPTERA

Suborder Blattodea
    Family Blaberidae (1 genus, 1 species)
    Family Blattellidae (3 genera, 5 species)
    Family Blattidae (2 genera, 3 species)
Suborder Isoptera
    Family Rhinotermitidae (1 genus, 1 species)
Suborder Mantodea
    Family Mantidae (1 genus, 1 species)

The Royal Entomological Society Book of British Insects, First Edition. Peter C. Barnard.
© 2011 Royal Entomological Society. Published 2011 by Blackwell Publishing Ltd.

SPECIES OF CONSERVATION CONCERN

No species have any legal protection and none is on the UKBAP list.

The Families of British Dictyoptera

SUBORDER BLATTODEA

All the British cockroaches, including the common introductions, can be identified using Marshall and Haes (1988); their distribution is covered by Haes and Harding (1997). There is a Blattodea Culture Group for those who regard cockroaches as pets, rather than pests (http://blattodea-culture-group.org). For a world overview of cockroaches see Cochran (2009).

Family Blaberidae (1 genus, 1 species)
The only species regularly found in Britain is the root-feeding *Pycnoscelus surinamensis* (Surinam cockroach). It is an occasional import on house-plants and may breed in commercial glasshouses. *Nauphoeta cinerea* (Cinereous cockroach) is a very occasional introduction, though some may escape from colonies bred as reptile food.

British genus: *Pycnoscelus*

Family Blattellidae (3 genera, 5 species)
This family contains the only three native species of cockroaches in Britain, all in the genus *Ectobius*. They are mainly confined to southern England, but can be locally common. The males of all three species can fly but among the females only *E. pallidus* (Tawny cockroach) has fully developed wings, which are reduced in *E. lapponicus* (Dusky cockroach, Fig. 10.1) and vestigial in *E. panzeri* (Lesser cockroach). Found in a wide variety of habitats, they all lay overwintering oothecae.

The so-called German cockroach, *Blattella germanica*, is a cosmopolitan species that can only survive outside in Britain in warm places such as landfill sites. It is very common in heated buildings where it feeds on a wide variety of materials. *Supella longipalpa* (Brown-banded Cockroach) is less often seen, but also lives in warm buildings.

British subfamilies and genera:
   Blattellinae: *Blattella*, *Supella*
   Ectobiinae: *Ectobius*

Family Blattidae (2 genera, 3 species)
This family contains two widely known pest species, *Blatta orientalis* (Common or Oriental cockroach)

Fig. 10.1 A native cockroach, *Ectobius lapponicus* (Blattellidae) (Photo: Roger Key)

Fig. 10.2 An introduced cockroach, *Pelmatosilpha larifuga* (Blattidae) (Photo: Roger Key)

and *Periplaneta americana* (American or Ship cockroach). Both are cosmopolitan, and can be a serious problem as they are difficult to eradicate. *B. orientalis* can form huge colonies that are often out of sight, hidden in cavities and crevices during daylight, but appearing at night to feed. *P. americana* is less of a problem in Britain than elsewhere in the world, because it requires consistently higher temperatures to thrive. There are scattered records of *P. australasiae* (Australian cockroach), but this species does not usually live in houses, preferring commercial glasshouses and warehouses. *P. brunnea* (Brown cockroach) is an occasional introduction. There are many accidental introductions of exotic species such as *Pelmatosilpha larifuga*, the Vagabond cockroach (Fig. 10.2).

Fig. 10.3 Praying Mantis, *Mantis religiosa* (Mantidae) (Photo: Roger Key)

British genera: *Blatta*, *Periplaneta*

SUBORDER ISOPTERA

Family Rhinotermitidae (1 genus, 1 species)

A colony of termites was found in a house in north Devon in 1994 (see Ridout, 2000: 73–5); they are thought to have been accidentally introduced from the Canary Islands. The species was eventually identified as *Reticulitermes grassei*, which is endemic to the Iberian peninsula, although several colonies have been found further north in France. An intensive eradication programme was thought to have destroyed the Devon colony, but recent news articles have reported that a few individuals have been found at the same site, emphasizing that these subterranean species are very persistent.

A brief overview of Isoptera is given by Lewis (2009).

SUBORDER MANTODEA

Family Mantidae (1 genus, 1 species)

The Praying mantis, *Mantis religiosa*, is native to southern and central Europe (Fig. 10.3). There are very few records from Britain, but the species is likely to be collected as a novelty by tourists and it is also bred in captivity by some enthusiasts (see http://mantodea.myspecies.info). There is always the likelihood that escaped specimens will be seen, and establishment becomes a possibility if the British climate consistently warms up. For a brief world overview of Mantodea see Hurd (2009).

REFERENCES

COCHRAN, D.G. 2009. Blattodea (Cockroaches). In: Resh, V.H. & Cardé, R.T. (eds.) *Encyclopedia of insects* (2nd edn.). Academic Press/Elsevier, San Diego & London, pp. 108–12.

HAES, E.C.M. & HARDING, P.T. 1997. *Atlas of grasshoppers, crickets and allied insects in Britain and Ireland*. ITE Research Publication no. 11. The Stationery Office, London.

HURD, L.E. 2009. Mantodea (Praying Mantids). In: Resh, V.H. & Cardé, R.T. (eds.) *Encyclopedia of insects* (2nd edn.). Academic Press/Elsevier, San Diego & London, pp. 597–9.

LEWIS, V.R. 2009. Isoptera (Termites). In: Resh, V.H. & Cardé, R.T. (eds.) *Encyclopedia of insects* (2nd edn.). Academic Press/Elsevier, San Diego & London, pp. 535–8.

MARSHALL, J.A. & HAES, E.C.M. 1988. *Grasshoppers and allied insects of Great Britain and Ireland*. Harley Books, Colchester.

RIDOUT, B. 2000. *Timber decay in buildings*. Spon, London.

WEBSITES

http://www.orthoptera.org.uk/
The Orthopteroid recording scheme website contains details of the British Dictyoptera.

http://blattodea-culture-group.org/
The Blattodea Culture Group.

http://mantodea.myspecies.info/
The Mantis Study Group.

11

Order Orthoptera: the grasshoppers, crickets and bush-crickets

33 species in 10 families

Even in their restricted sense, excluding the Dictyoptera, Mantodea and Phasmida, the Orthoptera still form the largest of the Polyneopteran or 'orthopteroid' orders; they are probably most closely related to the Phasmida. Included within the group are the locusts, which are among the most devastating crop pests in many parts of the world; hence the Orthoptera as a whole are relatively well studied, at least by economic entomologists.

Members of the Orthoptera are generally medium to large insects, with long hind-legs that are often modified for jumping; the enlarged hind femora have a distinctive 'herring-bone' pattern of musculature (see Fig. 11.2). Their mandibulate mouthparts are often cited as the classic example of the biting and chewing type (see Fig. 11.10). The fore wings are generally narrow but thickened to form wing-covers that protect the much broader and more delicate hind wings when folded at rest. In several families the male fore wings form stridulatory organs, sometimes in combination with the hind legs. Females in some families have highly developed external ovipositors with which they lay their eggs in crevices in bark, plant stems or in the soil. Those groups that stridulate have a variety of hearing organs, either at the base of the abdomen or on the front tibiae.

Being relatively large insects, with distinctive colour patterns and well-defined gross morphological characters, many species can be identified from photos, or at least with a hand-lens, but in some families there are many colour varieties which can be confusing (Figs. 11.1 & 11.2). In some groups the stridulatory calls, known as songs, can be useful for identification especially if the specimen cannot be seen clearly, and the experienced orthopterist can recognize many species in the field by their songs alone, rather like ornithologists.

The classic work on this group was by Ragge (1965) though inevitably this is now somewhat outdated. Nearly all the British species can be identified using Marshall and Haes (1990) apart from one species recently discovered in Britain, *Meconema meridionale* (Meconematidae), and a few migrant species that have now become established; their distributions are covered by Haes and Harding (1997). To identify the often characteristic songs, a set of cassette tapes was issued to accompany this book by Burton and Ragge (1987). For a more detailed account of the songs of western European species see the book and two CDs by Ragge and Reynolds (1988a, 1988b). A pocket guide to the group was provided by Pinchen (2006) and a brief fold-out guide by Marshall and Ovenden (1999). Evans and Edmondson (2007) is the most up-to-date photographic guide, and is particularly useful for illustrating many colour varieties of some species. All the native species have well-established common names. However, there are several introduced species, migrants, etc., and some other species are found only in the Channel Islands.

There are around 1000 species in 15 families in Europe, with 24,000 species in 29 families worldwide. A world overview of the Orthoptera is provided by Ingrisch and Rentz (2009), and a readable and well-illustrated account of the whole group is given by Preston-Mafham (1990).

The Royal Entomological Society Book of British Insects, First Edition. Peter C. Barnard.
© 2011 Royal Entomological Society. Published 2011 by Blackwell Publishing Ltd.

HIGHER CLASSIFICATION OF BRITISH ORTHOPTERA

Suborder Caelifera
  Superfamily Acridoidea
    Family Acrididae (6 genera, 11 species)
  Superfamily Tetrigoidea
    Family Tetrigidae (1 genus, 3 species)
Suborder Ensifera
  Superfamily Grylloidea
    Family Gryllidae (3 genera, 3 species)
    Family Gryllotalpidae (1 genus, 1 species)
    Family Mogoplistidae (1 genus, 1 species)
  Superfamily Rhaphidophoroidea
    Family Rhaphidophoridae (1 genus, 1 species)
  Superfamily Tettigonioidea
    Family Conocephalidae (2 genera, 3 species)
    Family Meconematidae (1 genus, 2 species)
    Family Phaneropteridae (2 genera, 2 species)
    Family Tettigoniidae (5 genera, 6 species)

Fig. 11.1 Lesser marsh grasshopper, *Chorthippus albomarginatus*, green colour form (Acrididae) (Photo: Roger Key)

SPECIES OF CONSERVATION CONCERN

Three species of Orthoptera are listed on the Wildlife and Countryside Act 1981: *Decticus verrucivorus* (Tettigoniidae), *Gryllotalpa gryllotalpa* (Gryllotalpidae) and *Gryllus campestris* (Gryllidae). All three are also on the UKBAP list, as is *Stethophyma grossum* (Acrididae).

Fig. 11.2 *Chorthippus albomarginatus*, brown colour form (Acrididae) (Photo: Roger Key)

The Families of British Orthoptera

SUBORDER CAELIFERA

This suborder is characterized by having short, thick antennae; a very short external ovipositor; stridulatory organs (when present) on the hind legs and fore wings; and hearing organs (when present) on the abdomen.

SUPERFAMILY ACRIDOIDEA

Family Acrididae (6 genera, 11 species)
The grasshoppers are perhaps the best known group of Orthoptera although to many people they are more often heard than seen; their cryptic colouring and habit of concealing themselves amongst vegetation can make them hard to spot, even at close range (Figs. 11.1–11.3). Acrididae produce their songs by rubbing pegs on the hind femur against a thickened vein on the fore wing. The hearing organs are on the sides of the abdomen.

Fig. 11.3 Nymph of *Chorthippus* sp., purple form (Acrididae) (Photo: Roger Key)

Fig. 11.4 Large marsh grasshopper, *Stethophyma grossum* (Acrididae) (Photo: Robin Williams)

Fig. 11.5 Common groundhopper, *Tetrix undulatus* (Tetrigidae) (Photo: Roger Key)

Grasshoppers are active during the day, especially in warm sunshine, and they feed on various plants. The genus *Oedipoda* is found only in the Channel Islands; on the UK mainland the only member of the Oedipodinae is *Stethophyma grossum* (Large marsh grasshopper, Fig. 11.4), which is on the UKBAP list.

Among the commonest grasshoppers in Britain are the usually flightless *Chorthippus parallelus* (Meadow grasshopper), *Chorthippus brunneus* (Field grasshopper) and *Omocestus viridulus* (Common green grasshopper).

British subfamilies and genera:

Gomphocerinae: *Chorthippus, Gomphocerippus, Myrmeleotettix, Omocestus, Stenobothrus*

Oedipodinae: *Stethophyma*

SUPERFAMILY TETRIGOIDEA

Family Tetrigidae (1 genus, 3 species)
The groundhoppers, or grouse-locusts, are similar to the Acrididae in having short antennae and a very short external ovipositor, but they are distinguished by the long, backward-projecting pronotum that covers the abdomen (Fig. 11.5). They do not produce a song, and therefore have no stridulatory or hearing organs. Like the grasshoppers they are most active during warm sunshine, and feed mainly on mosses, lichens and algae.

Two of the three British species of *Tetrix* are reasonably common, but *T. ceperoi* (Cepero's groundhopper) is found at only a few sites in southern England and Wales.

British genus: *Tetrix*

SUBORDER ENSIFERA

This suborder is characterized by the possession of long thread-like antennae; a long sword-shaped ovipositor; the male stridulatory organs are modified areas of the fore wings; and hearing organs (when present) are on the fore tibiae.

SUPERFAMILY GRYLLOIDEA

A world overview of crickets in the broad sense is given by Alexander (2009).

Family Gryllidae (3 genera, 3 species)
The true crickets are well known for their chirping song. Their thread-like antennae are generally longer than the body, and they have a flattened appearance because the wings are folded flat over the body. Their ovipositor is unique in being very thin and needle like, for laying eggs in the soil. The males produce their persistent chirping by rubbing the fore wings together, and these have a modified area known as the 'harp', which amplifies the sound. Most crickets are omnivorous and may be either diurnal or nocturnal.

Fig. 11.6 House cricket, *Acheta domesticus* (Gryllidae) (Photo: Roger Key)

Fig. 11.7 Wood cricket, *Nemobius sylvestris* (Gryllidae) (Photo: Roger Key)

Acheta domesticus (House cricket, Fig. 11.6) is commonly associated with humans in many parts of the world; it is also widely bred in captivity as food for pet reptiles. Of the two native crickets in Britain, *Nemobius sylvestris* (Wood cricket) is a rare species found in a few southern English counties (Fig. 11.7). Even scarcer is *Gryllus campestris* (Field cricket), which is listed on the Wildlife and Countryside Act 1981 and on the UKBAP list; efforts are being made to reintroduce it to the wild from captive-bred stock. A few other species are seen as occasional vagrants, including escapes from insect farms that produce live food for reptiles, an example being *Gryllus bimaculatus* (Fig. 11.8).

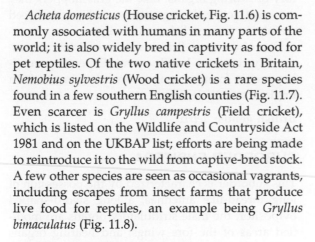

Fig. 11.8 *Gryllus bimaculatus* (Gryllidae) (Photo: Peter Barnard)

British subfamilies and genera:
Gryllinae: *Acheta*, *Gryllus*
Nemobiinae: *Nemobius*

Family Gryllotalpidae (1 genus, 1 species)
The mole-crickets have large, flattened fore legs, modified for burrowing; the antennae are thread-like though shorter than in the Gryllidae; the females have no external ovipositor. The eggs are laid in an underground burrow, and the female guards the eggs and young nymphs for some time.

The only British species is *Gryllotalpa gryllotalpa*, which is listed on the Wildlife and Countryside Act 1981, and is also on the UKBAP list (Fig. 11.9). Historically this seems to have been a common species in Britain, even reported as damaging root crops though it is essentially carnivorous, but it is now very scarce and any sightings may be of vagrants rather than residents.

Fig. 11.9 Mole cricket, *Gryllotalpa gryllotalpa* (Gryllotalpidae) (Photo: Roger Key)

British genus: *Gryllotalpa*

Family Mogoplistidae (1 genus, 1 species)

Some authors regard the scale-crickets as a sub-family of the Gryllidae; as their name suggests they are covered in small scales.

The sole British species is the Scaly cricket *Pseudomogoplistes vicentae* (previously misidentified as *Ps. squamiger*), known only from a few shingle beaches in southwest England and the Channel Islands, though it can be locally abundant. It is wingless and therefore cannot stridulate.

British genus: *Pseudomogoplistes*

SUPERFAMILY RHAPHIDOPHOROIDEA

Family Rhaphidophoridae (1 genus, 1 species)

The camel-crickets were formerly included within the superfamily Tettigonioidea. They are wingless, with no stridulatory or hearing organs.

The sole British species is *Tachycines asynamorus* (Greenhouse camel-cricket), easily distinguished by its extremely long legs. Although suspected of being an occasional plant pest it is mainly carnivorous. It survives only in heated greenhouses and similar situations, but is often eradicated before it can breed.

British genus: *Tachycines*

SUPERFAMILY TETTIGONIOIDEA

These are known generally as bush-crickets, but older texts may refer to them as long-horned grasshoppers.

Family Conocephalidae (2 genera, 3 species)

The cone-heads are regarded by some authors as a subfamily of the Tettigoniidae; their name comes from the acute angle between the top of the head and the frons, though several Tettigoniidae look rather similar. All three British species are uncommon and confined to southern England, though there is evidence that *Conocephalus discolor* (Long-winged conehead, Fig. 11.10) may be expanding its range.

British genera: *Conocephalus, Ruspolia*

Family Meconematidae (1 genus, 2 species)

The oak bush-crickets are regarded by some authors as a subfamily of the Tettigoniidae. The females have a long ovipositor that is slightly upturned (Fig. 11.11). The males do not stridulate, but drum on substrates such as leaves with a hind leg; they are

Fig. 11.10 Long-winged conehead, *Conocephalus discolor* (Conocephalidae) (Photo: Roger Key)

Fig. 11.11 Female oak bush-cricket, *Meconema thalassinum* (Meconematidae) (Photo: Roger Key)

mainly nocturnal and are attracted to lights. *Meconema thalassinum* (Oak bush-cricket) was long assumed to be the only British species, but in 2001 the short-winged *M. meridionale* (Southern oak bush-cricket) was found in southern England and it seems to be increasing its range. Both species are generally regarded as being carnivorous, but not exclusively (Fig. 11.12).

British genus: *Meconema*

Family Phaneropteridae (2 genera, 2 species)

This group of bush-crickets is treated as a subfamily of the Tettigoniidae by some authors but can be distinguished by the short and broad ovipositor, which is strongly upturned; both British species are covered in small brown spots.

Phaneroptera falcata (Sickle-bearing bush-cricket) was previously known only as an occasional

Fig. 11.12 Male *Meconema thalassinum* feeding on rotten apple (Meconematidae) (Photo: Peter Barnard)

Fig. 11.13 Speckled bush-cricket, *Leptophyes punctatissima* (Phaneropteridae) (Photo: Roger Key)

Fig. 11.14 Great green bush-cricket, *Tettigonia viridissima*, showing long wings (Tettigoniidae) (Photo: Peter Barnard)

Fig. 11.15 Roesel's bush-cricket, *Metrioptera roeselii* (Tettigoniidae) (Photo: Roger Key)

migrant, but a breeding colony now seems established in Sussex; recent records suggest that this continental species may be increasing its range. *Leptophyes punctatissima* (Speckled bush-cricket, Fig. 11.13) has very reduced wings; it is widespread through much of England, but scarcer towards the north.

British genera: *Leptophyes, Phaneroptera*

Family Tettigoniidae (5 genera, 6 species)
This is the main family of bush-crickets, which in its restricted sense now contains just six British species. Many, such as *Tettigonia viridissima* (Great green bush-cricket, Fig. 11.14) are confined to southern England, although at least one species, *Metrioptera roeselii* (Roesel's bush-cricket, Fig. 11.15) is expanding its range; *Pholidoptera griseoaptera* (Dark bush-cricket) is more widespread. Newly hatched tettigoniids can often be found in flowers (Fig. 11.16).

Decticus verrucivorus (Wart-biter) is listed on the Wildlife and Countryside Act 1981, and is also on the UKBAP list.

British subfamilies and genera:
Decticinae: *Decticus, Metrioptera, Pholidoptera, Platycleis*
Tettigoniinae: *Tettigonia.*

Fig. 11.16 Newly hatched Tettigoniid nymph on flower (Photo: Peter Barnard)

REFERENCES

ALEXANDER, R.D. 2009. Crickets. In: Resh, V.H. & Cardé, R.T. (eds.) *Encyclopedia of insects* (2nd edn.). Academic Press/Elsevier, San Diego & London, pp. 232–6.

BURTON, J.F. & RAGGE, D.R. 1987. *Sound guide to the grasshoppers and allied insects of Great Britain and Ireland* [Cassette]. Harley Books, Colchester.

EVANS, M. & EDMONDSON, R. 2007. *A photographic guide to the grasshoppers & crickets of Britain & Ireland*. WildGuides UK.

HAES, E.C.M. & HARDING, P.T. 1997. *Atlas of grasshoppers, crickets and allied insects in Britain and Ireland*. ITE Research Publication no. 11. The Stationery Office, London.

INGRISCH, S. & RENTZ, D.C.F. 2009. Orthoptera. In: Resh, V.H. & Cardé, R.T. (eds.) *Encyclopedia of insects* (2nd edn.). Academic Press/Elsevier, San Diego & London, pp. 732–43.

MARSHALL, J.A. & HAES, E.C.M. 1990. *Grasshoppers and allied insects of Great Britain and Ireland* (revised edn.). Harley Books, Colchester.

MARSHALL, J.A. & OVENDEN, D. 1999. *Guide to British grasshoppers and allied Insects*. Field Studies Council, Preston Montford, OP54.

PINCHEN, B.J. 2006. *Pocket guide to the grasshoppers, crickets and allied insects of Britain and Ireland*. Forficula Books, Lymington.

PRESTON-MAFHAM, K. 1990. *Grasshoppers and mantids of the world*. Blandford, London.

RAGGE, D.R. 1965. *Grasshoppers, crickets and cockroaches of the British Isles*. Warne, London.

RAGGE, D.R. & REYNOLDS, W.J. 1998a. *The songs of the grasshoppers and crickets of western Europe*. Harley Books, Colchester.

RAGGE, D.R. & REYNOLDS, W.J. 1998b. *A sound guide to the grasshoppers and crickets of western Europe* [Two CDs]. Harley Books, Colchester.

WEBSITES

http://www.orthoptera.org.uk/default.aspx
Orthopteroid recording scheme website.

http://orthoptera.speciesfile.org/HomePage.aspx
A taxonomic database of world Orthoptera.

http://www.brc.ac.uk/DBIF/homepage.aspx
A general site containing a database of the foodplants of British insects.

12

Order Phasmida: the stick-insects

4 species in 2 families

The Phasmida, also known as Phasmatodea or Phasmatoptera, are well known for their mimicry of twigs or leaves, and even their eggs resemble plant seeds, sometimes with sculptured surfaces (Fig. 12.1). They are clearly related to the Orthoptera; one group of Tettigoniidae in Australia shows a remarkable similarity to the stick-insects, but they also have many unique morphological and biological characters. They are generally confined to warmer parts of the world, with several species in southern Europe for example, so the few that can survive in Britain are confined to the south-west corners of England and Ireland.

The British species cannot be confused with any other group, except perhaps the aquatic bug *Ranatra linearis*, though the latter has a long terminal 'tail' and a piercing rostrum, unlike the mandibulate mouthparts of the stick insects. They are all wingless and long legged, with narrow bodies, and are basically nocturnal. Resting motionless on their foodplants during the day they can be very hard to see in the wild and are probably overlooked by the casual observer. Three of the four British species were introduced on plants from New Zealand, two in the early 20th century and the other apparently more recently. In their native country males are known in some species, but all the British populations reproduce parthenogenetically. It is not always easy to identify this group to species level, especially in the smaller instars, and two of the species were confused until relatively recently. Identification

can be made using the key by Marshall and Haes (1990) though this does not include *Bacillus*, or with the Orthopteroid Recording Scheme website. The distribution records in Haes and Harding (1997) have been updated on some of the websites listed below.

The Phasmida are a popular group of insects for breeding in captivity and many species are available commercially. Although some need specialist care, many are easy to rear on widely available British plants such as *Rubus* (bramble). One of their intriguing characteristics is the ease with which lost legs can be regenerated at successive moults (Fig. 12.2). More information can be found on websites such as the Phasmid Study Group; the two books by Brock (1991, 1999) give useful background information on the whole group.

There are 15 species in three families in Europe, with over 3000 species in 13 families worldwide.

HIGHER CLASSIFICATION OF BRITISH PHASMIDA

Family Bacillidae (1 genus, 1 species)
Family Phasmatidae (2 genera, 3 species)

SPECIES OF CONSERVATION CONCERN

No species have any legal protection and none is on the UKBAP list.

The Royal Entomological Society Book of British Insects, First Edition. Peter C. Barnard.
© 2011 Royal Entomological Society. Published 2011 by Blackwell Publishing Ltd.

Fig. 12.1 Phasmid eggs (Bacillidae) (Photo: Peter Barnard)

Fig. 12.3 Mediterranean stick-insect (*Bacillus rossius*, Bacillidae) is found in both brown and green forms (Photo: Peter Barnard)

Fig. 12.2 Shortened front leg of stick-insect showing regeneration after loss (Bacillidae) (Photo: Peter Barnard)

The Families of British Phasmida

Family Bacillidae (1 genus, 1 species)
The Mediterranean stick-insect, *Bacillus rossius*, survives on Tresco in the Isles of Scilly. As its common name suggests it is a native of southern Europe and north Africa along with other species in the same genus. Like the other species in Britain it feeds on Rosaceae such as *Rubus*. Superficially similar to the British species of Phasmatidae, *B. rossius* can be easily distinguished by its very short antennae, which are not much longer than the head; like several other species it can exist in green or brown forms (Fig. 12.3).

British genus: *Bacillus*

Family Phasmatidae (2 genera, 3 species)
Clitarchus hookeri (Smooth stick-insect) and *Acanthoxyla geisovii* (Prickly stick-insect) are known from Devon and Cornwall, as is *A. inermis* (Unarmed stick-insect), which has also been recorded in southwest Ireland. All three species were introduced from New Zealand. *A. inermis* was not recognized until the 1980s and it closely resembles *C. hookeri*. All can feed on a wide variety of rosaceous plants, with many kept on *Rubus* (bramble) in captivity. None of these species is found elsewhere in Europe.

Many other exotic species are kept in laboratories or as pets and may occasionally be seen as escapes though they cannot reproduce outdoors; for example *Carausius morosus* (Laboratory stick-insect) is occasionally found in the summer months on *Ligustrum* (privet) hedges.

British genera: *Acanthoxyla, Clitarchus*

REFERENCES

BROCK, P.D. 1991. *Stick insects of Britain, Europe and the Mediterranean*. Fitzgerald Publishing, London.

BROCK, P.D. 1999. *The amazing world of stick and leaf-insects*. Amateur Entomologists' Society, London.

HAES, E.C.M. & HARDING, P.T. 1997. *Atlas of grasshoppers, crickets and allied insects in Britain and Ireland*. ITE Research Publication no. 11. The Stationery Office, London.

MARSHALL, J.A. & HAES, E.C.M. 1990. *Grasshoppers and allied insects of Great Britain and Ireland* (revised edn.). Harley Books, Colchester.

WEBSITES

http://www.orthoptera.org.uk/default.aspx
Orthopteroid recording scheme website.

http://phasmida.speciesfile.org
A taxonomic database of the world species.

http://www.erccis.co.uk/
The site of the Environmental Records Centre for Cornwall and the Isles of Scilly has a downloadable information sheet on the established stick-insects in the area.

http://phasmid-study-group.org/
The phasmid study group.

13 Order Plecoptera: the stoneflies

34 species in 7 families

Although the Plecoptera clearly belong in the 'orthopteroid' group of orders, they are probably most closely related to the non-British Zoraptera and Embioptera, these three orders sometimes being placed in the superorder Plecopterida. However, they are apparently an early offshoot because their nymphs are always aquatic, unlike the entirely terrestrial nymphs of the other two orders.

The fore wings of stoneflies are narrow, and in several families there are three parallel veins joined by rows of cross-veins, giving a characteristic ladder-like appearance. Although the fore wings are not specially modified as wing covers all the wings tend to be rather shiny and leathery, giving rise to the anglers' alternative name of hard-winged flies. The adults sometimes retain the pair of long cerci always seen in the nymphs, but in some families (Taeniopterygidae, Nemouridae and Leuctridae) they are lost when the adult emerges. Adults feed very little, if at all, though some species are known to graze on lichens and algae; they are short-lived and are reluctant to fly, often found running around amongst stones and vegetation by the side of rivers. Their flight is slow and rather erratic, and in some families brachyptery is common, especially at higher altitudes (see Fig. 13.3). Stoneflies are generally nocturnal, and males communicate with females by drumming their abdomen on the substrate in a specific pattern of pulses. Although most species emerge in summer, and may be on the wing in autumn, there are several common species that are seen in early spring, even in February.

Plecopteran nymphs, known to some anglers as 'creepers', are usually confined to fast running water because they require high oxygen concentrations. They are therefore more commonly found in the upland areas of northern and western Britain, although some members of the Nemouridae are better adapted to slower currents and can be common in southern England. All are sensitive to organic pollution and the group as a whole is therefore important in water quality assessment. Nymphs are always herbivorous or detrivorous in early life, but in some families the later nymphs become more omnivorous, preying on small invertebrates. Most species are univoltine although the larger species can take two or three years to reach maturity, with up to twenty moults.

Identification of the British species is relatively straightforward; most nymphs can be identified by clear morphological, though microscopic, characters. A few adults have distinctive colour patterns although most are uniformly brown, but the genitalia in both sexes are specifically distinct. The only detailed identification key is by Hynes (1977), but a useful pictorial guide is provided by Pryce, Macadam and Brooks (2007), with some additional references by Brooks (1999). The better-known species are described in angling guides such as Goddard (1988, 1991). Many of the more common species have been given vernacular names by fly-fishermen; Pryce et al. (2007) have now assigned all the British species such names for consistency. The brachyptery seen in some males has given rise to confusion about the identity of certain species; for example, the British forms of *Perla bipunctata* and

The Royal Entomological Society Book of British Insects, First Edition. Peter C. Barnard.
© 2011 Royal Entomological Society. Published 2011 by Blackwell Publishing Ltd.

Taeniopteryx nebulosa are short-winged, while the populations in continental Europe vary. This has led some workers to suggest that the British populations are genetically distinct and may be different taxa, a problem that needs more investigation.

There are over 400 species in 7 families in Europe; over 3000 species in 16 families worldwide. A brief world overview of the Plecoptera is give by Stewart (2009).

Fig. 13.1 The Willow fly, *Euleuctra geniculata* (Leuctridae) (Photo: Stuart Crofts)

> ### HIGHER CLASSIFICATION OF BRITISH PLECOPTERA
>
> Superfamily Nemouroidea
>   Family Capniidae (1 genus, 3 species)
>   Family Leuctridae (2 genera, 6 species)
>   Family Nemouridae (4 genera, 11 species)
>   Family Taeniopterygidae (3 genera, 4 species)
> Superfamily Perloidea
>   Family Chloroperlidae (3 genera, 3 species)
>   Family Perlidae (2 genera, 2 species)
>   Family Perlodidae (4 genera, 5 species)

SPECIES OF CONSERVATION CONCERN

No species have any legal protection but two are on the UKBAP list: *Isogenus nubecula* (Perlodidae) and *Brachyptera putata* (Taeniopterygidae).

Fig. 13.2 Nemourid stonefly (Photo: Robin Williams)

The Families of British Plecoptera

SUPERFAMILY NEMOUROIDEA

Family Capniidae (1 genus, 3 species)
The three British species of *Capnia* emerge early in the spring; they are small dark insects, known generally as black stoneflies or winter stoneflies, and none is particularly common. Like the Chloroperlidae their fore wings lack the ladder-like venation seen in many other families.

British genus: *Capnia*

Family Leuctridae (2 genera, 6 species)
Needle flies (Fig. 13.1) get their common name from the way they wrap their wings round their body in a tubular shape, whereas other families usually hold them flat above the body. Some species can be seen flying into late autumn, whereas the aptly named Early needle fly (*Leuctra hippopus*) is on the wing from February. One species, *Euleuctra geniculata*, has the common name of the Willow fly, but probably the most common species in this family is *Leuctra inermis*. The nymphs of most Leuctridae need fast-flowing water.

British genera: *Euleuctra, Leuctra*

Family Nemouridae (4 genera, 11 species)
Known generally as the small brown stoneflies, these resemble Leuctridae but the wings are held flat over the body (Fig. 13.2). Because the nymphs of some members of this family can tolerate lower oxygen levels these may be the only species seen in slower rivers in southern England. Some species are on the wing in early spring, but others do not emerge until late summer. Several are common and abundant, including *Nemurella pictetii*, *Protonemura meyeri* and *Nemoura cinerea*.

British genera: *Amphinemura, Nemoura, Nemurella, Protonemura*

Family Taeniopterygidae (3 genera, 4 species)
These are the anglers' February reds, all confined to running water, and the adults of all four British species have at least one brownish band across the fore wings.

Brachyptera putata (Northern February red) is on the UKBAP list and is generally regarded as being endemic to Britain.

British genera: *Brachyptera, Rhabdiopteryx, Taeniopteryx*

SUPERFAMILY PERLOIDEA

Family Chloroperlidae (3 genera, 3 species)
The anglers' Small yellow sallies are small stoneflies, yellowish-green in colour, both as nymphs and adults. As in the Capniidae the adult fore wings do not show the ladder-like arrangement of cross-veins. Although *Xanthoperla apicalis* is currently on the British list, there are no confirmed records of this species, which is quite widespread in continental Europe.

British genera: *Chloroperla, Siphonoperla, Xanthoperla*

Family Perlidae (2 genera, 2 species)
This family has the two largest British species, with nymphs up to 35 mm in length, and adults up to 25 mm long, with a wingspan of around 50 mm. Both *Dinocras cephalotes* and *Perla bipunctata* are more common in northern Britain.

British genera: *Dinocras, Perla*

Family Perlodidae (4 genera, 5 species)
There are two distinct groups of genera within this family. The larger species in *Isogenus, Perlodes* and *Diura* are brown with characteristic yellowish markings on the head and prothorax, whereas members of *Isoperla* are smaller and green or yellowish in colour.

Diura bicaudata (Fig. 13.3) is a common upland species and *Perlodes microcephala* (Fig. 13.4) is the only large stonefly to be seen on southern chalk streams; it is only slightly smaller than the two species of Perlidae. Of the two species in *Isoperla, I. grammatica*, the anglers' Yellow Sally, is found throughout much of Britain except for south-east England, and its nymph has distinctive colour patterns (Fig. 13.5). There are no recent records of *I. obscura* in Britain, and it may be extinct in this country.

Isogenus nubecula (Scarce yellow Sally) is on the UKBAP list; it is currently known from only a few

Fig. 13.3 Male brachypterous *Diura bicaudata* (Perlodidae) (Photo: Roger Key)

Fig. 13.4 Male *Perlodes microcephalus* (Perlodidae) (Photo: Stuart Crofts)

Fig. 13.5 Yellow Sally nymph, *Isoperla grammatica* (Perlodidae) (Photo: Stuart Crofts)

sites on a single river and seems to be declining rapidly.

British genera: *Perlodes, Diura, Isogenus, Isoperla*

REFERENCES

BROOKS, S.J. 1999. Plecoptera: the stoneflies. In: Barnard, P.C. (ed.) *Identifying British insects and arachnids: an annotated bibliography of key works.* Cambridge University Press, Cambridge, pp. 33–5.

GODDARD, J. 1988. *John Goddard's waterside guide.* Unwin Hyman, London.

GODDARD, J. 1991. *Trout flies of Britain and Europe.* A & C Black, London.

HYNES, H.B.N. 1977. Adults and nymphs of British stoneflies (Plecoptera). *Scientific Publications of the Freshwater Biological Association* 17: 92 pp. [reprinted 1993].

PRYCE, D., MACADAM, C. & BROOKS, S. 2007. *Guide to the British stonefly (Plecoptera) families: adults and larvae.* Field Studies Council, Preston Montford, OP 113.

STEWART, K.W. 2009. Plecoptera. In: Resh, V.H. & Cardé, R.T. (eds.) *Encyclopedia of insects* (2nd edn.). Academic Press/Elsevier, San Diego & London, pp. 810–13.

WEBSITES

http://www.brc.ac.uk/schemes/RRS/plecoptera.htm
The Plecoptera Recording Scheme is currently hosted on the Riverflies Recording Schemes pages on the website of the Biological Records Centre.

http://plecoptera.speciesfile.org
The Plecoptera Species File has a catalogue of world species, together with many useful references.

PART 5 Paraneoptera

At first sight the Paraneoptera might appear to consist of a rather disparate group of insect orders, namely the true bugs, lice, book lice and thrips, but the monophyly of this group is generally accepted on both morphological and molecular characters. The exact relationships between the constituent orders are not as clear, though the Psocoptera and Phthiraptera probably form a monophyletic pair, the superorder Psocodea. There is even some controversial evidence that the Psocoptera may not be monophyletic, and that the Phthiraptera evolved from a psocopteran sub-group (summarized on the Tree of Life website; http://tolweb.org/Psocodea/8235). Whether the Thysanoptera are more closely related to these, or to the Hemiptera, is not certain, but some authors combine them with the Hemiptera in the super-order Condylognatha.

The Royal Entomological Society Book of British Insects, First Edition. Peter C. Barnard.
© 2011 Royal Entomological Society. Published 2011 by Blackwell Publishing Ltd.

14 Order Hemiptera: the true bugs

c. 1830 species in 63 families

Unfortunately there is an increasing tendency to call any insect a 'bug', so the term 'true bug' has to be used to signify a hemipteran. Although the affinities of the Hemiptera with other insect orders need further clarification, there is no doubt about the monophyly of the order. The formation of the mouthparts into the elongated rostrum is unique to the Hemiptera and is not found even in other groups with sucking mouthparts. The mandibles and maxillae are modified into long pointed stylets, and palps are entirely lacking; the labium forms an almost complete tube containing the stylets, with the labrum closing off the remaining gap in the rostrum, and the latter is usually directed backwards underneath the head. These extreme modifications are adapted for piercing and sucking liquids from plant tissue, nearly always vascular plants, although some families have become predatory. Such a limited feeding method might be expected to limit the diversity of the hemipteran evolution, both morphologically and biologically, but in fact the order is very diverse with members showing a wide variety of lifestyles. Bugs can be found in most terrestrial and freshwater habitats; some are even marine, and many are well known as horticultural pests, either causing direct damage to plants or transmitting virus diseases. The Hemiptera are clearly a hemimetabolous group, with nymphal stages showing a gradual adult-like appearance, but some of the Sternorrhyncha such as whiteflies and scale-insects have an abrupt change in morphology at the stage between the nymphs and the adult, effectively a pupa, which confused early authors trying to classify this group.

The higher classification of the group is reasonably stable within Europe but there are major differences in opinion on each side of the Atlantic. For anyone studying the Hemiptera on a wider scale it is important to understand how the two different systems relate to each other. Traditionally in Europe the order was divided into two suborders, the Heteroptera and Homoptera, with the latter divided into the Auchenorrhyncha and Sternorrhyncha. It was then recognized that the Homoptera did not form a monophyletic group so, for a time, three suborders were recognized: Heteroptera, Auchenorrhyncha and Sternorrhyncha. The latest view is that the Auchenorrhyncha are not monophyletic, so the system adopted here is to use four suborders: Heteroptera, Cicadomorpha, Fulgoromorpha and Sternorrhyncha. In the USA, the Heteroptera are combined with the southern hemisphere group Coleorrhyncha to form the suborder Prosorrhyncha; the Auchenorrhyncha are retained as a single unit, as are the Sternorrhyncha, though some of these are occasionally regarded as orders, not suborders. However, a glance through the various chapters on bugs in a work such as Resh and Cardé (2009) shows that there is still much disagreement on how these groups should be treated. Whatever system one adopts, the inner classification of the Hemiptera is complex, with a great range of hierarchical categories such as infraorders that are not found necessary in all orders.

The Hemiptera form the fifth largest group of insects in Britain, following the Hymenoptera,

The Royal Entomological Society Book of British Insects, First Edition. Peter C. Barnard.
© 2011 Royal Entomological Society. Published 2011 by Blackwell Publishing Ltd.

Diptera, Coleoptera and Lepidoptera. Not surprisingly for such a large group, there are no identification guides to cover the entire order; the larger and more colourful bugs such as the pentatomoid shield-bugs can be identified using well-illustrated books, whereas the smaller and technically more demanding groups such as aphids and coccids are often tackled only by expert taxonomists or economic entomologists. For the same reason, checklists of the different families vary in their usefulness, and some are in urgent need of updating; the only complete list of the British species (Kloet & Hincks, 1964) is no longer reliable. Identification literature is therefore listed under each group as appropriate, but there is one general book of importance to be noted here. Dolling (1991) gives an excellent overview of the whole order from a British perspective, and includes keys to families as well as good biological and morphological summaries, though a few groups have been added to the British fauna since the book appeared. A useful simplified key to families is provided by Unwin (2001). For a more informal view of the group worldwide, McGavin (1993) is well illustrated. Martin and Webb (1999) provide many specialized references that help to update earlier works. Although superseded in many ways, the early book by Butler (1923) is still useful for information on many immature stages. Wachmann's (1988) book gives many useful colour photos of British species.

There are nearly 8000 species in 94 families in Europe, and around 100,000 species in 104 families worldwide.

HIGHER CLASSIFICATION OF BRITISH HEMIPTERA

Suborder Heteroptera
  Infraorder Cimicomorpha
    Superfamily Cimicoidea
      Family Anthocoridae (12 genera, 32 species)
      Family Cimicidae (2 genera, 4 species)
      Family Nabidae (2 genera, 12 species)
    Superfamily Miroidea
      Family Microphysidae (2 genera, 6 species)
      Family Miridae (94 genera, 220 species)
    Superfamily Reduvioidea
      Family Reduviidae (4 genera, 7 species)
    Superfamily Tingoidea
      Family Tingidae (13 genera, 24 species)
  Infraorder Dipsocoromorpha
    Superfamily Dipsocoroidea
      Family Ceratocombidae (1 genus, 1 species)
      Family Dipsocoridae (1 genus, 2 species)
  Infraorder Leptopodomorpha
    Superfamily Saldoidea
      Family Aepophilidae (1 genus, 1 species)
      Family Saldidae (8 genera, 22 species)
  Infraorder Nepomorpha
    Superfamily Corixoidea
      Family Corixidae (9 genera, 35 species)
    Superfamily Gerroidea
      Family Gerridae (3 genera, 10 species)
      Family Veliidae (2 genera, 5 species)
    Superfamily Hebroidea
      Family Hebridae (1 genus, 2 species)
    Superfamily Hydrometroidea
      Family Hydrometridae (1 genus, 2 species)
    Superfamily Mesovelioidea
      Family Mesoveliidae (1 genus, 1 species)
    Superfamily Naucoroidea
      Family Aphelocheiridae (1 genus, 1 species)
      Family Naucoridae (1 genus, 1 species)
    Superfamily Nepoidea
      Family Nepidae (2 genera, 2 species)
    Superfamily Notonectoidea
      Family Notonectidae (1 genus, 4 species)
    Superfamily Pleoidea
      Family Pleidae (1 genus, 1 species)
  Infraorder Pentatomomorpha
    Superfamily Aradoidea
      Family Aradidae (2 genera, 7 species)
    Superfamily Coreoidea
      Family Alydidae (1 genus, 1 species)
      Family Coreidae (9 genera, 10 species)
      Family Rhopalidae (6 genera, 10 species)
      Family Stenocephalidae (1 genus, 2 species)
    Superfamily Lygaeoidea
      Family Berytidae (4 genera, 9 species)
      Family Lygaeidae (37 genera, 81 species)
      Family Pyrrhocoridae (1 genus, 1 species)
    Superfamily Pentatomoidea
      Family Acanthosomatidae (4 genera, 5 species)
      Family Cydnidae (6 genera, 7 species)
      Family Pentatomidae (19 genera, c. 20 species)
      Family Scutelleridae (2 genera, 4 species)
      Family Thyreocoridae (1 genus, 1 species)
    Superfamily Piesmatoidea
      Family Piesmatidae (2 genera, 2 species)
Suborder Cicadomorpha
  Superfamily Cercopoidea
      Family Aphrophoridae (3 genera, 9 species)
      Family Cercopidae (1 genus, 1 species)

Superfamily Cicadoidea
    Family Tibicinidae (1 genus, 1 species)
Superfamily Membracoidea
    Family Cicadellidae (*c.* 105 genera, *c.* 285 species)
    Family Membracidae (2 genera, 2 species)
    Family Ulopidae (2 genera, 2 species)
Suborder Fulgoromorpha
    Family Cixiidae (4 genera, 12 species)
    Family Delphacidae (40 genera, *c.* 75 species)
    Family Issidae (1 genus, 2 species)
    Family Tettigometridae (1 genus, 1 species)
Suborder Sternorrhyncha
    Superfamily Aleyrodoidea
        Family Aleyrodidae (13 genera, 19 species)
    Superfamily Aphidoidea
        Family Aphididae (*c.* 160 genera, *c.* 630 species)

Superfamily Coccoidea
    Family Aclerdidae (1 genus, 1 species)
    Family Asterolecaniidae (4 genera, 5 species)
    Family Coccidae (14 genera, 30 species)
    Family Diaspididae (18 genera, 28 species)
    Family Eriococcidae (7 genera, 15 species)
    Family Kermesidae (1 genus, 3 species)
    Family Margarodidae (2 genera, 2 species)
    Family Ortheziidae (4 genera, 5 species)
    Family Pseudococcidae (16 genera, 50 species)
Superfamily Phylloxeroidea
    Family Adelgidae (7 genera, 13 species)
    Family Phylloxeridae (2 genera, 3 species)
Superfamily Psylloidea
    Family Calophyidae (1 genus, 1 species)
    Family Homotomidae (1 genus, 1 species)
    Family Psyllidae (16 genera, 60 species)
    Family Triozidae (3 genera, 18 species)

SPECIES OF CONSERVATION CONCERN

One species of Hemiptera is listed on the Wildlife and Countryside Act 1981: *Cicadetta montana* (Tibicinidae). It is also on the UKBAP list along with nine other species: *Chlorita viridula*, *Doratura impudica*, *Euscelis venosus* and *Macrosteles cyane* (Cicadellidae); *Eurysa douglasi* and *Ribautodelphax imitans* (Delphacidae); *Hydrometra gracilenta* (Hydrometridae); *Saldula setulosa* (Saldidae); *Physatocheila smreczynskii* (Tingidae).

The following two species were removed from the UKBAP list in 2007: *Aphrodes duffieldi* (Cicadellidae) and *Orthotylus rubidus* (Miridae). A useful review of some threatened species was given by Kirby (1992).

The Families of British Hemiptera

SUBORDER HETEROPTERA

The Heteroptera are sometimes linked with the Coleorryncha (which includes the single southern hemisphere family Peloridiidae) to form the suborder Prosorrhyncha; for a world overview see Schaefer (2009). Amongst the Hemiptera the members of this suborder show the widest variety of morphology; they exploit a great range of food sources from fruits and leaves to mammal blood, and use the greatest range of habitats, from freshwater to stored products. Heteroptera are easily recognized by the form of the fore wings, which have a thickened opaque basal section, separated abruptly from the distal transparent section. Many families

have scent glands that produce a defensive acrid fluid, which has given rise to the alternative name of stink-bugs for the pentatomoid shield-bugs. As well as other forms of chemical communication some groups can produce sounds, either by stridulation or using abdominal tymbals, though this ability is better known in the Auchenorrhyncha. Stridulation is best developed in the aquatic bugs such the Gerridae (pondskaters) that live on the surface of the water, and in the Corixidae and Notonectidae (water boatmen) that swim under the surface. The truly aquatic groups breathe from an air bubble trapped in a layer of hairs or under the wings, or else using a siphon, as in the Nepidae (water scorpions).

Identifying most British species of Heteroptera is straightfoward; the single best book on the terrestrial families is Southwood and Leston (1959), out of print and somewhat out of date, but now available on CD-ROM. The freshwater groups are covered by Savage (1989, 1999), and Denton (2007) gives useful information and illustrations of many British species. Many of the British Heteroptera are illustrated on the British Bugs website (http://www.britishbugs.org.uk) together with much information on their distribution and biology. A useful world overview of the world Heteroptera is given by Schuh and Slater (1995).

Infraorder Cimicomorpha

SUPERFAMILY CIMICOIDEA

Family Anthocoridae (12 genera, 32 species)
Despite their common name of flower-bugs, the Anthocoridae are predators of soft-bodied

invertebrates (Fig. 14.1). As most species do not exceed 4mm in length they can only tackle small prey such as aphids, though some are specialized at detecting and attacking leaf-mining caterpillars. A few of the larger species such as *Anthocoris nemorum* have been known to bite people, though they cannot be considered as aggressive. This tendency does, however, indicate the close relationship between this family and the bed-bugs of the Cimicidae. Some species have a very narrow range of prey; for example *Dufouriellus ater* seems to prefer scolytid

Fig. 14.1 *Anthocoris* (Anthocoridae) (Photo: Colin Rew)

Fig. 14.2 The bed-bug, *Cimex lectularius*, slide preparation (Cimicidae) (Photo: Peter Barnard)

beetles, while *Brachysteles parvicornis* feeds almost exclusively on mites. Other species are associated with ants, often living deep within the nests. As well as *Anthocoris nemorum*, there are several other common species in this family, found feeding on insects on a wide variety of plants; members of the genus *Xylocoris* are often found in compost heaps and similar habitats, giving rise to the name Hot-bed bugs.

British subfamilies and genera:

Anthocorinae: *Acompocoris, Anthocoris, Elatophilus, Orius, Temnostethus, Tetraphleps*

Lyctocorinae: *Brachysteles, Cardiastethus, Dufouriellus, Lyctocoris, Xylocoridea, Xylocoris*

Family Cimicidae (2 genera, 4 species)

The bed-bugs are closely related to the anthocorid flower-bugs, and the two families have occasionally been treated as one. Cimicidae all live on avian or mammalian blood and their flattened bodies enable them to hide in crevices when not feeding. Unlike fleas and lice they have no means of clinging to fur or feathers so they are only found on their hosts while actually feeding. The well-known bed-bug, *Cimex lectularius* (Fig. 14.2), is less frequently seen in domestic houses than in former centuries, though some inner-city hotels have persistent populations. Their bites are irritating but bed-bugs do not transmit diseases. Most establishments are naturally eager to get rid of them, but their ability to hide in small cracks during the daytime can make them difficult to eradicate. Among the other species of *Cimex* in Britain, *C. pipistrelli* lives on bats. *Oeciacus hirundinis* is the Martin bug (Fig. 14.3), which seems to feed mainly on house martins (*Delichon urbica*); it

Fig. 14.3 *Oeciacus hirundinis* (Cimicidae) (Photo: Roger Key)

has been reported as entering domestic houses and occasionally biting people, though at around 3 mm it is the smallest member of the family in Britain.

A useful account of the biology of the common bed-bug is Anon (1973).

British genera: *Cimex, Oeciacus*

Family Nabidae (2 genera, 12 species)
Despite their delicate-sounding name of damsel-bugs, the Nabidae are aggressive predators on a variety of insects and spiders. Some species are flightless, and it is possible that the ability to disperse by means of flight is less important in such active predators that walk long distances in search of their prey. Many species live on the ground or on low vegetation, and only *Himacerus apterus* (Fig. 14.4) is reported as living on taller shrubs and trees. *Nabis rugosus* is commonly seen in gardens though there are other similar-looking species (Fig. 14.5).

Fig. 14.4 *Himacerus apterus* **(Nabidae) (Photo: Roger Key)**

Fig. 14.5 Juvenile Nabidae (Photo: Roger Key)

Identification: the European species were covered by Péricart (1987) in French; six of the British species are illustrated on the British Bugs website (http://www.britishbugs.org.uk).

British genera: *Himacerus, Nabis*

SUPERFAMILY MIROIDEA

Family Microphysidae (2 genera, 6 species)
The members of this small family are only 2–3 mm long and are commonly known as minute-bugs. The biology of such small bugs has not been well studied but most species seem to live among lichens and mosses on tree bark or walls, preying on small insects such as psocids, aphids and springtails. Some British species may be fairly common but their small size means that they are easily overlooked.

British genera: *Loricula, Myrmedobia*

Family Miridae (At Least 94 genera, 220 species)
The Miridae is not only the largest family of British Heteroptera, but it also includes some of the most important species, either as horticultural pests or as predators of other pest species. They are commonly known as capsid bugs because the family was previously known as the Capsidae. Perhaps a more appropriate name is leaf-bugs, as they seem well adapted to negotiating all kinds of vegetation. Two small species are notable as being the only British Heteroptera to feed on ferns: the Bracken bug, *Monalocoris filicis*, feeds mainly on the sporangia of bracken and the Fern bug, *Bryocoris pteridis*, on other species of ferns.

Members of the small subfamily Deraeocorinae are all predators, yet some are confined to particular tree species. Many of the more important plant pests are in the large subfamily Mirinae (Figs. 14.6 & 14.7), and these include *Lygocoris pabulinus* (Common green capsid) on soft fruits, *Adelphocoris lineolatus* (Lucerne bug) on several legumes and *Lygus rugulipennis* (Tarnished plant bug) on many field and garden crops. In the Orthotylinae *Blepharidopterus angulatus* (Black-kneed capsid) is a useful predator on fruit trees where it helps to control red spider mites. One striking species is *Myrmecoris gracilis*, a rare ant-mimic of sandy heathland (Fig. 14.8). Some lists include more genera than those listed here, but the generic limits are not always clear.

Identification: as well as Southwood and Leston (1959), the keys (in French) by Wagner and Weber (1964) on the French fauna can also be helpful.

Fig. 14.6 *Pantilius tunicatus* (Miridae) (Photo: Roger Key)

Fig. 14.7 Miridae (Photo: Colin Rew)

Fig. 14.8 *Myrmecoris gracilis* (Miridae) (Photo: Roger Key)

British subfamilies and genera:

Bryocorinae (including Dicyphinae): *Bryocoris, Campyloneura, Dicyphus, Macrolophus, Monalocoris, Tupiocoris*

Deraeocorinae: *Alloeotomus, Bothynotus, Deraeocoris*

Mirinae: *Acetropis, Adelphocoris, Agnocoris, Apolygus, Calocoris, Camptozygum, Capsodes, Capsus, Charagochilus, Closterotomus, Dichrooscytus, Grypocoris, Leptopterna, Liocoris, Lygocoris, Lygus, Megacoelum, Megaloceroea, Miridius, Miris, Myrmecoris, Neolygus, Notostira, Orthops, Pantilius, Phytocoris, Pinalitus, Pithanus, Polymerus, Rhabdomiris, Stenodema, Stenotus, Teratocoris, Trigonotylus, Zygimus*

Orthotylinae: *Blepharidopterus, Cyllecoris, Cyrtorhinus, Dryophilocoris, Fieberocapsus, Globiceps, Halticus, Heterocordylus, Heterotoma, Malacocoris, Mecomma, Orthocephalus, Orthotylus, Pachytomella, Platycranus, Pseudoloxops, Strongylocoris*

Phylinae: *Amblytylus, Asciodema, Atractotomus, Brachyarthrum, Campylomma, Chlamydatus, Compsidolon, Conostethus, Europiella, Hallodapus, Harpocera, Hoplomachus, Lopus, Macrotylus, Megalocoleus, Monosynamma, Oncotylus, Orthonotus, Parapsallus, Phoenicocoris, Phylus, Pilophorus, Placochilus, Plagiognathus, Plesiodema, Psallodema, Psallus, Salicarus, Sthenarus, Systellonotus, Tinicephalus, Tuponia, Tytthus*

SUPERFAMILY REDUVIOIDEA

Family Reduviidae (4 genera, 7 species)
Primarily a tropical group, the assassin-bugs form a relatively small family in Britain. They are easily recognized by the very short, thick and strongly curved rostrum; as well as being highly adapted for catching and feeding on their prey, the rostrum is also used to stridulate by rubbing it against the prosternum. *Reduvius personatus* is perhaps the best known species because it is synanthropic and has been widely transported around the world; its nymph covers its body with detritus, rather like some lacewing larvae, and this has given it the common name of Masked assassin-bug (Fig. 14.9). In contrast, the three species of *Empicoris* are unusually delicate-looking insects with raptorial fore legs (Fig. 14.10); they prey on small insects such as barklice. The Heath assassin-bug, *Coranus subapterus*, is widespread on sandy heaths and dunes; a further species in this genus is found only in the Channel Islands. There are no recent records of *Pygolampis bidentata*, which may no longer occur in Britain.

Fig. 14.9 *Reduvius personatus* (Reduviidae) (Photo: Roger Key)

Fig. 14.11 *Tingis cardui* (Tingidae) (Photo: Roger Key)

Fig. 14.10 *Empicoris vagabundus* (Reduviidae) (Photo: Roger Key)

British subfamilies and genera:
Emesinae: *Empicoris*
Harpactorinae: *Coranus*
Reduviinae: *Reduvius*
Stenopodainae: *Pygolampis*

SUPERFAMILY TINGOIDEA

Family Tingidae (13 genera, 24 species)
The lace-bugs get their common name from the reticulate sculpturing on the pronotum and fore wings; they superficially resemble the Piesmidae, though the latter are now placed in a separate superfamily. Most species are only around 4 mm long, and they feed on a variety of plants, some species being associated with mosses, others on broom and gorse, with a few confined to mature trees. One of the best known species is *Tingis cardui*, which lives on thistles (Fig. 14.11).

Physatocheila smreczynskii (Apple lace-bug) is on the UKBAP list.

Identification: the European species were covered by Péricart (1983) in French; around half the British species are illustrated on the British Bugs website (http://www.britishbugs.org.uk).

British genera: *Acalypta, Agramma, Campylosteira, Catoplatus, Derephysia, Dictyla, Dictyonota, Kalama, Lasiacantha, Oncochila, Physatocheila, Stephanitis, Tingis*

Infraorder Dipsocoromorpha
SUPERFAMILY DIPSOCOROIDEA

The few tiny bugs in this superfamily could easily be mistaken for Anthocoridae, though there are small morphological differences.

Family Ceratocombidae (1 genus, 1 species)
The only British species in this family, *Ceratocombus coleoptratus*, is a predator that lives in damp mosses. Very little is known of its detailed life history, though it is probably quite widespread in its distribution.

British genus: *Ceratocombus*

Family Dipsocoridae (1 genus, 2 species)
Both species in this family are predators and are found amongst mosses and gravels by the side of fast-flowing streams, often in upland areas; the adult stages have been collected during the winter months. *Cryptostemma alienum* is the more common species.

British genus: *Cryptostemma*

Infraorder Leptopodomorpha

SUPERFAMILY SALDOIDEA

Family Aepophilidae (1 genus, 1 species)
The only British species, *Aepophilus bonnairei*, known as the Marine bug, lives in crevices in rocks on west-facing coasts and is not easy to find: 'the collector will need a crow-bar and a pooter' (Southwood & Leston, 1959). This small wingless bug is not often seen but is probably quite widespread in suitable habitats, though its prey species are not certain. It was formerly included in the Saldidae. *A. bonnairei* is illustrated on the British Bugs website.

British genus: *Aepophilus*

Family Saldidae (8 genera, 22 species)
Shore-bugs is not a very appropriate name for this group; although some are found in coastal marshes, most species are associated with the margins of freshwater habitats, although *Saldula orthochila* is found on dry heaths. They are all large-eyed predators that actively pursue their prey, even though the front legs are not modified for grasping (as they are in the Nabidae, Reduviidae and the families of water-bugs). One of the commonest British species is *Saldula saltatoria*, the Common shore-bug, which is found around the edges of ponds and lakes (Fig. 14.12).
Saldula setulosa (Hairy shore-bug) is on the UKBAP list.

Identification: the European species were covered by Péricart (1990) in French.

British subfamilies and genera:
Chiloxanthinae: *Chiloxanthus*
Saldinae: *Chartoscirta, Halosalda, Macrosaldula, Micracanthia, Salda, Saldula, Teloleuca*

Infraorder Nepomorpha

This group includes all the water-bugs, formerly placed in two informal groupings; the surface-living families in the 'Amphibicorisae' and the fully aquatic ones in the 'Hydrocorisae', though these reflected ecological rather than taxonomic affinities. The surface-dwelling species all have long and conspicuous antennae, like the terrestrial families, whereas in the aquatic groups the antennae are short and hidden in pits under the head.

SUPERFAMILY CORIXOIDEA

Family Corixidae (9 genera, 35 species)
These are generally known as lesser water-boatmen, to distinguish them from the Notonectidae. It is a surprisingly large group, and identification is not always easy. Corixids spend most of their time at the bottom of still water bodies, only coming to the surface to renew the air-bubble with which they respire. They swim through the water using their fringed hind legs and, unlike the Notonectidae, keep their dorsal side up (Fig. 14.13). Many species can stridulate, often quite loudly, using a variety of structures on the legs and head; *Micronecta* males rotate their genitalia against the abdomen to produce a clearly audible sound. Although some Corixidae are predaceous, most species seem to

Fig. 14.12 *Saldula saltatoria* (Saldidae) (Photo: Roger Key)

Fig. 14.13 *Corixa punctata* (Corixidae) (Photo: Robin Williams)

feed on detritus sifted from the bottom sediments with their fringed anterior tarsi. Others scavenge on dead animal matter and sometimes algae, and it seems best to consider them as generalized feeders. Common species include *Corixa punctata* and *Sigara dorsalis*.

Identification: Savage (1989, 1999).

British subfamilies and genera:
　Corixinae: *Arctocorisa*, *Callicorixa*, *Corixa*, *Glaenocorisa*, *Hesperocorixa*, *Paracorixa*, *Sigara*
　　Cymatiainae: *Cymatia*
　　Micronectinae: *Micronecta*

SUPERFAMILY GERROIDEA

Family Gerridae (3 genera, 10 species)
The pondskaters, sometimes called water-striders, are a familiar sight on almost any area of still freshwater, down to the humblest garden pond. Their long middle legs are used to row themselves across the surface in rapid pursuit of their prey of small insects, which they grasp with their short raptorial fore legs. Because they rely on speed, they tend to prefer open areas of water, rather than ponds with much emergent vegetation, though floating duckweed is no obstacle (Fig. 14.14). The most common species, *Gerris lacustris*, shows a variety of wing-length from being totally apterous to fully macropterous; the fully winged forms are strong fliers and this accounts for the rapidity with which this species colonizes ponds every year.

Identification: Savage (1989).

British genera: *Aquarius*, *Gerris*, *Limnoporus*

Family Veliidae (2 genera, 5 species)
The surface-dwelling water-crickets are less familiar than the Gerridae, although at least two species, *Velia caprai* and *Microvelia reticulata*, are common and widespread. *Microvelia* species are up to 2 mm long (Fig. 14.15), compared with the much larger *Velia* species at around 6–8 mm. All are predators on a variety of small invertebrates and they can walk slowly on the surface or row themelves rapidly with their middle legs, like Gerridae.

Identification: Savage (1989).

British subfamilies and genera:
　Microveliinae: *Microvelia*
　Veliinae: *Velia*

SUPERFAMILY HEBROIDEA

Family Hebridae (1 genus, 2 species)
The two species of Sphagnum-bugs prey on small invertebrates amongst the damp moss by the side of freshwater, and are less than 2 mm long. For such small insects, the water surface can be a formidable obstacle, preventing them from walking up the curve of the meniscus to reach prey, for example. In all the surface-living bugs the body and legs are unwettable, except for the pretarsus of each leg, so that the body weight is easily supported by the surface film, with just six depressions where each pretarsus meets the water. *Hebrus* makes use of this phenomenon by raising some of its legs until the

Fig. 14.14 Pond-skaters (Gerridae) (Photo: Peter Barnard)

Fig. 14.15 *Microvelia reticulata* (Veliidae) (Photo: Roger Key)

surface film is lifted up, rather than depressed, and the surface forces then pull the bug up the curved meniscus in an apparently effortless gliding motion. This is clearly useful where the strongly curved water surfaces that cover damp moss or similar plants can create high surface tensions that would otherwise be impossible for small surface-dwellers to negotiate.

Identification: Savage (1989).

British genus: *Hebrus*

SUPERFAMILY HYDROMETROIDEA

Family Hydrometridae (1 genus, 2 species)
The water-measurers are slender surface-living bugs, up to 12 mm long, whose heads are particularly elongated (Fig. 14.16). They walk slowly across weedy ponds and are apparently capable of finding small invertebrate prey beneath the surface. The most common species is *Hydrometra stagnorum*.

Hydrometra gracilenta (Lesser water-measurer) is on the UKBAP list.

Identification: Savage (1989).

British genus: *Hydrometra*

SUPERFAMILY MESOVELIOIDEA

Family Mesoveliidae (1 genus, 1 species)
There is just one species of pondweed-bug in Britain, *Mesovelia furcata*. Unusually for surface-dwelling bugs it is greenish in colour, rather than the usual dull brown of the Gerridae and so on. Like *Hydrometra* it can catch prey through the surface

film. Although widely distributed across Britain it is probably not common anywhere.

Identification: Savage (1989).

British genus: *Mesovelia*

SUPERFAMILY NAUCOROIDEA

Family Aphelocheiridae (1 genus, 1 species)
The only British species, *Aphelocheirus aestivalis*, is quite widespread though locally distributed in well-oxygenated fast streams and rivers. It is brownish, flattened and oval-shaped, about 10 mm long, and superficially resembles a cimicid bed-bug. A coating of fine hairs over most of the body trap a silvery layer of air, a classic example of a plastron, and gaseous exchange between this air and the surrounding water is so effective that the bug does not need to renew the air supply at the water surface. Living at the bottom of rivers it feeds on various insect larvae.

Identification: Savage (1989).

British genus: *Aphelocheirus*

Family Naucoridae (1 genus, 1 species)
There is just one species of saucer-bug in Britain, *Ilyocoris cimicoides* (Fig. 14.17), though a second member of the family is found in the Channel Islands. This dark brown oval bug can be up to 15 mm long, and has distinctive stoutly curved fore legs that are used to trap prey. This is one of the water-bugs that can give an unwary person a sharp nip.

Fig. 14.16 *Hydrometra stagnorum* (Hydrometridae) (Photo: Roger Key)

Fig. 14.17 *Ilyocoris cimicoides* (Naucoridae) (Photo: Roger Key)

Identification: Savage (1989).

British genus: *Ilyocoris*

SUPERFAMILY NEPOIDEA

Family Nepidae (2 genera, 2 species)
The two species of water-scorpions look very different from each other, though both have a long respiratory siphon and raptorial fore legs. *Nepa cinerea* is a very flattened dark brown bug that can walks around on the bottom of shallow water; it is very common but can be hard to see (Fig. 14.18). It is a fierce predator, even catching small fish. The Water stick-insect, *Ranatra linearis*, is less common than *Nepa*; it is very elongated, living in deeper water with more vegetation, and is an equally active predator though it takes smaller-sized prey (Fig. 14.19).

Identification: Savage (1989).

British subfamilies and genera:
Nepinae: *Nepa*
Ranatrinae: *Ranatra*

SUPERFAMILY NOTONECTOIDEA

Family Notonectidae (1 genus, 4 species)
The true water-boatmen, as distinct from the Corixidae, are sometimes known as backswimmers from their habit of swimming with their ventral side uppermost (Fig. 14.20). They keep a bubble of air trapped on the body by hairs and must frequently renew this air supply at the surface, especially in warm weather (Fig. 14.21). They are large active swimmers, up to 15 mm long, and they will prey on anything up to their own size including tadpoles

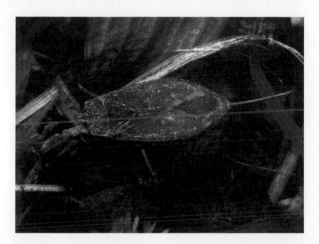

Fig. 14.18 *Nepa cinerea* (Nepidae) (Photo: Roger Key)

Fig. 14.20 *Notonecta* in its usual swimming position (Notonectidae) (Photo: Roger Key)

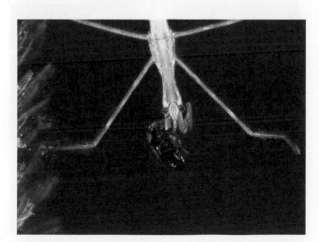

Fig. 14.19 *Ranatra linearis* (Nepidae) with *Laccophilus* beetle prey (Photo: Roger Key)

Fig. 14.21 *Notonecta* with its trapped air supply (Notonectidae) (Photo: Roger Key)

and juvenile fish; they will also bite an incautious human. *Notonecta glauca* is the most common of the four British species, though each has slightly different habitat preferences.

Identification: Savage (1989).

British genus: *Notonecta*

SUPERFAMILY PLEOIDEA

Family Pleidae (1 genus, 1 species)
The Pleidae resemble small Notonectidae, also swimming on their back. The only British species is *Plea minutissima* (*P. atomaria* or *P. leachi* in earlier literature), which is 2–3 mm long and found in weedy still or slow-flowing water where it preys on small invertebrates such as water-fleas.

Identification: Savage (1989).

British genus: *Plea*

Infraorder Pentatomomorpha

In some cases identification of bugs in this group is only possible using the rather outdated Southwood and Leston (1959), but more recent works are listed for some families.

SUPERFAMILY ARADOIDEA

Family Aradidae (2 genera, 7 species)
The members of this small family are known as flat-bugs or bark-bugs. Most are fungal feeders, gathering hyphae from the bark of living and dead trees. Their flattened body, strong legs and short antennae enable them to enter the spaces under the bark. The two species of *Aneurus*, formerly placed in a separate family (Aneuridae) are both quite common, whereas the five species of *Aradus* are more restricted. *Aradus cinnamoneus*, the Pine flat-bug, is unusual in being a sap-feeder on Scots pine trees.

British subfamilies and genera:
  Aneurinae: *Aneurus*
  Aradinae: *Aradus*

SUPERFAMILY COREOIDEA

Family Alydidae (1 genus, 1 species)
The only British species in this family is *Alydus calcaratus*, a rather local inhabitant of sandy heathlands and similar habitats. It is a very dark species that shows a flash of orange on its abdomen when it flies; the young nymphs are ant-mimics. *A. cal-*

caratus is illustrated on the British Bugs website (http://www.britishbugs.org.uk).

Identification: the European species were covered by Moulet (1995) in French.

British genus: *Alydus*

Family Coreidae (9 genera, 10 species)
Until recently this group had no common name, but the American name of squash-bugs is being used increasingly; in the USA they are common pests on squash plants, though the name is less appropriate in Britain where most species feed on the fruit and seeds of a various legumes and other plants. They are occasionally known as leather-bugs from their general appearance. *Coreus marginatus*, the Dock-bug, is one the largest and most common species (Fig. 14.22), feeding on Dock and other Polygonaceae; *Gonocerus acuteangulatus*, the Box-bug, is traditionally known only from Box trees on Box Hill in Surrey, but is now increasing its range; *Leptoglossus occidentalis*, the Western conifer seed-bug, was introduced to Europe from the USA and the many records in southern England suggest that it is becoming established on British pine trees. *Enoplops scapha* (Fig. 14.23) is a local coastal species that feeds on Asteraceae.

In popular identification guides the Coreidae are often regarded as honorary shieldbugs, which they resemble. The common British species are illustrated by Nau (2004) and all can be identified using Evans and Edmondson (2005) although *Leptoglossus* is not included.

Fig. 14.22 *Coreus marginatus* (Coreidae) (Photo: Peter Barnard)

Fig. 14.23 *Enoplops scapha* (Coreidae) (Photo: Roger Key)

Fig. 14.24 *Corizus hyoscyami* (Rhopalidae) (Photo: Roger Key)

British subfamilies and genera:

Coreinae: *Coreus, Enoplops, Gonocerus, Leptoglossus, Spathocera, Syromastus*

Pseudophloeinae: *Arenocoris, Bathysolen, Ceraleptus, Coriomeris*

Family Rhopalidae (6 genera, 10 species)
A little-known family of which only *Myrmus miriformis* could be considered as common. Most are broad-bodied bugs resembling Coreidae though with broader heads, but there are two species with narrow bodies: *Chorosoma schillingi* is particularly elongated, with *M. miriformis* less so. *Corizus hyoscami* (Fig. 14.24) resembles some of the Lygaeidae in its colouring.

Identification: the European species were covered by Moulet (1995) in French; all except one species are illustrated on the British Bugs website (http://www.britishbugs.org.uk)

British genera: *Chorosoma, Corizus, Liorhyssus, Myrmus, Rhopalus, Stictopleurus*

Family Stenocephalidae (1 genus, 2 species)
The two British species of spurge-bugs are moderate sized, from 12–14 mm long, and have a characteristic bifurcated front to the head (Fig. 14.25). Both species of *Dicranocephalus* feed on different species of spurges.

Identification: the European species were covered by Moulet (1995) in French. Both British species are illustrated on the British Bugs website (http://www.britishbugs.org.uk).

British genus: *Dicranocephalus*

SUPERFAMILY LYGAEOIDEA

Family Berytidae (4 genera, 9 species)
The stilt-bugs get their name from their very long legs; the femora all have enlarged apices, giving a distinct 'kneed' appearance. The antennae are also noticeably long. *Berytinus minor* is probably the commonest British species; *Neides tipularius* (Fig. 14.26) is rather local in southern Britain, and *Metatropis rufescens* (Fig. 14.27) is more widespread. All the species feed on low-growing plants, some on grasses.

Identification: the European Berytidae were covered by Péricart (1984) in French. Six of the British species are illustrated on the British Bugs website (http://www.britishbugs.org.uk).

Fig. 14.25 *Dicranocephalus agilis* (Stenocephalidae) (Photo: Roger Key)

Fig. 14.26 *Neides tipularius* (Berytidae) (Photo: Roger Key)

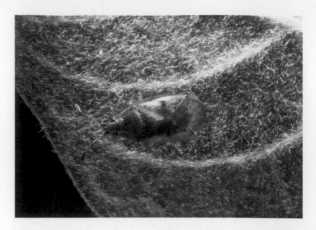

Fig. 14.28 *Kleidocerys resedae* (Lygaeidae) (Photo: Roger Key)

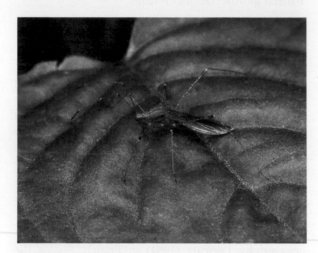

Fig. 14.27 *Metatropis rufescens* (Berytidae) (Photo: Roger Key)

British subfamilies and genera:
Berytinae: *Berytinus*, *Neides*
Gampsocorinae: *Gampsocoris*
Metacanthinae: *Metatropis*

Family Lygaeidae (37 genera, 81 species)
The Lygaeidae are a large family, second only to the Miridae within the Heteroptera. Commonly known as ground-bugs or seed-bugs they are frequently found on the ground, on low vegetation or amongst leaf-litter. Nearly all are phytophagous and elsewhere in the world some species are serious pests on plants such as cereals. A few species have common names, usually reflecting their preferred host plant, including *Ischnodemus sabuleti* (European chinchbug), *Gastrodes abietum* (Spruce cone bug), *G. grossipes* (Pine cone bug), *Heterogaster urticae* (Nettle ground-bug) and *Kleidocerys resedae* (Birch catkin bug, Fig. 14.28).

There are also several vagrant species, and two others are found only in the Channel Islands. Many species are illustrated on the British Bugs website (http://www.britishbugs.org.uk).

British subfamilies and genera:
Artheneinae: *Chilacis*
Blissinae: *Ischnodemus*
Cyminae: *Cymus*
Henestarinae: *Henestaris*
Heterogastrinae: *Heterogaster*
Ischnorhynchinae: *Kleidocerys*
Orsillinae: *Nysius*, *Orsillus*, *Ortholomus*
Oxycareninae: *Macroplax*, *Metopoplax*
Rhyparochrominae: *Acompus, Aphanus, Beosus, Drymus, Emblethis, Eremocoris, Gastrodes, Graptopeltus, Ischnocoris, Lamproplax, Lasiosomus, Macrodema, Megalonotus, Notochilus, Pachybrachius, Peritrechus, Pionosomus, Plinthisus, Pterotmetus, Raglius, Rhyparochromus, Scolopostethus, Sphragisticus, Stygnocoris, Taphropeltus, Trapezonotus, Tropistethus*

Family Pyrrhocoridae (1 genus, 1 species)
The bright red and black fire-bug, *Pyrrhocoris apterus*, is common in Europe and the Channel Islands, but is on the edge of its range in Britain, with only occasional sightings. It feeds on the fruit of tree-mallow and lime trees, and is illustrated on the British Bugs website (http://www.britishbugs.org.uk).

British genus: *Pyrrhocoris*

SUPERFAMILY PENTATOMOIDEA

Several families within this group go under the general name of shieldbugs, because of the large scutellum that covers much or all of the wings, as well as from their general shield-shaped bodies (see Fig. 14.33). Their habit of emitting pungent fluids when threatened has also given rise to the name of stink-bugs.

Family Acanthosomatidae (4 genera, 5 species)
Three members of this small family are common in Britain, including the well-known Hawthorn shield-bug, *Acanthosoma haemorrhoidale* (Fig. 14.29). Most species feed on various trees, including *Elasmucha grisea*, known as the Parent bug from the female's habit of guarding its eggs until they hatch, and even staying with the young nymphs. It lives on birch trees, as does *Elasmostethus interstinctus* (Fig. 14.30), which can be confused with *Acanthosoma*. One species, *Elasmucha ferrugata*, has not been seen in

Britain for some time, and its status is unclear. The other four can be identified using Nau (2004) or Evans and Edmondson (2005); all are illustrated on the British Bugs website (http://www.britishbugs. org.uk).

British genera: *Acanthosoma, Cyphostethus, Elasmostethus, Elasmucha*

Family Cydnidae (6 genera, 7 species)
The members of this family of shieldbugs have stout spiny legs, used for digging in the ground, and giving rise to the common name of burrower-bugs (Fig. 14.31). They are essentially ground-dwelling bugs that may burrow around the bases of their food-plants, though they often ascend the plant in order to feed. One of the commonest species is *Tritomegas bicolor* (formerly in *Sehirus*), the Pied shieldbug, which guards its eggs and young nymphs like *Elasmucha* in the Acanthosomatidae. *Adomerus biguttatus* (also formerly in *Sehirus*), the Cow-wheat shieldbug (Fig. 14.32) is a local species that seems

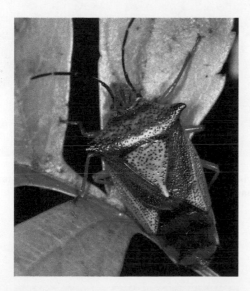

Fig. 14.29 *Acanthosoma* (Acanthosomatidae) (Photo: Peter Barnard)

Fig. 14.31 Spiny legs in the Cydnidae (Photo: Roger Key)

Fig. 14.30 *Elasmostethus interstinctus* (Acanthosomatidae) (Photo: Roger Key)

Fig. 14.32 *Adomerus biguttatus* (Cydnidae) (Photo: Roger Key)

to be declining. *Byrsinus flavicornis* has only been recorded once in Britain, though it is found in the Channel Islands and continental Europe.

Identification: Nau (2004) and Evans and Edmondson (2005).

British subfamilies and genera:
Cydninae: *Byrsinus*

Sehirinae: *Adomerus, Canthophorus, Geotomus, Legnotus, Sehirus, Tritomegas*

Family Pentatomidae (19 genera, *c.* 20 species)
Although this is the main family of shieldbugs (Fig. 14.33) in Britain , only a few species could be described as common, because the Pentatomidae as a whole prefer warmer climates. However, some species may become established, such as *Graphosoma lineatum*, which is spreading north through Europe as far as Denmark, and has been reported from the Channel Islands; there are possible sightings in the UK (Fig. 14.34). Members of the Asopinae such as *Picromerus bidens* (Fig. 14.35) are predatory on slow-moving insects such as caterpillars, but the other species are phytophagous. The more common species have vernacular names, including *Podops inuncta* (Turtle bug), *Aelia acuminata* (Bishop's mitre), *Pentatoma rufipes* (Forest bug), *Palomena prasina* (Green shieldbug), *Dolycoris baccarum* (Sloe bug) and the recently discovered *Nezara viridula*

(Southern green shieldbug). There are several introduced species of uncertain status, with others in the Channel Islands.

Identification: Nau (2004), for the common species, Evans and Edmondson (2005), and Pinchen (2009). Many species are illustrated on the British Bugs website (http://www.britishbugs.org.uk).

Fig. 14.34 *Graphosoma* (Pentatomidae) (Photo: Peter Barnard)

Fig. 14.33 Typical shield-shaped pentatomid (Photo: Colin Rew)

Fig. 14.35 *Picromerus bidens* (Pentatomidae) (Photo: Roger Key)

British subfamilies and genera:

Asopinae: *Jalla, Picromerus, Rhacognathus, Troilus, Zicrona*

Pentatominae: *Aelia, Carpocoris, Dolycoris, Eurydema, Eysarcoris, Holcostethus, Neottiglossa, Nezara, Palomena, Pentatoma, Piezodorus, Sciocoris*

Podopinae: *Graphosoma, Podops*

Family Scutelleridae (2 genera, 4 species)

In this family the scutellum is so large that it reaches the the tip of the abdomen, almost covering the wings entirely, and giving rise to the common name of tortoise-bugs. None of the four British species is common; the two species of *Eurygaster* feed on grasses and rushes, and this group can be serious pests of cereals elsewhere in Europe. The two species of *Odontoscelis* are found in sandy coastal areas.

Identification: Nau (2004) and Evans and Edmondson (2005); three of the four British species are illustrated on the British Bugs website (http://www.britishbugs.org.uk).

British subfamilies and genera:

Eurygastrinae: *Eurygaster*

Odontoscelinae: *Odontoscelis*

Family Thyreocoridae (1 genus, 1 species)

The only British species, *Thyreocoris scarabaeoides*, was formerly included in the Cydnidae. It is a small metallic black or bronze, rounded bug, only 4mm long, which superficially resembles a beetle. Locally distributed in southern England in dry and sunny locations, its food plants are species of *Viola*.

Identification: Nau (2004), Evans and Edmondson (2005), and the British Bugs website (http://www.britishbugs.org.uk).

British genus: *Thyreocoris*

SUPERFAMILY PIESMATOIDEA

Family Piesmatidae (2 genera, 2 species)

Formerly known as Piesmidae, the two species in this family are commonly called beet-bugs. Only around 2–3mm long, they have a reticulated pattern that makes them slightly resemble the lace-bugs, though the scutellum is not covered by the pronotum as it is in the Tingidae. *Piesma maculatum* is moderately common in southern Britain where it feeds on Chenopodiaceae. *Parapiesma quadratum* is also widely distributed, but more common on coastal saltmarshes; there is a subspecies *P. quadratum spergulariae* found only on the Scilly Isles. Both species are illustrated on the British Bugs website (http://www.britishbugs.org.uk).

British genera: *Parapiesma, Piesma*

SUBORDER CICADOMORPHA

The Cicadomorpha and Fulgoromorpha are sometimes linked as the Auchenorrhyncha, which was previously regarded as a suborder of the Hemiptera though there is little evidence for its monophyly. For a broad overview of these groups see Dietrich (2009). There are many useful photos of this group and notes (in German) in Remane and Wachmann (1993). Ossiannilsson (1981, 1983) has good figures and descriptions (in English) of many British species.

The Cicadomorpha includes the Cercopoidea (froghoppers or spittlebugs), Cicadoidea (cicadas) and Membracoidea (leafhoppers and treehoppers). All feed on the sap of terrestrial vascular plants, none being predatory or aquatic, and they constitute the major group of insects in grassland ecosystems, with many leafhoppers also found on trees and shrubs. Worldwide, some leafhoppers can be regarded as pests, though they are rarely a serious problem in Europe. Many have enlarged back legs that give them the ability to jump, hence the term 'hopper' in so many of their vernacular names.

SUPERFAMILY CERCOPOIDEA

The froghoppers, or spittlebugs, feed on xylem and this causes them to produce large quantities of dilute excreta, unlike phloem-feeders. Froghoppers get their name from the frog-like shape of their large heads.

Family Aphrophoridae (3 genera, 9 species)

The members of this family were previously placed in the Cercopidae. The nymphs of all species use the copious quantities of excreta to produce the foamy mass known commonly as cuckoo-spit (Fig. 14.36). The sticky froth not only hides the nymphs from view, but it also deters predators and helps to prevent dehydration on exposed plants. Most adults are brownish and inconspicuous (Fig. 14.37) though they may have exhibit many colour varieties. The more common species include *Neophilaenus lineatus*, found on grasses; the four species of *Aphrophora* on several trees and shrubs; and the Common froghopper, *Philaenus spumarius*, which is found on a wide variety of plants, frequently in gardens.

Identification: LeQuesne (1965); all except one species are illustrated on the British Bugs website (http://www.britishbugs.org.uk).

Fig. 14.36 *Philaenus spumarius* nymph and 'cuckoo-spit' (Aphrophoridae) (Photo: Roger Key)

Fig. 14.38 *Cercopis vulneratus* (Cercopidae) (Photo: Colin Rew)

Fig. 14.37 Aphrophoridae (Photo: Colin Rew)

Fig. 14.39 *Cicadetta montana* (Tibicinidae) (Photo: Roger Key)

British genera: *Aphrophora, Neophilaenus, Philaenus*

Family Cercopidae (1 genus, 1 species)
Since the Aphrophoridae were removed as a separate family, the only remaining British species in this family is *Cercopis vulnerata*. It is a striking red and black species (Fig. 14.38) found on a wide variety of plants, but often associated with trees. The nymphs are rarely seen as they feed underground on roots.

British genus: *Cercopis*

SUPERFAMILY CICADOIDEA

Family Tibicinidae (1 genus, 1 species)
The Tibicinidae are sometimes regarded as a subfamily of the Cicadidae, which includes the well-known tropical cicadas. The only British species is

the scarce *Cicadetta montana*, the New Forest cicada (Fig. 14.39), which is listed on the Wildlife and Countryside Act 1981 and is also on the UKBAP list.

British genus: *Cicadetta*

SUPERFAMILY MEMBRACOIDEA

Family Cicadellidae (*c.* 105 genera, *c.* 285 species)
The leafhoppers in the family Cicadellidae (previously known as Jassidae) form one of the largest families of British bugs; they are found on a wide variety of plants and sometimes cause feeding damage, though rarely of economic significance. Although some species have characteristic colour patterns, recognition of many species in such a large group requires microscopic examination. There are

Fig. 14.40 *Cicadella viridis*, pale colour variety (Cicadellidae) (Photo: Roger Key)

Fig. 14.41 *Cicadella viridis*, dark colour variety (Cicadellidae) (Photo: Roger Key)

many common species, including *Cicadella viridis* (Figs. 14.40 & 14.41) found on grasses, which has a number of colour varieties. Perhaps the most striking cicadellid is *Ledra aurita*. It is the largest British species and cannot jump but its unique shape and cryptic coloration make it very hard to spot on tree bark (Fig. 14.42).

Four species are on the UKBAP list: *Chlorita viridula* (Sea-wormwood leafhopper), *Doratura impudica* (Large dune leafhopper), *Euscelis venosus* (Carline thistle leafhopper) and *Macrosteles cyane* (Pondweed leafhopper).

Identification: LeQuesne (1965, 1969) and LeQuesne and Payne (1981). Other useful volumes for identification (in French) are Ribaut (1936, 1952), with the nomenclature updated by della Giustina (1989). Some species are illustrated on the British Bugs website (http://www.britishbugs.org.uk).

Fig. 14.42 *Ledra aurita* nymph (Cicadellidae) (Photo: Roger Key)

British subfamilies and genera:

Agalliinae: *Agallia, Anaceratagallia, Austroagallia*

Aphrodinae: *Anoscopus, Aphrodes, Planaphrodes, Stroggylocephalus*

Cicadellinae: *Cicadella, Graphocephala*

Deltocephalinae: *Adarrus, Allygidius, Allygus, Anoplotettix, Arocephalus, Arthaldeus, Athysanus, Balclutha, Cicadula, Colladonus, Conosanus, Cosmotettix, Deltocephalus, Doratura, Ebarrius, Elymana, Erotettix, Errastunus, Euscelidius, Euscelis, Graphocraerus, Grypotes, Hardya, Hesium, Idiodonus, Jassargus, Lamprotettix, Limotettix, Macrosteles, Macustus, Metalimnus, Mocuellus, Mocydia, Mocydiopsis, Ophiola, Opsius, Paluda, Paralimnus, Paramesus, Platymetopius, Psammotettix, Recilia, Rhopalopyx, Rhytistylus, Sagatus, Sardius, Sonronius, Sorhoanus, Speudotettix, Streptanus, Thamnotettix, Turrutus, Verdanus*

Dorycephalinae: *Eupelix*

Evacanthinae: *Evacanthus*

Iassinae [formerly spelled Jassinae]: *Batracomorphus, Iassus*

Idiocerinae: *Acericerus, Idiocerus, Metidiocerus, Populicerus, Rhytidodus, Stenidiocerus, Tremulicerus*

Ledrinae: *Ledra*

Macropsinae: *Hephathus, Macropsis, Oncopsis, Pediopsis*

Megophthalminae: *Megophthalmus*

Typhlocybinae: *Aguriahana, Alebra, Alnetoidea, Arboridia, Austroasca, Chlorita, Dikraneura, Edwardsiana, Emelyanoviana, Empoasca, Erythria, Eupterycyba, Eupteryx, Eurhadina, Fagocyba, Forcipata,*

Fig. 14.43 *Centrotus cornutus* (Membracidae) (Photo: Roger Key)

Fig. 14.44 Cixiidae (Photo: Colin Rew)

Hauptidia, Kybos, Lindbergina, Linnavuoriana, Notus, Ossiannilssonola, Ribautiana, Typhlocyba, Wagneripteryx, Zonocyba, Zygina, Zyginidia

Family Membracidae (2 genera, 2 species)
There are just two species of treehoppers in Britain; *Gargara genistae* is a scarce inhabitant of broom in southern England, and *Centrotus cornutus* (Fig. 14.43) is polyphagous, more widespread though still local. Both species have the large backward-pointing process on the pronotum, which is a main characteristic of this family; they also move slowly and cannot jump.

Identification: LeQuesne (1965). Both species are illustrated on the British Bugs website (http://www.britishbugs.org.uk).

British genera: *Centrotus, Gargara*

Family Ulopidae (2 genera, 2 species)
This family was formerly treated as a subfamily of the Cicadellidae. The two species of *Ulopa* are small, up to 4 mm, usually brachypterous and with coarse puncturing on the fore wings. *U. reticulata* is the most common species, associated with heather on heathland.

Identification: LeQuesne (1965); *U. reticulata* is illustrated on the British Bugs website (http://www.britishbugs.org.uk).

British genus: *Ulopa*

SUBORDER FULGOROMORPHA

The planthoppers are dominated by the large family Delphacidae. There are many useful photos of this group and notes (in German) in Remane and Wachmann (1993). Ossiannilsson (1978) has good figures and descriptions (in English) of many British species.

Family Cixiidae (4 genera, 12 species)
The Cixiidae (Fig. 14.44) resemble the Delphacidae, though they lack the tibial spurs. The nymphs of most species are subterranean root-feeders, often on grasses; the adult food-plant associations are not always known, but several species seem to have precise habitat requirements. *Cixius nervosus* is one of the most common species, found on a wide variety of trees and shrubs.

Identification: LeQuesne (1960) can be used, though it is somewhat out of date.

British genera: *Cixius, Oliarus, Tachycixius, Trigonocranus*

Family Delphacidae (*c.* 40 genera, *c.* 75 species)
This is by far the largest family of planthoppers in Britain and can be distinguished by a large articulated spur on the apex of each hind tibia. Brachyptery is quite common, and some species such as *Delphax pulchellus* (Figs. 14.45 & 14.46) show clear sexual dimorphism. Nearly all species are associated with low-growing plants such as grasses and no species are restricted to trees or taller shrubs. Common species include *Javesella pellucida* and *Muellerianella fairmairei*.

Two species are on the UKBAP list: *Eurysa douglasi* (Chalk planthopper) and *Ribautodelphax imitans* (Tall fescue planthopper).

Identification: LeQuesne (1960) can be used with care, but it is very out of date, especially with regard to nomenclature. Many species are illustrated on

Fig. 14.45 *Delphax pulchellus* male (Delphacidae) (Photo: Roger Key)

Fig. 14.47 *Issus muscaeformis* (Issidae) (Photo: Roger Key)

Fig. 14.46 *Delphax pulchellus* female (Delphacidae) (Photo: Roger Key)

the British Bugs website (http://www.britishbugs.org.uk).

British subfamilies and genera:
Asiracinae: *Asiraca*
Delphacinae: *Acanthodelphax*, *Calligypona*, *Chloriona*, *Conomelus*, *Criomorphus*, *Delphacinus*, *Delphacodes*, *Delphax*, *Dicranotropis*, *Ditropis*, *Euconomelus*, *Euides*, *Eurybregma*, *Eurysa*, *Eurysanoides*, *Eurysella*, *Eurysula*, *Florodelphax*, *Gravesteiniella*, *Hyledelphax*, *Javesella*, *Kosswigianella*, *Laodelphax*, *Megamelodes*, *Megamelus*, *Muellerianella*, *Muirodelphax*, *Nothodelphax*, *Oncodelphax*, *Paradelphacodes*, *Paraliburnia*, *Ribautodelphax*, *Scottianella*, *Stiroma*, *Struebingianella*, *Xanthodelphax*
Kelisiinae: *Anakelisia*, *Kelisia*
Stenocraninae: *Stenocranus*

Family Issidae (1 genus, 2 species)
This family includes only two British species, both in the genus *Issus*. They differ from the other Fulgoromorpha in having numerous crossveins in the fore wings. *Issus coleoptratus* is the more common species, found on a range of woody plants and trees. *I. muscaeformis* (Fig. 14.47) is known from only a few specimens.

Identification: *I. coleoptratus* is illustrated on the British Bugs website (http://www.britishbugs.org.uk).

British genus: *Issus*

Family Tettigometridae (1 genus, 1 species)
The only British species in this family is *Tettigometra impressopunctata*, a rare species confined to dry habitats such as chalk or dunes in southern England.

Identification: LeQuesne (1960).

British genus: *Tettigometra*

SUBORDER STERNORRHYNCHA

This suborder includes the most important plant pests in the Hemiptera. All are sap-feeders, all cause some direct damage by their feeding, and some also transmit viruses and other diseases of agriculturally or horticulturally important plant species. Coupled with this is their ability to reproduce rapidly, especially the aphids, so that populations increase very quickly. Other groups like scale insects have a waxy coating that repels contact insecticides, and the rapid rates of reproduction also enable the development of insecticide resistance. The whole group is well defined by morphological characters, including the rostrum apparently arising on the prosternum between the front coxae, giving the Sternorrhyncha their name. However, there is great variability in morphological development, ranging from the typical bug shape of Psylloidea to the extreme reduction in female scale insects. There are also complex lifecycles in some groups; aphids have alternation of host plants, coupled with many morphological forms, parthenogenesis and vivipary. Whiteflies and scale insects have a complete change in morphology between the nymphal and adult stages, known as the puparium, which is analogous with the pupal stage of endopterygote insects.

For some groups in this suborder it is necessary to consult specialist literature, not listed here, and the account by Martin and Webb (1999) should be examined for such papers. For a world overview of the Sternorrhyncha see Gullan and Martin (2009).

SUPERFAMILY ALEYRODOIDEA

All the British whiteflies are in the family Aleyrodidae. This is predominantly a tropical group, hence the relatively small British fauna of 19 species, which is in complete contrast to the aphids where the British fauna is relatively large. Adult whiteflies are very small insects, only 1–2 mm long, and are covered with white wax, even on the wings. They usually reproduce sexually but are not viviparous. Despite their small size, the eggs can be conspicuous because they are often arranged in arcs or circles. The nymphal stages are very complex; the newly hatched nymph is mobile and seeks a suitable feeding site, but the next three instars are entirely sessile so cannot relocate if the conditions become less than favourable. The fifth instar is functionally a pupa, and is termed a puparium.

Identification of whiteflies is not easy; the taxonomy is not based on the adults, which are difficult to identify when not associated with other stages of the life cycle, or with a particular host plant. The young nymphs are also difficult to identify, and the classification of the group is based on the puparial stage. Not only does this stage show useful taxonomic characters, but it is always associated with the appropriate host plant, thus automatically providing host data. Even the deserted puparial 'case' is sufficient to recognize most species (Fig. 14.48). A complication arises from the fact that puparia can in fact vary in response to their immediate surroundings, such as leaf texture. Fortunately this is less of a problem with the restricted British fauna, though it can make the identification of trop-

Fig. 14.48 *Aleurotuba jelinekii* puparial skin (Aleyrodidae) (Photo: Peter Barnard)

Fig. 14.49 Outdoor whitefly (Aleyrodidae) (Photo: Roger Key)

ical pest species a highly skilled process.

Worldwide there are over 1500 described species of whiteflies, with no doubt many more yet to be recognized, and the British fauna of less than 20 species is therefore very small; even in southern Europe the number of species is only around 50 or so.

Family Aleyrodidae (13 genera, 19 species)
Only a dozen or so of the British species can survive outdoors (Fig. 14.49), with the others confined to glasshouses. The two species of *Aleurochiton, A.*

Fig. 14.50 *Trialeurodes vaporarium*, adults and nymphs (Aleyrodidae) (Photo: Roger Key)

SUPERFAMILY APHIDOIDEA

The aphids are probably the best known of the plant-feeding bugs, if only because the domestic gardener spends much time trying to eradicate them from vegetables and flowers; some species are of course serious pests in agricultural and horticultural establishments. Although some feed on the mature leaves of plants, sometimes through the bark, or even on roots, most species prefer new shoots and young leaves, and this method of feeding can cause much damage to the overall growth of the plant. Many species are restricted to certain plant genera, so that host data can be helpful for identification of the aphids, though others are widely polyphagous. Not all species are like the familiar garden greenfly; some are brown, pink or red; a few produce waxy coatings like the mealybug scale-insects, which in some cases is so extensive that it hides the insects completely. Even more confusing is that the complex lifecycle of aphids leads to the development of morphologically different forms, either at different times of the year and on different host plants, or else co-existing with each other. Many species reproduce parthenogenetically during the main feeding season and are also viviparous, giving birth to live young; these stages are usually apterous. Those species that have an alternative winter host plant will produce winged sexual forms in the autumn, which lay overwintering eggs. Thus, the aphids have the benefit of rapid population growth during the summer months, coupled with the genetic advantages of sexual reproduction. There are many variations on this theme, some groups never reproducing sexually, with just the two phases of apterous, viviparous females alternating with alate, oviparous ones. Further information on the many complexities of aphid life cycles is given in books such as Dixon (1973, 1985) and Blackman (1974).

aceris and *A. acerinus*, are both found on maple trees, and both are bivoltine, but most other outdoor species are univoltine. Species that live on deciduous trees usually overwinter as puparia, whereas those on evergreen trees remain as nymphs through the winter. One of the best-known species is the Cabbage whitefly, *Aleyrodes proletella*, which flies up in dense clouds when infested cabbage plants are disturbed; it can have several generations in the summer months. The other common species is *Trialeurodes vaporariorum*, the Greenhouse whitefly, which cannot survive outdoors but can pass through several generations each year in heated glasshouses, and even in domestic houses (Fig. 14.50).

Occasional pest species are intercepted by plant quarantine agencies, but these can rarely become established in Britain even in glasshouses. Many tropical pests are now cosmopolitan, and the relatively small numbers of these are covered in specialist literature. The notorious pest *Bemisia tabaci* is occasionally found in Britain under glass, but does not seem to persist.

The British species were covered by Mound (1966), though a few species have been found since, and there is useful data on host plants and distributions in the world catalogue by Mound and Halsey (1978). A more recent world catalogue is by Martin and Mound (2007). For other specialized references see Martin and Webb (1999).

British genera: *Aleurochiton, Aleuropteridis, Aleurotuba, Aleurotulus, Aleyrodes, Asterobemisia, Bemisia, Dialeurodes, Filicaleyrodes, Pealius, Siphoninus, Tetralicia, Trialeurodes*

Many species of aphid are associated with ants, which feed on the sugary 'honeydew' excreted from the anus of the aphids (Fig. 14.51); in return the ants help to protect the aphids from predators, and they also move them to more suitable locations, a process known aptly as 'farming'. Gardeners who see numbers of ants trailing up and down an unhealthy plant often mistakenly assume that the ants are responsible for the damage, not realizing that aphids are the true cause. Even those aphids that feed underground on roots may be attended by ants (Fig. 14.52) and it is common to see ants, predators such as ladybird larvae and parasitic wasps, all part of an aphid-based community (Fig. 14.53). Nearly

Fig. 14.51 Aphids with attendant ants (Aphididae) (Photo: Peter Barnard)

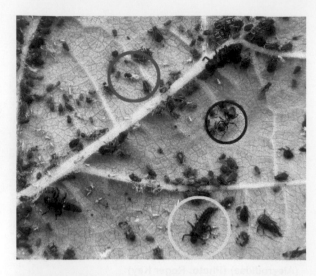

Fig. 14.53 Aphid community with ant (black circle), ladybird larva (yellow circle) and parasitic wasp (red circle) (Photo: Peter Barnard)

Fig. 14.52 Root aphids (*Trama*) with ants (Aphididae) (Photo: Peter Barnard)

Fig. 14.54 Prominent cornicles (siphunculi) on aphids (Photo: Colin Rew)

all aphids have a unique morphological character, a pair of siphunculi or cornicles on the dorsal surface of the abdomen, which may be short processes or elongated tubes, sometimes of a different colour to the rest of the body (Fig. 14.54). Some early texts asserted that the cornicles secrete honeydew, but it is now known that they can secrete a waxy fluid that deters potential predators, as well as producing alarm pheromones; thus there is communication between the numerous individuals in an aphid colony.

There are around 5000 species of aphids known worldwide, so the British fauna of over 600 species is a remarkably high percentage of the world species. This reflects the fact that aphids are principally a temperate group, poorly represented in tropical regions, in complete contrast to the

96

Aleyrodidae. Inevitably, such a successful group has been carried round much of the world on plant hosts, but there are relatively few species that can be considered as serious cosmopolitan pests; nonetheless the damage caused by these is out of all proportion to the small number of species. The works by Blackman and Eastop (1994, 2000) should be consulted for detailed information.

Remaudière and Remaudière (1997) produced a world catalogue of aphid species, but this is now largely superseded by the on-line world list (http://aphid.speciesfile.org).

Identification of such a complex group is not easy, and there are no recent guides to the whole family in Britain. The last attempt to cover the whole fauna was by Theobald (1926, 1927, 1929) but this work is now of historical interest only. However, there are several handbooks covering the major subgroups, including Stroyan (1977, 1984), Blackman (2010) and Carter (1982). The earlier book by Blackman (1974) includes a key to the main genera, together with a guide to over 100 of the most common species. The Scandinavian fauna is quite similar to that of Britain, so the keys by Heie (1980, 1982, 1986, 1992, 1994, 1995) are also useful. Dixon and Thieme (2007) is a comprehensive guide to the aphids on deciduous trees. A brief world overview of aphids is given by Sorensen (2009).

Family Aphididae (*c.* 160 genera, *c.* 630 species)
The higher classification of the aphids, including the familiar greenfly and blackfly, is complex and contunally changing. Currently all species are placed in the family Aphididae, and groups such as Anoeciidae, Callaphididae, Chaitophoridae, Hormaphididae, Lachnidae, Mindaridae, Pemphigidae, Phloeomyzidae and Thelaxidae are now subgroups within the Aphididae.

Inevitably in such a large and important group there are numerous significant species, often best known as pests on agricultural crops or on vegetables and ornamental plants in domestic gardens. Some of the better known species are *Eriosoma lanigerum*, the Woolly aphid on apple trees (Fig. 14.55); *Brevicoryne brassicae*, the Cabbage aphid; *Aphis fabae*, the Black bean aphid or Blackfly (not to be confused with the dipteran family Simuliidae); *Myzus persicae*, the Peach–potato aphid; *Metopolophium dirhodum*, the Rose–grain aphid; *Macrosiphum rosae*, the Rose aphid; *Rhopalosiphum padi*, the Bird cherry aphid; several willow aphids (Fig. 14.56), and many more.

Because of the ever-changing groupings within this family, no attempt is made here to place genera

Fig. 14.55 Woolly aphid, *Eriosoma lanigerum* (Aphididae) (Photo: Peter Barnard)

Fig. 14.56 Giant willow aphid, *Tuberolachnus salignus* (Aphididae) (Photo: Roger Key)

into subfamilies, but further information can be found on the Aphid Species File website (http://aphid.speciesfile.org).

British genera: *Acaudella, Acaudinum, Acyrthosiphon, Aloephagus, Amphorophora, Anoecia, Anthracosiphon, Anuraphis, Aphis, Aploneura, Appendiseta, Aspidaphis, Atheroides, Aulacorthum, Baizongia, Betulaphis, Brachycaudus, Brachycolus, Brevicoryne, Calaphis, Callipterinella, Capitophorus, Caricosipha, Cavariella, Cerataphis, Ceruraphis, Chaetosiphella, Chaetosiphon, Chaitophorus, Chromaphis, Cinara, Clethrobius, Clypeoaphis, Colopha, Coloradoa, Corylobium, Cryptaphis, Cryptomyzus, Cryptosiphum, Crypturaphis, Ctenocallis, Decorosiphon, Delphiniobium, Diuraphis, Drepanosiphum, Dysaphis, Elatobium, Ericaphis, Eriosoma, Eucallipterus, Eucarazzia, Euceraphis, Eulachnus, Forda, Geoica,*

Glyphina, Gootiella, Hamamelistes, Hayhurstia, Hyadaphis, Hyalopteroides, Hyalopterus, Hydaphias, Hyperomyzus, Idiopterus, Illinoia, Impatientinum, Iziphya, Jacksonia, Juncobia, Kaltenbachiella, Lachnus, Laingia, Linosiphon, Liosomaphis, Lipamyzodes, Lipaphis, Longicaudus, Macrosiphoniella, Macrosiphum, Maculolachnus, Megoura, Megourella, Melanaphis, Melaphis, Metopeurum, Metopolophium, Microlophium, Microsiphum, Mimeuria, Mindarus, Monaphis, Muscaphis, Myzaphis, Myzocallis, Myzodium, Myzotoxoptera, Myzus, Nasonovia, Nearctaphis, Neotoxoptera, Ovatomyzus, Ovatus, Pachypappa, Pachypappella, Paczoskia, Panaphis, Paoliella, Paracletus, Paramyzus, Patchiella, Pemphigus, Pentalonia, Periphyllus, Phloeomyzus, Phorodon, Phyllaphis, Pleotrichophorus, Plocamaphis, Prociphilus, Protaphis, Protrama, Pseudacaudella, Pseudobrevicoryne, Pterocallis, Pterocomma, Rhodobium, Rhopalomyzus, Rhopalosiphoninus, Rhopalosiphum, Schizaphis, Schizolachnus, Semiaphis, Sipha, Sitobion, Smynthurodes, Staegeriella, Staticobium, Stomaphis, Subacyrthosiphon, Subsaltusaphis, Symydobius, Takecallis, Tetraneura, Thecabius, Thelaxes, Therioaphis, Thripsaphis, Tinocallis, Toxoptera, Toxopterina, Trama, Trichosiphonaphis, Tubaphis, Tuberculatus, Tuberolachnus, Uroleucon, Utamphorophora, Vesiculaphis, Wahlgreniella

SUPERFAMILY COCCOIDEA

The members of this subfamily are arguably the most highly modified of all the plant-feeding bugs, as many of the females are sessile with their legs reduced or absent. Although known generally as scale insects, they exhibit a wide range of morphological forms; in some groups the females are mobile, though still always wingless. Some species reproduce parthenogenetically, but sexual reproduction may be more common than is apparent, because male coccids are rarely seen, and are unknown for many species. The males are always small, delicate insects with only the front wings present, which makes them superficially resemble small Diptera. They cannot feed, cannot fly far, and are generally short-lived. Males are hard to identify, and the classification of the group is based on the adult female stage.

Newly hatched nymphs are known as crawlers because they are mobile and can seek out suitable feeding sites but, once they have settled, the nymphs secrete a waxy coating, which can vary greatly between different families. In the Pseudococcidae, Margarodidae and most Eriococcidae the wax is loose and powdery; in the Asterolecaniidae the wax

forms a hard translucent covering that is separate from the body; the Ortheziidae secrete characteristic flat plates of wax; and in the Coccidae the 'scale' is formed from the dorsal surface of the body itself.

The morphological variety is so great in this group that it is not easy to define them, but when legs are present they have only a single claw, unlike the paired claws seen in all other Hemiptera. The scale-like forms can only be confused with nymphal Aleyrodoidea or a few highly modified aphids. As in the aphids there some economically significant plant pests among the Coccoidea; the waxy coating makes them highly resistant to contact insecticides and their inconspicuous appearance makes them difficult to see, even during rigorous plant quarantine procedures, which has enabled some species to become cosmopolitan pests. On a more positive side, a few species are useful in other parts of the world, where their secretions are used to create products such as dyestuffs like cochineal, the classic example being the Lac insect in Asia whose wax is the raw ingredient of shellac. Like aphids, associations with ants are often important, as scale insects excrete large amounts of watery honeydew (Fig. 14.57).

Like the Aleyrodoidea, the Coccoidea are principally a tropical group, and the British fauna is thus relatively small, with up to about 150 species recorded out of the world total of around 8000. However, several so-called British species can only

Fig. 14.57 Scale insects with attendant ants (Photo: Peter Barnard)

surive in glasshouses or similar protected environments, and many species are intercepted each year by plant quarantine services.

The higher classification of the Coccoidea is a matter of some debate, with some small families not being recognized as distinct by all authors. There are no recent guides to cover all the British groups; there is a key to families in Dolling (1991) and some families are covered by appropriate guides, listed below, but the student of this group will need to become familiar with many European and worldwide publications such as Danzig (1967, 1993), Kosztarab and Kozár (1988) and Williams and Watson (1988a, 1988b, 1990). The only complete British checklist (Boratynski & Williams, 1964) is very out of date.

Family Aclerdidae (1 genus, 1 species)
The sole British species in this family is *Aclerda tokionis*, a grass scale, and it is only found under glass. It is a cosmopolitan species that feeds on various bamboos.

British genus: *Aclerda*

Family Asterolecaniidae (4 genera, 5 species)
The pit scales have a hard, translucent covering with peripheral glands all around the body that secrete glassy filamentous fringes. The three native species of *Asterodiaspis* all feed on oak twigs where their presence creates a pit with a raised rim, a kind of rudimentary gall. *Planchonia arabidis* is also indigenous, and there are two introduced species, *Bambusaspis bambusae* and *Russellaspis pustulans*.

British genera: *Asterodiaspis, Bambusaspis, Planchonia, Russellaspis*

Family Coccidae (14 genera, 30 species)
The so-called soft scales are the best known of the scale insects in general, and this group includes many of the pest species, important both in domestic gardens and in commercial horticulture. The actual scale is formed by a thickened dorsal surface of the body, but some species secrete powdery or translucent wax coverings as well. If legs are present beyond the crawler stage they are always hidden beneath the scale, most adults remain sessile though walking is possible in some species at a laboriously slow pace. The cryptic nature of soft scales can make them difficult to see until they build up to large numbers, by which time control is more difficult. Many are attended by ants, and this can provide a clue to their presence.

Worldwide this is a large family, with over 1100 species known, and around 30 are regarded as

Fig. 14.58 *Chloropulvinaria floccifera*, a cushion scale (Coccidae) (Photo: Roger Key)

British, although there are probably a similar number of exotic species recorded as introductions. Identification of the British species is not easy because there are no recently published keys. Hodgson (1994) provided a generic key, but the student of this group will need to become familiar with works such as Kosztarab and Kozár (1988) and Danzig (1967). Some of the more common species are illustrated on horticultural websites, but identification based on general appearance, even when coupled with host-plant data, is not always reliable.

Among the more common species, *Parthenolecanium corni*, the Brown scale, is a pest on hazel, as well as some other trees and soft fruits. *Pulvinaria vitis*, the Woolly vine scale, lives on many fruit trees as well as grape vines; and *Eriopeltis* species can be common on grasses. *Chloropulvinaria floccifera*, the Cushion scale (Fig. 14.58) and *Pulvinaria regalis* are among the exotic species that have become well established in glasshouses as well as outside. *P. regalis* can be particularly common on garden or street trees.

British genera: *Chloropulvinaria, Coccus, Eriopeltis, Eulecanium, Exaeretopus, Lecanopsis, Lichtensia, Luzulaspis, Palaeolecanium, Parafairmairia, Parasaissetia, Parthenolecanium, Pulvinaria, Saissetia, Vittacoccus*

Family Diaspididae (18 genera, 28 species)
Armoured scales are rather different from other coccoids in that the secreted scale is separate from the body of the insect, surrounding it loosely, and with its own distinct shape; moreover, the scales of early instars are incorporated into those of later stages, sometimes forming overlapping, concentric

structures. The diaspids can be the most difficult group to recognize on plants because the scales are easily taken to be markings on stems or fruits. They are unusual in feeding on parenchyma tissue rather than on sap; hence they do not excrete honeydew and are not attended by ants.

There are well over 2000 species of diaspids known worldwide, so the British fauna is very small, with around a dozen native species and more exotic ones that have become established under glass, or rarely outside. Many more species are regularly intercepted by quarantine authorities.

The native species include *Carulaspis carueli* and *C. juniperi*, which are found on juniper and cypress, and the two most common species are probably *Chionaspis salicis* and *Lepidosaphes ulmi*, both feeding on deciduous trees and shrubs. *L. ulmi* is known as the Mussel scale, from its shape, and can be a serious pest of apple trees; *Aulacaspis rosae*, the Rose scale, can be a pest on rose bushes in nurseries.

As with the Coccidae there is no recent key to the British species. For the exotic species Williams and Watson (1988a) is invaluable, and Kosztarab and Kozár (1988) or Danzig (1967) are the most useful works for the native species.

British genera: *Acutaspis, Aonidia, Aspidiotus, Aulacaspis, Carulaspis, Chionaspis, Chrysomphalus, Diaspidiotus, Diaspis, Dynaspidiotus, Furchadaspis, Gymnaspis, Hemiberlesia, Ischnaspis, Kuwanaspis, Lepidosaphes, Pinnaspis, Unaspis*

Family Eriococcidae (7 genera, 15 species)
The felted scales are usually covered in powdery wax and therefore resemble the pseudococcid mealybugs though the females are usually shorter and more rounded. The native British species are not well studied, but many seem to feed on grasses, though there are important exceptions. *Eriococcus devoniensis* feeds entirely on *Erica tetralix*, and causes the shoots to curve round with the insect on the inside. *Pseudochermes fraxini* can be common on branches of ash trees, and *Cryptococcus fagisuga* lives on the trunks of beech, where it may form large conspicuous colonies that can render the trees susceptible to fungal damage. Cochineal insects of the genus *Dactylopius* occasionally appear on potted cactus plants, and there are numerous other exotic species recorded, most intercepted during quarantine.

The native species of eriococcids were covered by Williams (1985).

British genera: *Acanthococcus, Cryptococcus, Gossyparia, Greenisca, Kuwanina, Noteococcus, Ovaticoccus, Pseudochermes*

Family Kermesidae (1 genus, 3 species)
Just three species of *Kermes* are found in Britain, and all live on oak trees, where the mature females swell up into hard globular structures up to 5 mm across that act as egg-chambers. Dolling (1991) provides references to the British species.

British genus: *Kermes*

Family Margarodidae (2 genera, 2 species)
The margarodids have a powdery wax coating like that of mealybugs. There are just two British species: *Matsucoccus pini* lives under the bark of Scots pine and is parthenogenetic; *Steingelia gorodetskia* is a bisexual species that lives on birch trees. A few exotic species are occasionally introduced to Britain, and one significant pest is the Cottony cushion scale, *Icerya purchasi*, a native of Australia that has spread to other countries. Its control in America by the introduced ladybird *Rodolia cardinalis* is one of the classic examples of biological control; it is a serious pest on citrus trees and can easily survive under glass in Britain, so vigilance is needed against its introduction.

British genera: *Matsucoccus, Steingelia*

Family Ortheziidae (4 genera, 5 species)
The members of this small family have the colourful name of ensign scales, from the distinctive large overlapping flat plates of wax that cover most or all of the body. In adult females these plates extend far behind the body to produce an ovisac, or brood pouch, in which the eggs are laid. Despite this characteristic appearance, ortheziids are actually the least modified of the coccoid families; all stages are fully mobile with normal legs and antennae and most are bisexual, although facultative parthenogenesis occurs in some species. The best-known species is *Orthezia urticae* (Fig. 14.59), which is common on several plants and has even been found on salt-marshes with periodic submersion in seawater. There are a few occasionally introduced species.

British genera: *Arctorthezia, Newsteadia, Orthezia, Ortheziola*

Family Pseudococcidae (16 genera, *c.* 50 species)
Mealybugs get their common name from the powdery or floury covering of wax that they secrete (Fig. 14.60). Many species are mobile, with well-developed legs but others, especially the subterranean species are apodous. Of around 30 native British species, about two-thirds live on grasses, sometimes on the roots where they are associated

Fig. 14.59 *Orthezia urticae* (Ortheziidae) (Photo: Roger Key)

Fig. 14.60 Mealybugs (Pseudococcidae) (Photo: Roger Key)

with ants. As a primarily tropical group it seems likely that such species are able to survive by exploiting sheltered habitats either underground or in leaf-sheaths. No British species can be considered a pest, but many introduced species can cause much damage to cultivated plants under glass, where they attack a wide range of plants from cacti to bulbs and can also transmit virus diseases. Although most native species live on herbaceous plants, one of the commonest British species is *Phenacoccus aceris*, which lives on many trees and woody shrubs, especially gorse.

The most recent guide to the British species was by Williams (1962), who included 42 species in the British fauna, of which 13 were regarded as established glasshouse species. This work is still usable,

although a further half-dozen species are now considered to be established, and there are many more exotic species regularly intercepted by quarantine authorities.

British genera: *Antonina, Atrococcus, Balanococcus, Brevennia, Dysmicoccus, Euripersia, Heterococcus, Peliococcus, Phenacoccus, Planococcus, Pseudococcus, Rhizoecus, Spilococcus, Trionymus, Trochiscococcus, Vryburgia*

SUPERFAMILY PHYLLOXEROIDEA

The two families in this group are sometimes included in the Aphidoidea though they lack cornicles (siphunculi). All species are oviparous, even if no sexual reproduction has occurred, and they overwinter as nymphs, not as eggs.

Family Adelgidae (7 genera, 13 species)
The conifer woolly aphids are an entirely Holarctic group that have a complex lifecycle on conifers. Over a two-year period there are different generations, both sexual and asexual, on spruce trees, where the second asexual generation produces galls; there is also a series of asexual generations on other conifers such as *Pinus*, *Larix* and *Pseudotsuga*, depending on the species of adelgid.

Confusingly, there are two generic classifications in common use within the adelgids; some authors place all species the two genera *Adelges* and *Pineus*; others use a more complex system with many more genera recognized. The differences arise because each system is based on different sets of morphological data. This is not an easy group to identify, and the host plant and gall morphology can be useful. Carter (1971) enables the identification of the British species, though the nomenclature is now out of date.

British genera: *Adelges, Cholodkovskya, Dreyfusia, Eopineus, Gilletteella, Pineus, Sacchiphantes*

Family Phylloxeridae (2 genera, 3 species)
Unlike the adelgids all the members of this family are found only on deciduous trees. *Phylloxera glabra* is probably the only common species in Britain, where it lives on deciduous oak trees. There are also supposedly records of *Phylloxera quercus*, though these two species have not always been carefully separated, and each uses different primary hosts. *Moritziella corticalis* is an introduced species that also lives on oak. Two other species, *Phylloxerina populi* and *Phylloxerina salicis* are occasionally found on Lombardy poplar and willows respectively, but both are probably introductions. The most notorious

member of this family is the Grape phylloxera, *Daktulosphaera* [= *Viteus*] *vitifoliae*, a serious pest of grapevines in the 19th century that was only controlled by selective breeding of resistant vine varieties. Part of its lifecycle is an asexual phase that feeds on vine roots, severely stunting the growth of the plant. There are occasional reports of the species in Britain in recent years, but most modern vines are grafted on to American rootstocks, which are resistant.

British genera: *Moritziella, Phylloxera*

SUPERFAMILY PSYLLOIDEA

Although the jumping plant lice, or psyllids, are classified in the Sternorrhyncha, they can be mistaken for cicadellid leafhoppers, both from their general appearance and from their ability to jump. However, a close inspection shows that even the nymphs have the typical rostrum arising between the front coxae as in all Sternorrhyncha. The Psylloidea have many distinguishing characters, including more complex wing venation, two-segmented tarsi (three-segmented in the rest of the suborder) and long, multi-segmented antennae (Fig. 14.61). They can also be mistaken for Psocoptera, but again a closer look will show that the Psocopteran biting mouthparts, with long palps, are quite different. All species reproduce sexually, and eggs are laid on the food plants, on which the nymphs are phloem-feeders, with some species causing gall formation.

Identification of the British Psylloidea is relatively straightforward, using the keys of Hodkinson and White (1979) for the adults, and White and Hodkinson (1982) for fifth instar nymphs. The latter is particularly useful because nymphs can often be found unassociated with adults at certain times of the year. A few species have been added to the British list since these two handbooks were published, and the higher classification has also changed. The former six families have now been reduced to four, with the Aphalaridae, Liviidae and Spondyliaspididae now included in the Psyllidae, and the family Calophyidae created for *Calophya*. Another useful work for identification is Ossiannilsson (1992), which covers most of the British species. Many species are illustrated on the British Bugs website (http://www.britishbugs.org.uk).

Family Calophyidae (1 genus, 1 species)
The only British species in this family is *Calophya rhois*, known from a single record in the Hebrides. It feeds on ornamental shrubs, and is an accidental introduction.

British genus: *Calophya*

Family Homotomidae (1 genus, 1 species)
The sole British species is *Homotoma ficus*, introduced from southern Europe; it is the only British psylloid to feed on fig trees, and is confined to southern Britain and the Channel Islands. The family was formerly known as Carsidaridae.

British genus: *Homotoma*

Family Psyllidae (16 genera, 60 species)
The main family of psyllids in Britain includes a few pest species. Some have complex lifecycles, overwintering on different plants from their host plants, and hibernating as either eggs or adults. Many species are host-specific and the more common species include *Livia juncorum*, whose nymphs cause red 'tassel galls' on rushes (*Juncus*). *Livia* was formerly placed in the separate family Liviidae. Two species in *Psylla* are pests of fruit trees: *Psylla mali*, the Apple psyllid or Apple sucker, and *Ps. pyricola*, the Pear sucker. The feeding of *Ps. buxi* often causes distorted shoots on box trees, *Ps. alni* (Fig. 14.62) is common on alder, and the nymphs of *Psyllopsis* spp. cause leaf-edge galls on ash trees. One small species, *Ctenarytaina eucalypti*, was introduced from Australia, and is now widely established on ornamental *Eucalyptus* spp. It was formerly placed in the family Spondyliaspididae.

Some species of Psyllidae are illustrated on the British Bugs website (http://www.britishbugs.org.uk).

British genera: *Aphalara, Aphorma, Arytaina, Arytainilla, Baeopelma, Cacopsylla, Chamaepsylla, Craspedolepta, Ctenarytaina, Livia, Livilla, Psylla,*

Fig. 14.61 Psyllidae (Photo: Colin Rew)

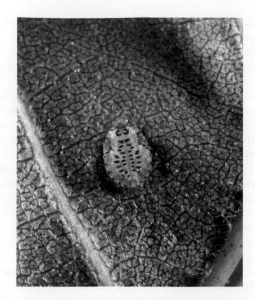

Fig. 14.62 *Psylla alni* nymph (Psyllidae) (Photo: Roger Key)

Psyllopsis, Rhinocola, Spanioneura, Strophingia

Family Triozidae (3 genera, 18 species)
The Triozidae are a distinctive family of psyllids, characterized by the three main veins in the fore wing, R, M and Cu all arising from the same point. The best-known species is probably *Trioza urticae*, a common insect on nettles, where its feeding causes some leaf distortion. Some other species cause galls to form on the host plants, including *T. alacris* on Bay laurel, and *Trichochermes walkeri* on Buckthorn.

A few species are illustrated on the British Bugs website (http://www.britishbugs.org.uk).

British genera: *Bactericera, Trichochermes, Trioza*

REFERENCES

ANON. 1973. *The bed-bug*. British Museum (Natural History) Economic series 5: 17 pp.

BLACKMAN, R.L. 1974. *Aphids*. Ginn & Co., London & Aylesbury.

BLACKMAN, R.L. 2010. Aphids – Aphidinae (Macrosiphini). *Handbooks for the identification of British insects* 2(7): 420 pp.

BLACKMAN, R.L. & EASTOP, V.F. 1994. *Aphids on the world's trees: an identification and information guide*. CAB International, Wallingford [updated digital version at http://www.aphidsonworldsplants.info].

BLACKMAN, R.L. & EASTOP, V.F. 2000. *Aphids on the world's crops: an identification and information guide* (2nd edn.). Wiley, Chichester.

BORATYNSKI, K.L. & WILLIAMS, D.J. 1964. Coccoidea, pp. 87–94. In: Kloet, G.S. & Hincks, W.D. A check-list of British insects (2nd edn., revised). Part 1: Small orders

and Hemiptera. *Handbooks for the identification of British insects* 11(1): 119 pp.

BUTLER, E.A. 1923. *A biology of the British Hemiptera-Heteroptera*. Witherby, London.

CARTER, C.I. 1971. Conifer woolly aphids (Adelgidae) in Britain. *Forestry Commission Bulletin* 42: 51 pp. HMSO, London.

CARTER, C.I. 1982. Conifer lachnids. *Forestry Commission Bulletin* 58: 75 pp. HMSO, London.

DANZIG, E.M. 1967. Coccinea. In: Bei-Bienko, G.Y. (ed.) Keys to the insects of the European USSR. Vol. I. Apterygota, Palaeoptera, Hemimetabola, pp. 800–850. English translation by J. Salkind. Israel Program for Scientific Translations, Jerusalem.

DANZIG, E.M. 1993. [Fauna of Russia and neighbouring countries. Rhynchota, volume X. Scale insects (Coccinea), families Phoenicococcidae and Diaspididae.] Nauka, St. Petersburg [in Russian].

DELLA GIUSTINA, W. 1989. Homoptères Cicadellidae Volume 3 (Compléments aux ouvrages d'Henri Ribaut). *Faune de France* 73: 350 pp.

DENTON, J. 2007. *Water bugs and water beetles of Surrey*. Surrey Wildlife Trust, Pirbright.

DIETRICH, C.H. 2009. Auchenorrhyncha (Cicadas, Spittlebugs, Leafhoppers, Treehoppers, and Planthoppers). In: Resh, V.H. & Cardé, R.T. (eds.) *Encyclopedia of insects* (2nd edn.). Academic Press/Elsevier, San Diego & London, pp. 56–64.

DIXON, A.F.G. 1973. *Biology of aphids*. Arnold, London.

DIXON, A.F.G. 1985. *Aphid ecology*. Blackie, Glasgow & London.

DIXON, T. & THIEME, T. 2007. *Aphids on deciduous trees*. Naturalists' Handbooks 29, Richmond Publishing, Slough.

DOLLING, W.R. 1991. *The Hemiptera*. Oxford University Press, Oxford.

EVANS, M. & EDMONDSON, R. 2005. *A photographic guide to the shieldbugs and squashbugs of the British Isles*. WildGuides UK.

GULLAN, P.J. & MARTIN, J.H. 2009. Sternorrhyncha. In: Resh, V.H. & Cardé, R.T. (eds.) *Encyclopedia of insects* (2nd edn.). Academic Press/Elsevier, San Diego & London, pp. 957–67.

HEIE, O.E. 1980. The Aphidoidea (Hemiptera) of Fennoscandia and Denmark. I. General part: the families Mindaridae, Hormaphididae, Thelaxidae, Anoeciidae, and Pemphigidae. *Fauna Entomologica Scandinavica* 9: 236 pp.

HEIE, O.E. 1982. The Aphidoidea (Hemiptera) of Fennoscandia and Denmark. II. The family Drepanosiphidae. *Fauna Entomologica Scandinavica* 11: 176 pp.

HEIE, O.E. 1986. The Aphidoidea (Hemiptera) of Fennoscandia and Denmark. III. Family Aphididae: subfamily Pterocommatinae and tribe Aphidini of subfamily Aphidinae. *Fauna Entomologica Scandinavica* 17: 314 pp.

HEIE, O.E. 1992. The Aphidoidea (Hemiptera) of Fennoscandia and Denmark. IV. Family Aphididae: part 1 of tribe Macrosiphini of subfamily Aphidinae. *Fauna Entomologica Scandinavica* 25: 190 pp.

HEIE, O.E. 1994. The Aphidoidea (Hemiptera) of Fennoscandia and Denmark. V. Family Aphididae: part 2 of tribe Macrosiphini of subfamily Aphidinae. *Fauna Entomologica Scandinavica* 28: 242 pp.

HEIE, O.E. 1995. The Aphidoidea (Hemiptera) of Fennoscandia and Denmark. VI. Family Aphididae: part 3 of tribe Macrosiphini of subfamily Aphidinae, and family Lachnidae. *Fauna Entomologica Scandinavica* 31: 222 pp.

HODGSON, C.J. 1994. *The scale insect family Coccidae: an identification manual to genera.* CAB International, Wallingford.

HODKINSON, I.D. & WHITE, I.M. 1979. Homoptera: Psylloidea. *Handbooks for the identification of British insects* 11(5a): 98 pp.

KIRBY, P. 1992. A review of the scarce and threatened Hemiptera of Great Britain. *UK Nature Conservation* 2. Joint Nature Conservation Committee, Peterborough.

KLOET, G.S. & HINCKS, C. 1964. A check-list of British insects (2nd edn., completely revised). Part 1. *Handbooks for the identification of British insects* 11(1): 119 pp.

KOSZTARAB, M. & KOZÁR, F. 1988. *Scale insects of central Europe.* Akadémiai Kiadó, Budapest.

LEQUESNE, W.J. 1960. Hemiptera (Fulgoromorpha). *Handbooks for the identification of British insects* 2(3): 68 pp.

LEQUESNE, W.J. 1965. Hemiptera Cicadomorpha (excluding Deltocephalinae and Typhlocybinae). *Handbooks for the identification of British insects* 2(2a): 64 pp.

LEQUESNE, W.J. 1969. Hemiptera Cicadomorpha Deltocephalinae. *Handbooks for the identification of British insects* 2(2b): 148 pp.

LEQUESNE, W.J. & PAYNE, K.R. 1981. Cicadellidae (Typhlocybinae) with a check list of the British Auchenorhyncha (Hemiptera, Homoptera). *Handbooks for the identification of British insects* 2(2c): 95 pp.

MARTIN, J.H. & MOUND, L.A. 2007. An annotated check list of the world's whiteflies (Insecta: Hemiptera: Aleyrodidae). *Zootaxa* 1492: 1–84.

MARTIN, J.H. & WEBB, M.D. 1999. Hemiptera: the true bugs. In: Barnard, P.C. (ed.) *Identifying British insects and arachnids: an annotated bibliography of key works.* Cambridge University Press, Cambridge, pp. 54–75.

McGAVIN, G.C. 1993. *Bugs of the world.* Blandford, London.

MOULET, P. 1995. Hémiptères Coreoidea (Coreidae, Rhopalidae, Alydiidae, Pyrrhocoridae, Stenocephalidae) euro-méditerranéens. *Faune de France* 81: 336 pp.

MOUND, L.A. 1966. A revision of the British Aleyrodidae (Hemiptera–Homoptera). *Bulletin of the British Museum (Natural History) (Entomology)* 17: 397–428.

MOUND, L.A. & HALSEY, S.H. 1978. *Whitefly of the world: a systematic catalogue of the Aleyrodidae (Homoptera) with host plant and natural enemy data.* Wiley, Chichester.

NAU, B. 2004. *Guide to shieldbugs of the British Isles.* Field Studies Council, Preston Montford, OP85.

OSSIANNILSSON, F. 1978. The Auchenorrhyncha (Homoptera) of Fennoscandia and Denmark. Part 1: Introduction, Infraorder Fulgoromorpha. *Fauna Entomologica Scandinavica* 7(1): 1–222.

OSSIANNILSSON, F. 1981. The Auchenorrhyncha (Homoptera) of Fennoscandia and Denmark. Part 2: The families Cicadidae, Cercopidae, Membracidae and Cicadellidae (excl. Deltocephalinae). *Fauna Entomologica Scandinavica* 7(2): 223–593.

OSSIANNILSSON, F. 1983. The Auchenorrhyncha (Homoptera) of Fennoscandia and Denmark. Part 3: The family Cicadellidae, Deltocephalinae, catalogue, literature and index. *Fauna Entomologica Scandinavica* 7(3): 594–979.

OSSIANNILSSON, F. 1992. The Psylloidea (Homoptera) of Fennoscandia and Denmark. *Fauna Entomologica Scandinavica* 26: 346 pp.

PÉRICART, J. 1983. Hémiptères Tingidae euro-méditerranéens. *Faune de France* 69: 618 pp.

PÉRICART, J. 1984. Hémiptères Berytidae euro-méditerranéens. *Faune de France* 70: 180 pp.

PÉRICART, J. 1987. Hémiptères Nabidae d'Europe occidentale et du Maghreb. *Faune de France* 71: 185 pp.

PÉRICART, J. 1990. Hémiptères Saldidae et Leptopodidae d'Europe occidentale et du Maghreb. *Faune de France* 77: 238 pp.

PINCHEN, B. 2009. *A pocket guide to the shieldbugs and leatherbugs of Britain and Ireland.* Forficula Books, Lymington

REMAUDIÈRE, G. & REMAUDIÈRE, M. 1997. *Catalogue des Aphididae du monde (Homoptera Aphidoidea).* INRA Editions, Paris, 473 pp.

REMANE, R. & WACHMANN, E. 1993. *Zikaden kennenlernen beobachten.* Naturbuch Verlag, Augsburg.

Resh, V.H. & Cardé, R.T. (eds.) 2009. *Encyclopedia of insects* (2nd edn.). Academic Press/Elsevier, San Diego & London.

RIBAUT, H. 1936. Homoptères Auchénorhynques. I (Typhlocybidae). *Faune de France* 31: 321 pp.

RIBAUT, H. 1952. Homoptères Auchénorhynques. II (Jassidae). *Faune de France* 57: 474 pp.

SAVAGE, A.A. 1989. Adults of the British aquatic Hemiptera Heteroptera: a key with ecological notes. *Scientific Publications of the Freshwater Biological Association* 50: 173 pp.

SAVAGE, A.A. 1999. Keys to the larvae of British Corixidae. *Scientific Publications of the Freshwater Biological Association* 57: 56 pp.

SCHAEFER, C.W. 2009. Prosorrhyncha (Heteroptera and Coleorrhyncha). In: Resh, V.H. & Cardé, R.T. (eds.) *Encyclopedia of insects* (2nd edn.). Academic Press/ Elsevier, San Diego & London, pp. 839–55.

SCHUH, R.T. & SLATER, J.A. 1995. *True bugs of the world (Hemiptera: Heteroptera).* Cornell University, New York.

SORENSON, J.T. (2009) Aphids. In: Resh, V.H. & Cardé, R.T. (eds.) *Encyclopedia of insects* (2nd edn.). Academic Press/ Elsevier, San Diego & London, pp. 27–31.

SOUTHWOOD, T.R.E. & LESTON, D. 1959. *Land and water bugs of the British Isles.* Warne, London [also available on CD-ROM and as a facsimile reprint, from Pisces Conservation].

STROYAN, H.L.G. 1977. Aphidoidea – Chaitophoridae and Callaphididae. *Handbooks for the identification of British insects* 2(4a): 130 pp.

STROYAN, H.L.G. 1984. Aphids – Pterocommatinae and Aphidinae (Aphidini). *Handbooks for the identification of British insects* 2(6): 232 pp.

THEOBALD, F.V. 1926, 1927, 1929. *The plant lice or Aphididae of Great Britain*. Vols. 1–3. Headley Brothers, Ashford & London.

UNWIN, D. 2001. *A key to families of British bugs (Insecta, Hemiptera)*. AIDGAP guide 269, Field Studies Council, Preston Montford.

WACHMANN, E. 1988. *Wanzen beobachten – kennenlernen.* J. Neumann-Neudamm, Melsungen.

WAGNER, E. & WEBER, H. 1964. Hétéroptères Miridae. *Faune de France* 67: 592 pp.

WHITE, I.M. & HODKINSON, I.D. 1982. Psylloidea (nymphal stages): Hemiptera, Homoptera. *Handbooks for the identification of British insects* 11(5b): 50 pp.

WILLIAMS, D.J. 1962. The British Pseudococcidae (Homoptera: Coccoidea). *Bulletin of the British Museum (Natural History)* (Entomology) 12(1): 1–79.

WILLIAMS, D.J. 1985. The British and some other European Eriococcidae (Homoptera: Coccoidea). *Bulletin of the British Museum (Natural History)* (Entomology) 51(4): 347–93.

WILLIAMS, D.J. & WATSON, G.W. 1988a. *The scale insects of the tropical south Pacific region. Part 1. The armoured scales (Diaspididae)*. CAB International, Wallingford.

WILLIAMS, D.J. & WATSON, G.W. 1988b. *The scale insects of the tropical south Pacific region. Part 1. The mealybugs (Pseudoccidae)*. CAB International, Wallingford.

WILLIAMS, D.J. & WATSON, G.W. 1990. *The scale insects of the tropical south Pacific region. Part 1. The soft scales (Coccidae) and other families*. CAB International, Wallingford.

WEBSITES

http://www.britishbugs.org.uk
British Bugs website, with much useful information about distribution, as well as many excellent photos to aid identification.

http://www.hetnews.org.uk
Useful site on the British Heteroptera.

http://www.brc.ac.uk/DBIF
A general site on the foodplants of British insects.

http://www.aphidsonworldsplants.info
Contains updated, digitized versions of various books by Blackman and Eastop, starting with *Aphids on the world's trees* (1994).

http://aphid.speciesfile.org
The world list of aphid species.

15 Order Phthiraptera: the sucking and biting lice

c. 540 species in 17 families

Lice are unique among the insects in being ectoparasitic in both the juvenile and adult stages; indeed all stages of the lifecycle are spent on the host's body. Although they are very common, few people study the group and many people never see them at all unless they study the bird or mammal hosts. Traditionally the group was divided into two separate orders: the Anoplura, or sucking lice (also known as Siphunculata), which are all blood-feeders on mammals; and the Mallophaga, the chewing or biting lice, which feed on feathers, fur, blood, etc., on both birds and mammals. The apparently radical differences in mouthparts delayed their combination in a single order, but the monophyly of the Phthiraptera is now accepted. Most authorities now recognize four suborders; the Anoplura are unchanged, but the former Mallophaga are now divided into three, the Amblycera (on birds and mammals), the Ischnocera (also on birds and mammals) and the non-British Rhyncophthirina (on elephants and wart-hogs). There is now evidence that Phthiraptera may be a subgroup of the Psocoptera, which would make the latter a paraphyletic group; this is discussed on the 'Psocodea' section on the Tree of Life website (http://tolweb.org/Psocodea/8235).

All adult lice are wingless and dorsoventrally flattened; the eyes are reduced or absent and they usually have well-developed claws for clinging to their host. Most lice are just a few millimetres long. The mouthparts are of two distinct kinds: in the more generalized 'chewing' suborders the mandibulate mouthparts are used to bite and chew sections of feather, fur or hair, though feeding on skin is also recorded. The resultant irritation may cause the host to scratch the site, leading to small wounds and the lice then feed on the blood; some will also eat sebaceous secretions. In the more specialized Anoplura the mouthparts are highly modified into piercing stylets, which are normally retracted into the head but are everted to pierce the mammalian host skin, inject anticoagulants and feed on the blood directly.

Most species reproduce sexually, though parthenogenesis is known in some cases. The eggs are glued to the hair or feathers of the host and will only hatch if the host's body temperature is sufficiently high. There are usually three nymphal stages, and the whole lifecycle may take only a few weeks in favourable conditions. Young or unfit hosts may carry larger populations of lice, and there is evidence that some hosts can develop immunity, which helps to control the numbers of parasites.

Most lice species are highly host-specific, often restricted to a single host species, and even confined to certain body areas of the host. Thus a bird might have three or four species of lice, each living at a different site such as the neck, back and wings. A familiar example is the human louse, *Pediculus humanus*, formerly divided into subspecies: *P. humanus humanus* on the body and *P. humanus capitis* on the head, though these are often regarded as distinct species. This restricted movement of species, coupled with the fact that lice cannot fly or jump, means that transfer from one host to another usually takes place only when the hosts are in close contact. That this works quite efficiently is shown

The Royal Entomological Society Book of British Insects, First Edition. Peter C. Barnard.
© 2011 Royal Entomological Society. Published 2011 by Blackwell Publishing Ltd.

by the frequent epidemics of head-lice in schools, where children put their heads together, or perhaps share combs. Some lice, especially those living on birds, can be carried phoretically from one host to another by flying insects such as hippoboscid flies, thus aiding their dispersal.

There are a few species of lice that seem able to breed on a wide range of hosts. For example, *Saemundssonia lari*, is parasitic on every species of gull (*Larus*) in Britain and it has also been found on other groups of birds, though possibly as accidental 'stragglers'. Strangely, the human body-louse can also breed on the domestic pig (and some fleas are also found on both hosts) though this must be the result of an accidental similarity in the physiology of the two mammals.

The often close association between lice species and their hosts has made them a popular group for studying coevolution, and many phylogenetic trees have been drawn up to show how lice groups must have evolved along with the relevant bird or mammalian groups. There are, however, several anomalies in such theories that can be hard to explain. Although nearly all bird species have lice, some mammalian groups have none, and these include monotremes, bats and whales. The sensitivity of lice to changes in temperature might explain why they have not spread to bats that hibernate; also one might assume that marine mammals would be safe from lice because of the frequent immersion in sea-water. In fact, lice do occur on some hibernating mammals, and seals also have specialized lice, which can only breed when the seals come on land.

The number of species of lice in Britain is widely quoted as being around 540, and this figure is derived from the last complete checklist (Kloet & Hincks, 1964). The way that this list was generated is recorded in the introduction to that work, where it was pointed out that because students of the lice are more interested in the hosts than in the distribution of the insects, occurrences in Britain are not necessarily published: 'The only way, therefore, to produce a comprehensive list is to include all the species regularly associated with animals and birds that are considered as British. This has been done here.' Thus, this list should perhaps be regarded as a theoretical one; it may or may not be accurate, although many lice species have of course been confirmed as British. The only mammals and birds not included in the potential host list were those from circuses or zoos, but several pet species such as guinea-pigs seem to have good representation. A few of the lice found on domestic animals and birds

have been given common names, usually based on the host species.

This means that there are no faunistic guides to British lice, and the student of this group will need to become familiar with the literature on a world scale. Coupled with the fact that most specimens need to be prepared for microscopic examination, this makes the study of this group a somewhat specialized activity. The most useful works are listed at the end of the chapter; specialized publications with notes on their coverage are noted by Lyal (1999). A brief world overview of the Phthiraptera is given by Hellenthal and Price (2009).

There are around 800 species in 17 families in Europe, and over 5000 species in 24 families worldwide.

HIGHER CLASSIFICATION OF BRITISH PHTHIRAPTERA

Suborder Amblycera
  Superfamily Gyropoidea
    Family Gyropidae (1 genus, 1 species)
    Family Gliricolidae (2 genera, 2 species)
  Superfamily Laemobothrioidea
    Family Laemobothriidae (1 genus, 8 species)
  Superfamily Menopodoidea
    Family Menopodidae (23 genera, 147 species)
  Superfamily Ricinoidea
    Family Ricinidae (1 genus, 11 species)
  Superfamily Trimenoponoidea
    Family Trimenoponidae (1 genus, 1 species)
Suborder Anoplura
  Superfamily Echinophthirioidea
    Family Echinophthiriidae (1 genus, 1 species)
  Superfamily Linognathoidea
    Family Enderleinellidae (1 genus, 2 species)
    Family Hoplopleuridae (2 genera, 5 species)
    Family Linognathidae (2 genera, 5 species)
    Family Polyplacidae (3 genera, 7 species)
    Family Phthiridae (1 genus, 1 species)
  Superfamily Pediculoidea
    Family Haematopinidae (1 genus, 3 species)
    Family Pediculidae (1 genus, 2 species)
Suborder Ischnocera
  Superfamily Goniodoidea
    Family Goniodidae (5 genera, 25 species)
  Superfamily Philopteroidea
    Family Philopteridae (46 genera, 296 species)
  Superfamily Trichodectoidea
    Family Trichodectidae (4 genera, 21 species)

SPECIES OF CONSERVATION CONCERN

No species have any legal protection and none is on the UKBAP list. However, being obligate parasites, any lice that are specific to threatened vertebrate hosts would automatically become at risk themselves.

The Families of British Phthiraptera

SUBORDER AMBLYCERA

SUPERFAMILY GYROPOIDEA

Family Gyropidae (1 genus, 1 species)
The only British species is *Gyropus ovalis*, a parasite of guinea-pigs, and now cosmopolitan following the popularity of their hosts as domestic pets.

British genus: *Gyropus*

Family Gliricolidae (2 genera, 2 species)
This family was formerly included within the Gyropidae. Like *Gyropus*, *Gliricola porcelli* is found on guinea-pigs. *Pitrufquenia coypus* lives on the coypu, presumably now eradicated from Britain.

British genera: *Gliricola, Pitrufquenia*

SUPERFAMILY LAEMOBOTHRIOIDEA

Family Laemobothriidae (1 genus, 8 species)
The species of *Laemobothrion* are parasites of falconiform birds.

British genus: *Laemobothrion*

SUPERFAMILY MENOPODOIDEA

Family Menopodidae (23 genera, 147 species)
A large family of bird parasites, including some large genera like *Actornithophilus* and *Menacanthus*.

Menopon gallinae, the Shaft louse, is one of many cosmopolitan species found on poultry and some game birds; it is a small species, up to 2 mm long, that feeds on feathers. *Menacanthus stramineus*, also found on chickens and turkeys, feeds by chewing through developing feathers and taking blood. *Trinoton anserinum* is a common parasite of swans and some ducks (Fig. 15.1).

British genera: *Actornithophilus, Amyrsidea, Ancistrona, Ardeiphilus, Austromenopon, Ciconiphilus, Colpocephalum, Cuculiphilus, Dennyus, Eidmanniella, Eucolpocephalum, Eureum, Hohorstiella, Holomenopon, Kurodaia, Machaerilaemus, Menacanthus, Menopon,*

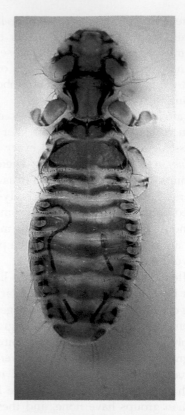

Fig. 15.1 *Trinoton anserinum* (Menopodidae) a parasite of swans and ducks (Photo: Lajos Rozsa, Creative Commons Licence)

Myrsidea, Nosopon, Pseudomenopon, Somaphantus, Trinoton

SUPERFAMILY RICINOIDEA

Family Ricinidae (1 genus, 11 species)
The species of *Ricinus* live on passerine birds, and are known to bite through the host's skin in order to feed on the blood.

British genus: *Ricinus*

SUPERFAMILY TRIMENOPONOIDEA

Family Trimenoponidae (1 genus, 1 species)
The only British species is *Trimenopon hispidum*, a parasite of guinea-pigs.

British genus: *Trimenopon*

SUBORDER ANOPLURA

The sucking lice.

SUPERFAMILY ECHINOPHTHIRIOIDEA

Family Echinophthiriidae (1 genus, 1 species)
The only British species is *Echinophthirius horridus*, the Seal louse, which is a parasite on *Phoca vitulina*, the Harbour or Common seal. Some members of this family trap an air-layer in hairs on the body and respire using this plastron, like some water-bugs; however, *E. horridus* does not breathe in this way. It can only reproduce and spread from one host to another while the seals are on dry land.

British genus: *Echinophthirius*

SUPERFAMILY LINOGNATHOIDEA

Family Enderleinellidae (1 genus, 2 species)
The two British species of *Enderleinellus* were formerly included in the Hoplopleuridae; they are parasites of squirrels.

British genus: *Enderleinellus*

Family Hoplopleuridae (2 genera, 5 species)
The members of this small family are found on rodents and squirrels.

British genera: *Hoplopleura, Schizophthirus*

Family Linognathidae (2 genera, 5 species)
The members of this family parasitize ungulate mammals, especially sheep and cattle, and different species of *Linognathus* have been reported on distinct areas of sheep, such as the legs or head.

British genera: *Linognathus, Solenopotes*

Family Polyplacidae (3 genera, 7 species)
This small family was formerly regarded as a subgroup of the Hoplopleuridae. *Haemodipsus* species are found on rabbits and hares, *Polyplax* species on rodents, and *Neohaematopinus sciuri* on squirrels.

British genera: *Haemodipsus, Neohaematopinus, Polyplax*

Family Pthiridae (1 genus, 1 species)
This family name was previously spelled Phthiridae, based on the incorrect spelling of the genus *Phthirus*, and was previously regarded as a subgroup of the Pediculidae.

The only species is *Pthirus* [= *Phthirus*] *pubis*, the notorious Pubic louse or Crab louse, found only in humans, and usually transmitted sexually, although occasionally via infested towels or clothes. This species does not transmit any diseases, but the anticoagulant saliva causes prolonged irritation.

British genus: *Pthirus*

SUPERFAMILY PEDICULOIDEA

Family Haematopinidae (1 genus, 3 species)
The best-known member of *Haematopinus* is *H. suis*, the Pig or Hog louse. This species is cosmopolitan, living on wild pigs and domesticated breeds. The nymphs are usually found on the warmer parts of the host body, such as inside the ears or folds of skin on the neck, but the adults spread over the rest of the body. All stages feed by cutting through the host skin and penetrating a blood vessel, usually a vein, with their stylets. The adults are relatively large, 5–6 mm long with reduced eyes, and they become sexually mature after about ten days. The whole life cycle lasts around a month, so that untreated infestations of pigs can increase in numbers very rapidly. The presence of *Haematopinus* is more threatening to piglets because of blood loss, but even in adult pigs they can produce skin irritations and loss of appetite; they also carry certain diseases. On pig farms these parasites are controlled by a variety of insecticides, but they are still considered a serious problem in the USA. Other species of *Haematopinus* can be found on cattle.

British genus: *Haematopinus*

Family Pediculidae (1 genus, 2 species)
Human lice are probably the best known of all the Phthiraptera and they are often the only species that people ever encounter. There are two forms of this louse, and there is still disagreement over whether they are distinct species or merely subspecies. They have distinct behaviours, can be separated by size and colour (at least statistically) and have different abilities to transmit disease; on the other hand they can interbreed. The Body louse, *Pediculus humanus*, generally feeds on the trunk of the human body, though it is an interesting exception to most of the lice in that it lays its eggs on clothing, not on the body itself. Also, the insects rest on the clothing between feeding sessions rather than remaining on the host body all the time. Presumably the relative lack of body hair in people means that the louse regards clothing as a substitute fur. In unhygienic conditions the numbers of body lice can reach many hundreds on a single person's clothing, and the general malaise resulting from such extensive feeding gave rise to the popular term of feeling 'lousy'. In some parts of the world this louse transmits typhus, a rickettsial disease,

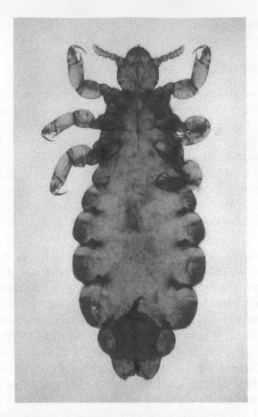

Fig. 15.2 *Pediculus capitis*, the Human head-louse (Pediculidae), microscope preparation (Photo: Peter Barnard)

and this can be a problem anywhere that people are crowded in unsanitary conditions, such as after a large-scale disaster. The control of typhus by DDT during the second World War was one of the major successes of a chemical pesticide.

The Head louse, *Pediculus capitis* (or *P. humanus capitis*, Fig. 15.2) is rather different in its habits. Usually confined to the hair on the human head, it behaves more like a 'normal' louse. The eggs, known commonly as nits, are attached to the hair shafts, and the insects themselves spend most of their time clinging to the hair. Both kinds of human lice are blood-feeders and control depends on their different habits: body lice are transmitted by infected clothing or bedding, whereas head lice are often transmitted by direct contact, or by sharing combs and hairbrushes. It is reported that *P. humanus* tends to leave the bodies of people suffering from a high fever, which may be an additional way of spreading to new hosts.

British genus: *Pediculus*

SUBORDER ISCHNOCERA
SUPERFAMILY GONIODOIDEA

Family Goniodidae (5 genera, 25 species)
The members of this family of chewing lice were formerly included within the Philopteridae, though the exact relationships of the genera are still the subject of debate. They are feather lice that live on various galliforms (hens and game birds) and columbiforms (pigeons and doves).

British genera: *Campanulotes, Chelopistes, Coloceras, Goniocotes, Goniodes*

SUPERFAMILY PHILOPTEROIDEA

Family Philopteridae (46 genera, 296 species)
This, the largest family of (mainly) bird parasites in the Ischnocera, is probably paraphyletic, even though some genera have already been placed in different families; it is likely that further research will divide the Philopteridae even more. Typically the members of this group are elongate in shape, which enables them to hide amongst feathers and to avoid being dislodged by the host's preening. One of the best known species is *Columbicola claviformis*, which is commonly found on pigeons. *Philopterus passerinus* and related species are found on sparrows and other passerine birds. Extreme host specificity is shown by some species, as in the various parasites of the crow genus *Corvus*. For example, *Philopterus atratus* is confined to the rook (*C. frugilegus*); *Ph. corvi* to the raven (*C. corvus*); *Ph. ocellatus* to the carrion crow (*C. corone*) and *Ph. guttatus* to the jackdaw (*C. monedulus*).

British genera: *Acidoproctus, Alcedoffula, Anaticola, Anatoecus, Aquanirmus, Ardeicola, Brueelia, Capraiella, Carduiceps, Cirrophthirius, Columbicola, Craspedonirmus, Craspedorrhynchus, Cuclotogaster, Cuculicola, Cuculoecus, Cummingsiella, Degeeriella, Falcolipeurus, Fulicoffula, Halipeurus, Haffneria, Ibidoecus, Incidifrons, Lagopoecus, Lipeurus, Lunaceps, Meropoecus, Mulcticola, Naubates, Ornithobius, Otidoecus, Oxylipeurus, Pectinopygus, Penenirmus, Perineus, Philopterus, Quadraceps, Rallicola, Rhynonirmus, Saemundssonia, Strigiphilus, Sturnidoecus, Syrrhaptoecus, Trabeculus, Upupicola*

SUPERFAMILY TRICHODECTOIDEA

Family Trichodectidae (4 genera, 21 species)
Although the Trichodectidae are closely related to

the Philopteridae, they are all parasites of mammals. *Damalinia bovis* is the principal chewing louse to be found on cattle; although controllable with various insecticides it is always a greater risk when cows are housed closely together. Other species in this family are found on horses, sheep, badgers and many other hosts.

British genera: *Damalinia, Felicola, Lutridia, Trichodectes*

REFERENCES

DURDEN, L.A. & MUSSER, G.G. 1994. The sucking lice (Insecta, Anoplura) of the world: a taxonomic checklist with records of mammalian hosts and geographical distributions. *Bulletin of the American Museum of Natural History* 218: 90 pp.

EMERSON, K.C. & PRICE, R.D. 1981. A host–parasite list of the Mallophaga on mammals. *Miscellaneous Publications of the Entomological Society of America* 12(1): 72 pp.

FERRIS, G.F. 1951. The sucking lice. *Memoirs of the Pacific Coast Entomological Society* 1: 320 pp.

HELLENTHAL, R.A. & PRICE, R.D. 2009. Phthiraptera. In: Resh, V.H. & Cardé, R.T. (eds.) *Encyclopedia of insects* (2nd edn.). Academic Press/Elsevier, San Diego & London, pp. 777–80.

HOPKINS, G.H.E. & CLAY, T. 1952. *A checklist of the genera and species of Mallophaga*. British Museum (Natural History), London.

KIM, K.C., PRATT, H.D. & STOJANOVICH, C.J. 1986. *The sucking lice of North America: an illustrated manual for identification*. Pennsylvania State University Press, University Park.

KLOET, G.S. & HINCKS, W.D. 1964. A check list of British insects (2nd edn.). Part 1: small orders and Hemiptera. *Handbooks for the identification of British insects* 11(1): 119 pp.

LYAL, C.H.C. 1999. Phthiraptera: the lice. In: Barnard, P.C. (ed.) *Identifying British insects and arachnids: an annotated bibliography of key works*. Cambridge University Press, Cambridge, pp. 47–50.

PRICE, R.D., HELLENTHAL, R.A., PALMA, R.L. JOHNSON, K.P. & CLAYTON, D.H. 2003. The chewing lice: world checklist and biological overview. *Illinois Natural History Survey Special Publication* 24: 501 pp.

TIMMERMANN, G. 1965. Die Federlingsfauna der Sturmvögel und die Phylogenese des procellariiformen Vogelstammes. *Abhandlungen und Verhandlungen des Naturwissenschaftlichen Vereins in Hamburg (N.F.)* 8 (suppl.): 249 pp.

WERNECK, F.L. 1948. *Os Malóphagos de mamiferos. Parte I. Amblycera e Ischnocera (Philopteridae e parte de Trichodectidae)*. Rio de Janeiro (Revista Brasileira de Biologia).

WERNECK, F.L. 1950. *Os Malóphagos de mamiferos. Parte II. Ischnocera (continuacao de Trichodectidae) e Rhyncophthirina*. Rio de Janeiro (Instituto Oswaldo Cruz).

WEBSITES

http://tolweb.org/Phthiraptera/8237
A useful page from the Tree of Life project.

http://phthiraptera.info
Website of the International Society of Phthirapterists.

http://tolweb.org/Psocodea/8235
For the latest thoughts on relationships between Psocoptera and Phthiraptera.

16 Order Psocoptera: the booklice and barklice

100 species in 19 families

The Pscocoptera have long been regarded as phylogenetically close to the Phthiraptera, but recent research has suggested that the Phthiraptera are actually a subgroup of the Psocoptera, within the wider group of Psocodea. Not surprisingly there is some reluctance to accept this view fully, because it would have implications for the integrity of the Psocoptera. Such a paraphyletic group could no longer retain its status as an order, a rather similar situation to that of the Isoptera and Dictyoptera. Further morphological and molecular evidence should clarify this situation. For the latest ideas on relationships within this group see the Psocodea section on the Tree of Life website (http://tolweb.org/Psocodea/8235).

Apart from their similarities with Phthiraptera, the Psocoptera are most likely to be confused with the jumping plant-lice in the hemipteran Psylloidea, as they can be found in similar situations. However, the Psocoptera all have biting mouthparts, quite different to the rostrum of the Psylloidea; the wing venation is different, and Psocoptera do not jump if disturbed. Most species are just a few millimetres long and the wings are well-developed, longer than the body, and sometimes distinctively marked (Fig. 16.1). The wing venation is of considerable use in the classification of many families. However, brachyptery and aptery are frequent in some groups and the common domestic pests such as *Liposcelis* and *Trogium* are always wingless. Occasionally the short fore wings in brachypterous species are easily shed, giving the appearance of complete aptery. The adult antennae are long and slender but a distinctive feature of the head is the enlarged, bulbous postclypeus, which often has a characteristic pattern of striations (Fig. 16.2). The eyes are generally well developed, but in many of the wingless species the eyes are greatly reduced.

There is no single common name for this group, and they are often simply called psocids. The indoor species are often termed booklice or dustlice, neither of which is particularly appropriate or accurate. The outdoor species are sometimes called barklice but some recent authors prefer the newly coined name of barkflies, to avoid the negative connotations of the term 'lice'. Interestingly, some can produce sounds audible to humans; some species of *Lepinotus* make a quiet chirping noise, and *Trogium pulsatorium* gets its specific name from the ticking sound it produces. These sounds are probably important as part of courtship rituals; most species reproduce bisexually, though facultative or obligatory parthenogenesis is widespread in some families. Psocid nymphs usually have six instars, though this number is reduced in some polymorphic species, especially associated with brachyptery or aptery.

Although most people will encounter the relatively few species that occur indoors, the majority of British species are found outside on a wide variety of plants. They feed on algae, lichens and fungi, often on the bark of trees or woody shrubs and their mouthparts are adapted to prising up these micro-epiphytes, biting off pieces and chewing them. Many species are generalized feeders, but others have very specific needs so that species that feed on algae cannot transfer to lichens, for example. Thus the association of psocids with particular trees is based on the habitat preferred by their food plant,

The Royal Entomological Society Book of British Insects, First Edition. Peter C. Barnard.
© 2011 Royal Entomological Society. Published 2011 by Blackwell Publishing Ltd.

Fig. 16.1 Patterned fore wings of macropterous psocid (Photo: Colin Rew)

Fig. 16.2 Swollen and patterned postclypeus of psocids (Photo: Colin Rew)

Fig. 16.3 Common domestic psocid (Photo: Peter Barnard)

rather than on direct requirements of the insects. Although some species are found on the leaves and small twigs, others are confined to the trunks of trees where they may even live under the bark. Thus, the group is not widely known or studied; standard 'beating' techniques will often dislodge huge numbers from the tree canopy but the entomologist has to search carefully for those found only on trunks or large branches. Being small and soft-bodied, specimens are usually collected and stored in alcohol, though for critical examination of some groups it is necessary to make microscope slide preparations. In the Lepidopsocidae, which have wing-patterns formed from scales, it can be useful to keep some dry, pinned or card-mounted specimens, because the scales are dislodged in fluid preservatives.

The domestic species of psocids that are found in houses or industrial food stores were initially adapted to feeding on the moulds found on farinaceous foodstuffs or on damp books and paper (Fig. 16.3). Some of these have adapted to feed directly on food such as flour and they have become independent of dampness because they can absorb water vapour from the atmosphere. This clearly accounts for the success in these few species being able to colonize and exploit conditions that would be hostile to most members of the group. Although they are not often serious pests in domestic houses, their presence in foodstuffs is clearly unacceptable to most people, and the contamination by faecal material is unpleasant. It can be a different matter in industrial scale food storage, because infestations may go unnoticed and therefore build up to large numbers very quickly. Although only a few cosmopolitan species fall in this category, their control is important in many countries.

There are currently 100 species on the British list, but this undoubtedly includes many casual introductions, particularly of species that can only live in artificially heated conditions. There is thus the same difficulty in defining a British list as with other groups that include cosmopolitan pest species such as aphids and coccids. The most important work for identifying the species currently known from Britain is New (2005), which also has extensive sections on the morphology and biology of the whole group. Even though New included notes on European species that were predicted to occur in

Britain, there have been one or two recent additions to our fauna that were not foreseen. For a wider view of the entire European fauna the standard work is Lienhard (1998). Because this is a little-studied group, the published distributions of several British species are likely to be incomplete. It is noteworthy that some species have recently been rediscovered after a century or more, e.g. *Peripsocus parvulus* and *Hyalopsocus morio*. There have been many changes in the family-level classification since the Kloet and Hincks (1964) checklist, mainly because groups such as Mesopsocidae, Polypsocidae and Pseudocaeciliidae are either no longer used or else have been redefined.

Over 200 species in 25 families have been recorded in Europe, and over 5000 species in 17 families are known worldwide. The last published world checklist was by Lienhard and Smithers (2002) and a brief review of the group is provided by Mockford (2009).

HIGHER CLASSIFICATION OF BRITISH PSOCOPTERA

Suborder Psocomorpha
  Infraorder Caeciliusetae
    Family Amphipsocidae (1 genus, 1 species)
    Family Caeciliusidae (4 genera, 8 species)
    Family Stenopsocidae (2 genera, 4 species)
  Infraorder Epipsocetae
    Family Epipsocidae (1 genus, 1 species)
  Infraorder Homilopsocidea
    Family Ectopsocidae (1 genus, 7 species)
    Family Elipsocidae (5 genera, 10 species)
    Family Lachesillidae (1 genus, 3 species)
    Family Mesopsocidae (1 genus, 3 species)
    Family Peripsocidae (1 genus, 7 species)
    Family Philotarsidae (2 genera, 3 species)
    Family Trichopsocidae (1 genus, 3 species)
  Infraorder Psocetae
    Family Psocidae (9 genera, 14 species)
Suborder Troctomorpha
  Infraorder Nanopsocetae
    Family Liposcelididae (2 genera, 18 species)
    Family Pachytroctidae (2 genera, 2 species)
    Family Sphaeropsocidae (1 genus, 1 species)
Suborder Trogiomorpha
  Infraorder Atropetae
    Family Lepidopsocidae (4 genera, 4 species)
    Family Psoquillidae (2 genera, 3 species)
    Family Trogiidae (3 genera, 6 species)
  Infraorder Psocathropetae
    Family Psyllipsocidae (2 genera, 2 species)

SPECIES OF CONSERVATION CONCERN

No species have any legal protection and none is on the UKBAP list.

The Families of British Psocoptera

SUBORDER PSOCOMORPHA

Infraorder Caeciliusetae

Family Amphipsocidae (1 genus, 1 species)
The members of this family were formerly included in the Polypsocidae or Caeciliidae (now Caeciliusidae). The only British species is *Kolbia quisquiliarum*, a scarce species that seems confined to grassland rather than trees. *Kolbia* was previously spelled *Kolbea*.

British genus: *Kolbia*

Family Caeciliusidae (4 genera, 8 species)
This family was formerly called Caeciliidae. Two common species are *Valenzuela burmeisteri*, found mainly on conifers and other evergreens, and the bright yellow *V. flavidus*, which is found on a variety of deciduous trees. Both species were previously placed in *Caecilius*.

British genera: *Caecilius, Enderleinella, Epicaecilius, Valenzuela*

Family Stenopsocidae (2 genera, 4 species)
Graphopsocus cruciatus has strongly marked wing patterns and is commonly found on a wide range of shrubs and trees, both evergreen and deciduous; it is a cosmopolitan species. *Stenopsocus immaculatus* is another common species found on a wide variety of trees. It is common across Europe and very variable; within Britain its morphological range seems to include specimens previously considered as *St. lachlani*, so the latter name may need to be removed from British list.

British genera: *Graphopsocus, Stenopsocus*

Infraorder Epipsocetae

Family Epipsocidae (1 genus, 1 species)
The only British species is *Bertkauia lucifuga*, previously in the genus *Epipsocus*. It is a scarce, dark brown, usually wingless species, found in leaf-litter and often associated with conifers.

British genus: *Bertkauia*

Infraorder Homilopsocidea

Family Ectopsocidae (1 genus, 7 species)
Ectopsocus briggsi is commonly found on the branches of a wide range of trees and shrubs; the wings have very dark spots at the apex of each main vein. Brachypterous forms may be confused with *E. petersi*, which was described relatively recently, and seems to be just as common and widespread though possibly more confined to tree trunks.

British genus: *Ectopsocus*

Family Elipsocidae (5 genera, 10 species)
The Elipsocidae were previously treated as a subfamily of the Mesopsocidae. The members of this family are found on the bark of a wide variety of trees and woody shrubs. *Elipsocus hyalinus* is common, though males are unknown in this parthenogenetic species, and the keys to this genus use the morphology of the females. The recognition of species in this group was very confused until recent keys were published, so early records are not always reliable.

British genera: *Cuneopalpus, Elipsocus, Propsocus, Pseudopsocus, Reuterella*

Family Lachesillidae (1 genus, 3 species)
Formerly in the Mesopsocidae, the British species of *Lachesilla* are usually found in leaf-litter, hay and straw, as well as tree trunks. None is common, though *L. pedicularia* is the most widespread and is probably the only native species; worldwide it is sometimes known as the Cosmopolitan grain psocid, though it is rarely found indoors in Britain.

British genus: *Lachesilla*

Family Mesopsocidae (1 genus, 3 species)
With the separation of families such as Lachesillidae and Elipsocidae, this family now includes only the genus *Mesopsocus*. They are relatively large psocids, with a body length up to 5 mm, found on the bark of a variety of trees. Two species, *M. immunis* and *M. unipunctatus* are widespread in Britain, though rarely in large numbers, and they are often found together.

British genus: *Mesopsocus*

Family Peripsocidae (1 genus, 7 species)
Three of the seven British species in this family are known only from females, and brachyptery occurs quite frequently. The wings of all species are either uniformly brown, or have strong patterns. The two most common species in Britain are *Peripsocus phaeopterus*, found on the branches of deciduous and evergreen trees and shrubs, but rarely on the trunk; and *P. subfasciatus*, found in similar habitats but also on the trunk. Throughout Europe *P. subfasciatus* is entirely parthenogenetic. *P. parvulus* was until recently known only from a single 19th-century record from the New Forest, but it has been rediscovered at three further sites.

British genus: *Peripsocus*

Family Philotarsidae (2 genera, 3 species)
Until very recently, the two species of *Philotarsus* were the only members of this family recorded in Britain; both have dark markings on the wings and usually live on bark. *Ph. parviceps* is most common on the branches of deciduous trees, whereas *Ph. picicornis* is not so widespread and is usually confined to conifers. *Aaroniella badonneli* was recently added to the British list.

British genera: *Aaroniella, Philotarsus*

Family Trichopsocidae (1 genus, 3 species)
Trichopsocus is the only genus in this family, and the three British species have few or no markings on the wings. None is particularly common, and all three are associated with the branches of trees and shrubs.

British genus: *Trichopsocus*

Infraorder Psocetae

Family Psocidae (9 genera, 14 species)
This is the second largest family of British Psocoptera, but it includes a few rare species of uncertain status. Most species are among the larger psocids, often 5 or 6 mm long, and they are usually found on tree bark. Several have well-defined markings on the fore wings. *Psococerastis gibbosa* has the distinction of being the largest British psocid, up to 7 mm long and has small but distinct fore wing markings; it is widespread but not usually found in large numbers (Fig. 16.4). One of the most common species is *Metylophorus nebulosus*, a widespread western Palaearctic species, found on tree branches; it is notable in showing clear sexual dimorphism, with the female fore wing being much darker and more strongly marked than that of the male.

Hyalopsocus morio is a very distinctive species with shiny dark brown or blackish fore wings; until recently the last known record was from the 19th century, but it has recently been rediscovered in Gloucestershire. Similarly, *Psocus bipunctatus* has

Fig. 16.4 *Psococerastis gibbosa* (Psocidae), the largest British species (Photo: Roger Key)

recently been reported from an oak tree in Kent, the first record since 1837. *Atlantopsocus adustus* was recently added to British list and seems well established along the south coast.

British subfamilies and genera:
   Amphigerontiinae: *Amphigerontia, Blaste*
   Psocinae: *Atlantopsocus, Hyalopsocus, Loensia , Metylophorus, Psococerastis, Psocus, Trichadenotecnum*

SUBORDER TROCTOMORPHA

Infraorder Nanopsocetae

Family Liposcelididae (2 genera, 18 species)
This is the largest family of British psocids, though several of its species are probably casual imports. The family was previously known as Liposcelidae. All species are apterous, with rare winged forms in *Embidopsocus*, the eyes are very reduced and the hind femora swollen. Members of the genus *Liposcelis* are known as booklice, and *L. bostrychophila* is probably the most common in domestic situations, rarely found outdoors; it is a cosmopolitan species and is parthenogenetic. The species in this genus are difficult to identify accurately, and older records are generally unreliable. Although a few species are found outdoors, the indoor species can become pests on dry foods such as flour and cereals, whether in commercial food-stores or in domestic kitchens. They can absorb water from a humid atmosphere, which clearly aids their persistence under adverse conditions, but the populations are often larger in damp situations; hence their frequent discovery in unventilated bookcases, where they feed on moulds or on the glues in book-

bindings. It should be noted that booklice have no connection with 'book-worms', which are various kinds of beetle larvae.

British genera: *Embidopsocus, Liposcelis*

Family Pachytroctidae (2 genera, 2 species)
The status of the two species in this family reported from Britain is very doubtful. *Tapinella castanea* is assumed to be an introduction on bananas from the Canary Islands, while an unknown species of *Pachytroctes* is also assumed to be introduced.

British genera: *Pachytroctes, Tapinella*

Family Sphaeropsocidae (1 genus, 1 species)
The only British species in this family is *Badonnelia titei*. This may be a native species, or it may be a frequent introduction, often in domestic environments; it is quite widespread in western Europe. This species is noteworthy for the thickened and sculptured fore wings in the female, which are very like coleopteran elytra; males are apterous.

British genus: *Badonnelia*

SUBORDER TROGIOMORPHA

Infraorder Atropetae

Family Lepidopsocidae (4 genera, 4 species)
The members of this family have flattened scales over much of the body, including the wings, where they often produce distinctive patterns, though some species are brachypterous. Only one species, *Pteroxanium kelloggi*, is believed to be native; it is an uncommon but cosmopolitan species found mainly in leaf-litter. The other three species recorded in Britain are known only from single records on stored products.

British genera: *Lepolepis, Nepticulomima, Pteroxanium, Soa*

Family Psoquillidae (2 genera, 3 species)
The three species in this family are all rare imports on stored products. The most frequently seen is *Psoquilla marginepunctata*, a cosmopolitan species with a striking fore-wing pattern.

British genera: *Psoquilla, Rhyopsocus*

Family Trogiidae (3 genera, 6 species)
This family contains a mixture of outdoor barklice and domestic species, usually wingless. The most common outdoor species is *Cerobasis guestfalica*, almost entirely parthenogenetic, and found on the bark of a wide range of trees and shrubs. The three species of *Lepinotus* are all found in domestic envi-

ronments, though *L. patruelis* is occasionally found outdoors. Similarly, *Trogium pulsatorium* is rare outside but is a common domestic species. The female of *T. pulsatorium* taps the sclerotized tip of her abdomen on the substrate to produce a regular ticking sound, which is clearly audible to humans in quiet environments. In former centuries this ominous sound was frequently heard by sick people lying in quiet rooms, which gave rise to name of 'death-watch' (not to be confused with the Death watch beetle in the coleopteran family Anobiidae).

British genera: *Cerobasis, Lepinotus, Trogium*

Infraorder Psocathropetae

Family Psyllipsocidae (2 genera, 2 species)
The two British species in this family are both found in domestic situations. Their wings are always reduced to some extent, and both are very small, up to 2mm in length.

British genera: *Dorypteryx, Psyllipsocus*

REFERENCES

KLOET, G.S. & HINCKS, W.D. 1964. A check list of British insects (2nd edn.). Part 1: small orders and Hemiptera. *Handbooks for the identification of British insects* 11(1): 119 pp.

LIENHARD, C. 1998. Psocoptères euro-méditerranéens. *Faune de France* 83: 517 pp.

LIENHARD, C. & SMITHERS, C.N. 2002. Psocoptera (Insecta): world catalogue and bibliography. Muséum d'histoire naturelle, Genève.

MOCKFORD, E.L. 2009. Psocoptera (psocids, booklice). In: Resh, V.H. & Cardé, R.T. (eds.) *Encyclopedia of insects* (2nd edn.). Academic Press/Elsevier, San Diego & London, pp. 858–60.

NEW, T.R. 2005. Psocids. Psocoptera (booklice and barklice). *Handbooks for the identification of British insects* (2nd edn) 1(7): 146 pp.

WEBSITES

http://www.brc.ac.uk/schemes/barkfly/homepage.htm
Although this covers only the outdoor species, there are many excellent photos, literature updates such as new records, and an on-line key.

http://www.psocodea.org/
Psoco Net is an international site for psocidologists, with the downloadable Psocid News.

http://tolweb.org/Psocodea/8235
This discusses the latest thoughts on the relationships between Psocoptera and Phthiraptera.

17 Order Thysanoptera: the thrips

179 species in 3 families

The thrips are probably most closely related to the Hemiptera, based on certain mouthpart structures and other characters. Their minute size means that they are a little known group, and many people are surprised to learn how many species there are. Some feed on fungi, others on higher plants, and a small number of these can be agricultural or horticultural pests.

Because of its ancient Greek origin, the name thrips is both singular and plural: there is no such word as 'thrip'. Most species are only 1–3 mm long, and the largest British species reaches 7 mm, but their extremely narrow bodies make even the larger ones inconspicuous, though they can be present in huge numbers in favourable conditions. The wings are extremely narrow with a very reduced venation, and they carry fringes of long cilia. The antennae are slender and unmodified, and the eyes are usually well developed, though they are often reduced in size and complexity in wingless species. The mouthparts are perhaps the most remarkable because of their extreme modification, and they form a cone on the underside of the head. Only the left mandible is present; it is needle-like, and is used to perforate the host plant to get access to the food source. The maxillae form narrow stylets that link together to form a tube, which is inserted into the hole made by the mandible so that liquid can be pumped up, often after saliva has been injected. There are slight variations on this theme, because a few species feed on fungal spores, and others are predatory on small insects and mites, but the right mandible is always absent.

Many species of thrips feed on fungi, but they are inconspicuous and often concealed from view under their host fungus, tree bark or leaf-litter. Most are in the family Phlaeothripidae and members of the subfamily Idolothripinae, including *Megathrips*, have wider maxillary stylets with which they eat fungal spores whole. Another large group of thrips feed on leaves, and these include most of the species that cause damage to plants, either by direct feeding in large numbers or by transmitting plant viruses; the best known is *Limothrips cerealium* the Grain thrips. Probably the easiest group to find are the flower thrips, which can be numerous and easily visible when their dark bodies stand out against brightly coloured flowers (Fig. 17.1). They feed on nectar, parts of the flower tissue and especially pollen grains; however, they are often covered in pollen and may help to pollinate some plant species, so their overall effect is not always clear. Some species are predators on very small insects and other arthropods such as mites, and this behaviour is sometimes seen among the flower thrips. Elsewhere in the world some species cause gall formation though none of these are found in Britain.

Just how many thrips are specific to particular host plants or fungi is not known, again because so little research has been done on this group, but a table of common associations of flower thrips with plant species was given by Kirk (1996). The handbook by Mound et al. (1976) gives the known host associations for each species.

Although the behaviour of only a few species has been studied in detail, it seems that thrips often have complex social structures, with communal activities, and male fighting associated with repro-

The Royal Entomological Society Book of British Insects, First Edition. Peter C. Barnard.
© 2011 Royal Entomological Society. Published 2011 by Blackwell Publishing Ltd.

Fig. 17.1 The small size of a typical thrips on a flower-head (Photo: Colin Rew)

Fig. 17.2 Microscope slide preparation of a thrips (Photo: Luis Fernández García, Creative Commons Licence)

duction. Most reproduce bisexually though parthenogenesis is found in a few species. As in the hemipteran family Coccidae there is a distinct 'pupal' stage before the adult; for this reason the juvenile stages tend to be known as larvae rather than nymphs, though the Thysanoptera are clearly 'exopterygote'. There are two larval instars, then a prepupa, followed by one (in the Terebrantia) or two (Tubulifera) pupal stages. Although the prepupae and pupae are non-feeding, they are mobile to some extent. Reproduction in the thrips is haplodiploid; the males develop from unfertilized eggs and have only half the chromosome numbers found in the females, which develop from fertilized eggs. A useful general account of thrips biology was given by Lewis (1973) though this is now somewhat out of date.

Because of their small size and specialized habitats, thrips are not always well represented in general insect collections. The specialist collector will 'beat' flowers, vegetation or dead wood (for the fungal feeding species). Being small and soft-bodied they are preserved in fluid, but any critical identification needs microscopic examination, so specimens are usually mounted permanently on slides (see Fig. 17.2).

The Thysanoptera are divided into two suborders: the Terebrantia is the larger of the two in Britain, containing the families Aeolothripidae and Thripidae. The other suborder, Tubulifera, contains only the family Phaeothripidae, which is the largest in the world, though not in Britain, reflecting the greater diversity of thrips in tropical regions. The main work for identifying the British species is the handbook by Mound et al. (1976), but Kirk's (1996) book is a very useful introduction to the group, which also contains keys to the more common species. The European Terebrantia were covered by Strassen (2003), and the economically important species were treated by Lewis (1997) and Mound and Kibby (1998).

There are around 570 species in 6 families recorded in Europe, and over 6000 species in 9 families described worldwide. A brief world overview of the Thysanoptera is given by Mound (2009) and a discussion of the higher classification of the group by Mound and Morris (2007).

HIGHER CLASSIFICATION OF BRITISH THYSANOPTERA

Suborder Terebrantia
    Family Aeolothripidae (3 genera, 14 species)
    Family Thripidae (44 genera, 117 species)
Suborder Tubulifera
    Family Phlaeothripidae (19 genera, 48 species)

SPECIES OF CONSERVATION CONCERN

No species have any legal protection and none is on the UKBAP list.

The Families of British Thysanoptera

SUBORDER TEREBRANTIA

Members of this suborder can be recognized by the blunt or slightly pointed apex of the abdomen. The fore wings have two longitudinal veins, in addition to the costa, and are covered with microsetae (Fig. 17.2).

Family Aeolothripidae (3 genera, 14 species)
Most members of this family are flower-feeders, though some are predatory. Only the species of *Aeolothrips* are at all well known; they are all very small, up to 2 mm in length, and are found in a wide variety of habitats. In the macropterous species the wings are banded black and white.

British genera: *Aeolothrips, Melanthrips, Rhipidothrips*

Family Thripidae (44 genera, 117 species)
Most members of this family are flower-feeders, though there are also many important leaf-feeders that can cause damage to crops or to cultivated flowers. This is by far the largest family of British thrips and many species seem to be highly host-specific, though several are known from only a few records.

Perhaps the best known is *Limothrips cerealium*, called the Thunderfly, though this name is often applied to any species with similar behaviour associated with cereals. Like all members of this genus it feeds on grasses, but it can become a major pest on cereal crops. The populations build up in cool weather, but in warm and humid conditions, which are often accompanied by thunderstorms, it takes to the air in countless numbers. These migrating swarms settle on people, causing some irritation, and they also enter houses through the tiniest of cracks around windows. They often settle on the inside of the windows, but they also manage to creep behind the glass in picture frames and similarly unlikely places. The overall effect is to create some distress to local inhabitants, although thrips are harmless, if persistent, invaders. In the past they have also entered smoke detectors, setting off false alarms, until the alarm manufacturers came up with new designs to keep the thrips out.

Another well-known pest in this family is *Kakothrips pisivorus*, the Pea thrips, which feeds on pea flowers and later causes unsightly damage to pea pods. *Thrips major* is the Rubus thrips, a very common species found in wide variety of flowers, particularly on bindweed (*Calystegia*). *Frankliniella occidentalis*, the Western flower thrips, is a North American native that has become a cosmopolitan pest; it spread rapidly across Europe and is now well established in Britain. Although little more than 1 mm in length it can be a major pest on cultivated plants in heated glasshouses. Direct feeding damage makes plants unsightly and unsaleable commercially, and it also transmits plant viruses.

British genera: *Anaphothrips, Apterothrips, Aptinothrips, Aurantothrips, Baliothrips, Belothrips, Bolacothrips, Caliothrips, Ceratothrips, Chaetanaphothrips, Chirothrips, Dendrothrips, Dichromothrips, Drepanothrips, Echinothrips, Euchaetothrips, Frankliniella, Helionothrips, Heliothrips, Hemianaphothrips, Hercinothrips, Iridothrips, Kakothrips, Leucothrips, Limothrips, Mycterothrips, Neohydatothrips, Odontothrips, Oxythrips, Parthenothrips, Platythrips, Psydrothrips, Rhaphidothrips, Rubiothrips, Scirtothrips, Scolothrips, Sericothrips, Stenchaetothrips, Stenothrips, Taeniothrips, Tameothrips, Tenothrips, Thrips, Tmetothrips*

SUBORDER TUBULIFERA

The members of this suborder are recognized by the tubular apex (segment 10) of the abdomen (Fig. 17.3); the fore wings have no longitudinal veins or

Fig. 17.3 Adults and nymphs of *Suocerathrips linguis* (Phlaeothripidae) on *Sansevieria trifasciata* (Photo: PaulT, Creative Commons Licence)

covering of microsetae. The Phlaeothripidae is the only family in this group.

Family Phlaeothripidae (19 genera, 48 species)
Although some members of this family are flower- or leaf-feeders, many live on fungi. *Phlaeothrips annulipes* is a common species on birch fungi, and the classic way to find it is to examine the birch 'brooms' used for fire-fighting in forest areas. *Hoplothrips pedicularius* is common on the dead wood fungus *Stereum*, which is found on many deciduous trees. *Megathrips lativentris* is another widespread species in which the males have lateral tubercles on the abdomen together with a pair of large abdominal horns, though the precise function of these structures is not clear. It feeds on fungal spores in leaf-litter. *Haplothrips leucanthemi*, the Ox-eye daisy thrips, is one of the common flower-feeding species in this family; although confined to one species of plant (*Leucanthemum vulgare*), this is such a common host that the thrips itself is also widespread. *Suocerathrips linguis* is a recent addition to the British list; it feeds on fungi on the leaves of cultivated *Sansevieria* (see Fig. 17.3) and was probably introduced from Africa.

British subfamilies and genera:

Idolothripinae: *Bacillothrips, Bolothrips, Cryptothrips, Megalothrips, Megathrips*

Phlaeothripinae: *Cephalothrips, Gynaikothrips, Haplothrips, Holothrips, Hoplandrothrips, Hoplothrips, Liothrips, Neoheegeria, Phlaeothrips, Poecilothrips, Sinuothrips, Suocerathrips, Xylaplothrips*

REFERENCES

KIRK, W.D.J. 1996. *Thrips*. Naturalists' handbook no. 25. Richmond Publishing, Slough.

LEWIS, T. 1973. *Thrips: their biology, ecology, and economic importance*. Academic Press, London.

Lewis, T. (ed.) 1997. *Thrips as crop pests*. CAB International, Wallingford.

MOUND, L.A. 2009. Thysanoptera. In: Resh, V.H. & Cardé, R.T. (eds.) *Encyclopedia of insects* (2nd edn.). Academic Press/Elsevier, San Diego & London, pp. 999–1003.

MOUND, L.A. & KIBBY, G. 1998. *Thysanoptera: an identification guide* (2nd edn.). CAB International, Wallingford.

MOUND, L.A., MORISON, G.D., PITKIN, B.R. & PALMER, J.M. 1976. Thysanoptera. *Handbooks for the identification of British insects* 1(11): 79 pp.

MOUND, L.A. & MORRIS, D.C. 2007. The insect Order Thysanoptera: classification versus systematics. *Zootaxa* 1668: 395–411 [downloadable at http://www.mapress.com/zootaxa/2007f/zt01668p411.pdf].

STRASSEN, R. ZUR 2003. Die terebranten Thysanopteren Europas und des Mittelmeer-Gebietes. *Die Tierwelt Deutschland* 74: 277 pp.

WEBSITES

http://www.ento.csiro.au/thysanoptera/worldthrips.html
A catalogue of the world species of Thysanoptera.

http://www.brc.ac.uk/DBIF/homepage.aspx
A general database of the foodplants of British insects, which has information on several species of thrips.

PART 6 Endopterygota

All these higher insects have a well-defined metamorphosis from larva via a pupa to adult, hence the alternative name of Holometabola, meaning a complete metamorphosis. Each stage of the lifecycle is morphologically (and usually biologically) quite distinct, and there is no sign of the wings developing in the larvae; this explains the name Endopterygota, meaning internal wings. All the other winged insects, now in the Palaeoptera, Polyneoptera and Paraneoptera, were formerly grouped as the Hemimetabola or Exopterygota on the grounds that they have no metamorphosis, the wings gradually developing externally throughout the nymphal stages. However, this is clearly a plesiomorphy and the paraphyletic group Exopterygota is no longer used; its inadequacy is emphasized by the pupal stages seen in groups such as the Coccidae (Hemiptera) and the Thysanoptera.

There are several subgroups within the Endopterygota, though not all are accepted universally. The Neuroptera, Megaloptera and Raphidioptera are usually grouped within the Neuropterida, and these may form the sister group of the Coleoptera (and possibly Strepsiptera). The Hymenoptera may have a similar relationship with the Panorpoid orders (Diptera, Lepidoptera, Mecoptera, Siphonaptera and Trichoptera). The Mecoptera and Siphonaptera (Mecopterida) may be combined with the Diptera in the Antliophora, and the Lepidoptera and Trichoptera are often united within the Amphiesmenoptera, despite their very different life histories. It is clear that these supraordinal groupings need more rigorous examination and definition before they can be adopted as formal taxa.

The Royal Entomological Society Book of British Insects, First Edition. Peter C. Barnard.
© 2011 Royal Entomological Society. Published 2011 by Blackwell Publishing Ltd.

18 Order Coleoptera: the beetles

c. 4000 species in 112 families

Adult beetles have a distinctive appearance, with the hard elytra covering most of the body and meeting down the mid-line, which makes them easy to recognize but this apparent uniformity hides an enormous variety of life histories. There are numerous variations in larval forms and feeding habits, and beetles can be found in a wide range of habitats. In turn this makes them important in ecological research and environmental surveys. The group has always been popular with general naturalists and insect collectors, second only to the Lepidoptera, and not least because most specimens are robust and easy to keep in collections.

In common with other endopterygote insects, the larvae are the main feeding stage but in many families the adults are long-lived and also feed. Sometimes larvae and adults occupy similar habitats with the same food requirements, but in other cases the adults have different habits from the larvae. The hard elytra have undoubtedly played a part in adult longevity because the well-protected insects can burrow under the ground or in wood without damage; most other adult endopterygotes have fragile wings that limit their activity

During the 19th century the entire British Coleoptera fauna was covered by monographic works like Fowler (1886–91) supplemented by Fowler and Donisthorpe (1913). The last attempt to cover all the British species was by Joy (1932), whose 'practical handbook' is still referred to by Coleoptera experts, but the keys are not easy for the novice to use. Its scarcity led to a reprint in 1976 though in a smaller format, meaning that the scale lines of the illustrations in volume 2 are slightly inaccurate. A facsimile of Joy's original book is also available on CD-ROM for those who still want to tackle the original text. An attempt was made to update this seminal work by Hodge and Jones (1995) though this book is also hard to find outside specialist libraries. A useful summary of the value of Joy's handbook to the modern Coleoptera is provided on this website: http://markgtelfer.co.uk/beetles/do-i-need-joy. Even when Joy's book is still useful for identification it must be remembered that the nomenclature of many species has changed considerably in recent years, and the checklist of British beetles on *The Coleopterist* website (http://www.coleopterist.org.uk) should always be consulted.

Linssen's (1959) books are widely quoted as a more recent account of the British species, but the work is not completely reliable for all families and must be used with care. The only modern book that gives a well-illustrated generic overview of the beetles is Harde's book on the central European fauna; the English translation (Harde & Hammond, 1984) includes an indication of which species are found in Britain.

The latest edition of *The Coleopterist's Handbook* (Cooter & Barclay, 2006) is essential reading for anyone interested in the British beetles, but the systematically minded may find it a little arduous to discover information about particular families because of the very basic contents list and an index that includes only genera. It is an excellent summary of the biology of all groups and contains many useful illustrations. Some families of British beetles are covered by recent handbooks and there are many more useful papers; for details see Hammond

The Royal Entomological Society Book of British Insects, First Edition. Peter C. Barnard.
© 2011 Royal Entomological Society. Published 2011 by Blackwell Publishing Ltd.

and Hine (1999). For some families it is still necessary to consult the monographic series of the European species *Die Käfer Mitteleuropas,* published in parts from 1964 onwards (in German), and referred to under individual authors in each section below. A useful key to families was provided by Unwin (1984); a more rigorous but specialist-level key was by Crowson (1956). For general texts on the biology of the whole group Crowson (1981) is still invaluable.

There is a projected new series on the Coleoptera of Europe, to be edited by T. Wagner and published by Apollo Books, which is due to start appearing in 2011. Also useful is the recently begun series of catalogues of the Palaearctic Coleoptera (Löbl & Smetana, 2003–10).

The identification of larvae is not always easy; some keys to adults also include larvae, but in many groups only a few larvae are known. Klausnitzer (1978) is a useful guide (in German) to the European groups, and the volumes of *Die Käfer Mitteleuropas* are always worth consulting (Klausnitzer, 1991,

1994, 1996). Lawrence et al. (1993) covers the families of the world. There is a useful overview of the British larvae in Cooter and Barclay (2006, pp. 178–99), though it is not listed on the contents page.

There are several works covering species of economic importance, such as wood-boring beetles and those associated with stored products. Some useful books are Booth et al. (1990), Hickin (1963) and Hinton (1945); there are many extra references in Cooter and Barclay (2006).

There are several websites of interest to the coleopterist and these are listed at the end of this chapter. These also include information about the several specialist journals, which contain many new records of species, updates to exisiting keys, and so on. Several beetle groups also have recording schemes, and their websites are always worth consulting for recent information.

There are around 12,500 species in 144 families in Europe, and at least 350,000 species in 175 families worldwide. For a world overview of the Coleoptera see McHugh and Liebherr (2009).

HIGHER CLASSIFICATION OF BRITISH COLEOPTERA

Suborder Adephaga
  Superfamily Caraboidea
    Family Carabidae (88 genera, 350 species)
    Family Dytiscidae (28 genera, 118 species)
    Family Gyrinidae (2 genera, 12 species)
    Family Haliplidae (3 genera, 19 species)
    Family Hygrobiidae (1 genus, 1 species)
    Family Noteridae (1 genus, 2 species)
Suborder Myxophaga
  Superfamily Sphaeriusoidea
    Family Sphaeriusidae (1 genus, 1 species)
Suborder Polyphaga
  Infraorder Bostrichiformia
    Superfamily Bostrichoidea
      Family Anobiidae (26 genera, 53 species)
      Family Bostrichidae (3 genera, 3 species)
      Family Dermestidae (13 genera, 40 species)
      Family Lyctidae (1 genus, 2 species)
    Superfamily Derodontoidea
      Family Derodontidae (1 genus, 1 species)
  Infraorder Cucujiformia
    Superfamily Chrysomeloidea
      Family Cerambycidae (49 genera, 65 species)
      Family Chrysomelidae (63 genera, 273 species)
      Family Megalopodidae (1 genus, 3 species)
      Family Orsodacnidae (1 genus, 2 species)
    Superfamily Cleroidea
      Family Cleridae (10 genera, 14 species)
      Family Dasytidae (4 genera, 9 species)

      Family Malachiidae (10 genera, 16 species)
      Family Phloiophilidae (1 genus, 1 species)
      Family Trogossitidae (5 genera, 5 species)
    Superfamily Cucujoidea
      Family Alexiidae (1 genus, 1 species)
      Family Biphyllidae (2 genera, 2 species)
      Family Bothrideridae (3 genera, 5 species)
      Family Byturidae (1 genus, 2 species)
      Family Cerylonidae (2 genera, 5 species)
      Family Coccinellidae (30 genera, 53 species)
      Family Corylophidae (4 genera, 11 species)
      Family Cryptophagidae (11 genera, 103 species)
      Family Cucujidae (1 genus, 2 species)
      Family Endomychidae (5 genera, 8 species)
      Family Erotylidae (3 genera, 7 species)
      Family Kateretidae (3 genera, 9 species)
      Family Laemophloeidae (4 genera, 11 species)
      Family Languriidae (1 genus, 1 species)
      Family Latridiidae (13 genera, 56 species)
      Family Monotomidae (3 genera, 23 species)
      Family Nitidulidae (16 genera, 89 species)
      Family Phalacridae (3 genera, 15 species)
      Family Silvanidae (10 genera, 12 species)
      Family Sphindidae (2 genera, 2 species)
    Superfamily Curculionoidea
      Family Anthribidae (8 genera, 9 species)
      Family Apionidae (33 genera, 87 species)
      Family Attelabidae (2 genera, 2 species)
      Family Curculionidae (152 genera, 475 species)

Family Dryophthoridae (2 genera, 4 species)
Family Erirhinidae (6 genera, 12 species)
Family Nanophyidae (2 genera, 2 species)
Family Nemonychidae (1 genus, 1 species)
Family Platypodidae (1 genus, 1 species)
Family Raymondionymidae (1 genus, 1 species)
Family Rhynchitidae (7 genera, 18 species)
Superfamily Lymexyloidea
  Family Lymexylidae (2 genera, 2 species)
Superfamily Tenebrionoidea
  Family Aderidae (3 genera, 3 species)
  Family Anthicidae (6 genera, 13 species)
  Family Ciidae (7 genera, 22 species)
  Family Melandryidae (10 genera, 17 species)
  Family Meloidae (3 genera, 10 species)
  Family Mordellidae (5 genera, 17 species)
  Family Mycetophagidae (6 genera, 13 species)
  Family Mycteridae (1 genus, 1 species)
  Family Oedemeridae (4 genera, 10 species)
  Family Pyrochroidae (2 genera, 3 species)
  Family Pythidae (1 genus, 1 species)
  Family Ripiphoridae (1 genus, 1 species)
  Family Salpingidae (6 genera, 11 species)
  Family Scraptiidae (2 genera, 14 species)
  Family Tenebrionidae (33 genera, 47 species)
  Family Tetratomidae (2 genera, 4 species)
  Family Zopheridae (9 genera, 12 species)
Infraorder Elateriformia
  Superfamily Buprestoidea
    Family Buprestidae (5 genera, 14 species)
  Superfamily Byrrhoidea
    Family Byrrhidae (7 genera, 13 species)
    Family Dryopidae (2 genera, 9 species)
    Family Elmidae (8 genera, 12 species)
    Family Heteroceridae (2 genera, 9 species)
    Family Limnichidae (1 genus, 1 species)
    Family Psephenidae (1 genus, 1 species)
    Family Ptilodactylidae (1 genus, 1 species)
  Superfamily Dascilloidea
    Family Dascillidae (1 genus, 1 species)

Superfamily Elateroidea
  Family Cantharidae (7 genera, 41 species)
  Family Drilidae (1 genus, 1 species)
  Family Elateridae (38 genera, 73 species)
  Family Eucnemidae (5 genera, 6 species)
  Family Lampyridae (3 genera, 3 species)
  Family Lycidae (3 genera, 4 species)
  Family Throscidae (2 genera, 5 species)
Superfamily Scirtoidea
  Family Clambidae (2 genera, 10 species)
  Family Eucinetidae (1 genus, 1 species)
  Family Scirtidae (7 genera, 20 species)
Infraorder Scarabaeiformia
  Superfamily Scarabaeoidea
    Family Aegialiidae (1 genus, 3 species)
    Family Aphodiidae (34 genera, 55 species)
    Family Bolboceratidae (1 genus, 1 species)
    Family Cetoniidae (4 genera, 6 species)
    Family Geotrupidae (4 genera, 7 species)
    Family Lucanidae (4 genera, 4 species)
    Family Melolonthidae (5 genera, 7 species)
    Family Rutelidae (3 genera, 3 species)
    Family Scarabaeidae (2 genera, 9 species)
    Family Trogidae (1 genus, 3 species)
Infraorder Staphyliniformia
  Superfamily Hydrophiloidea
    Family Georissidae (1 genus, 1 species)
    Family Helophoridae (1 genus, 20 species)
    Family Histeridae (20 genera, 51 species)
    Family Hydrochidae (1 genus, 7 species)
    Family Hydrophilidae (18 genera, 70 species)
    Family Spercheidae (1 genus, 1 species)
    Family Sphaeritidae (1 genus, 1 species)
  Superfamily Staphylinoidea
    Family Hydraenidae (5 genera, 30 species)
    Family Leiodidae (21 genera, 93 species)
    Family Ptiliidae (18 genera, 75 species)
    Family Scydmaenidae (9 genera, 32 species)
    Family Silphidae (7 genera, 21 species)
    Family Staphylinidae (264 genera, *c.* 1000 species)

SPECIES OF CONSERVATION CONCERN

Eight species of Coleoptera are listed on the Wildlife and Countryside Act 1981: *Chrysolina cerealis* (Chrysomelidae); *Curimopsis nigrita* (Byrrhidae); *Graphoderus zonatus* (Dytiscidae); *Hydrochara caraboides* and *Paracymus aeneus* (Hydrophilidae); *Hypebaeus flavipes* (Malachiidae); *Limoniscus violaceus* (Elateridae) and *Lucanus cervus* (Lucanidae). Two further species are on the Bern Convention: *Cerambyx cerdo* (Cerambycidae) and *Graphoderus bilineatus* (Dytiscidae).

There are 78 species on the UKBAP list: *Exapion genistae* and *Melanapion minimum* (Apionidae);

Curimopsis nigrita (Byrrhidae); *Agonum scitulum, Amara famelica, Amara fusca, Anisodactylus nemorivagus, Anisodactylus poeciloides, Bembidion humerale, Bembidion quadripustulatum, Bembidion testaceum, Brachinus sclopeta, Bracteon argenteolum, Calosoma inquisitor, Carabus intricatus, Carabus monilis, Chlaenius tristis, Cicindela hybrida, Cicindela sylvatica, Cylindera germanica, Harpalus froelichii, Harpalus honestus, Harpalus melancholicus, Lebia cyanocephala, Ophonus laticollis, Ophonus melletii, Ophonus puncticollis, Ophonus stictus, Panagaeus cruxmajor, Philorhizus quadrisignatus, Philorhizus vectensis, Poecilus kugelanni* and *Pogonus luridipennis* (Carabidae); *Oberea oculata* (Cerambycidae);

127

Gnorimus nobilis and *Gnorimus variabilis* (Cetoniidae); *Chrysolina graminis, Cryptocephalus coryli, Cryptocephalus decemmaculatus, Cryptocephalus exiguus, Cryptocephalus nitidulus, Cryptocephalus primarius, Cryptocephalus punctiger, Cryptocephalus sexpunctatus, Donacia aquatica, Donacia bicolora* and *Psylliodes luridipennis* (Chrysomelidae); *Bagous nodulosus* and *Orchestes testaceus* (Curculionidae); *Agabus brunneus, Bidessus minutissimus, Bidessus unistriatus, Graphoderus zonatus, Hydroporus necopinatus* subsp. *roni, Hydroporus rufifrons* and *Laccophilus poecilus* (Dytiscidae); *Ampedus rufipennis, Anostirus castaneus, Lacon querceus, Limoniscus violaceus, Megapenthes lugens, Melanotus punctolineatus* and *Synaptus filiformis* (Elateridae); *Helophorus laticollis* (Helophoridae); *Ochthebius poweri* (Hydraenidae); *Hydrochus nitidicollis* (Hydrochidae); *Lucanus cervus* (Lucanidae); *Platycis cosnardi* (Lycidae); *Malachius aeneus* (Malachiidae); *Melandrya barbata* (Melandryidae); *Meloe proscarabaeus, Meloe rugosus* and *Meloe violaceus* (Meloidae); *Byctiscus populi* (Rhynchitidae); *Meotica anglica, Stenus longitarsis, Stenus palposus* and *Thinobius newberyi* (Staphylinidae).

The following species were removed from the UKBAP list in 2007 for various reasons: *Gastrallus immarginatus* (Anobiidae); *Aphodius niger* (Aphodiidae); *Protapion ryei* (Apionidae); *Amara strenua, Badister collaris, Badister peltatus, Bembidion nigropiceum, Cicindela maritima, Dromius sigma, Harpalus cordatus, Harpalus dimidiatus, Harpalus parallelus, Lionychus quadrillum, Perileptus areolatus, Pterostichus aterrimus, Tachys edmondsi* and *Tachys micros* (Carabidae); *Chrysolina cerealis* and *Psylliodes sophiae* (Chrysomelidae); *Cathormiocerus britannicus, Ceutorhynchus insularis, Ernoporus tiliae* and *Pachytychius haematocephalus* (Curculionidae); *Dryophthorus corticalis* (Dryophthoridae); *Ampedus nigerrimus, Ampedus ruficeps* and *Elater ferrugineus* (Elateridae); *Procas granulicollis* (Erirhinidae); *Eucnemis capucina* (Eucnemidae); *Hydrochara caraboides* and *Paracymus aeneus* (Hydrophilidae); *Hypebaeus flavipes* (Malachiidae).

The two volumes by Hyman and Parsons (1992, 1994) are still of value in assessing conservation needs.

The Families of British Coleoptera

SUBORDER ADEPHAGA

This suborder is generally regarded as the most 'primitive' within the Coleoptera and most members are active predators as both larvae and adults. Five of the six British families are aquatic in all stages, but the ground beetles in the Carabidae are entirely terrestrial.

All the 'water beetles' have traditionally been treated together in identification guides, but this means that at least two taxonomically distinct groups are artificially combined: the aquatic groups of the Caraboidea, and the Hydrophiloidea in the Polyphaga, together with various other families that are either partly or wholly aquatic. The traditional series for identifying these groups was Balfour-Browne (1940, 1950, 1958), though most can be identified using Friday (1988). Denton (2007) is a useful photographic guide, which covers most of the common UK species despite being a county fauna. For the more critical groups it is necessary to use some European works such as Tachet et al. (2000).

All the British families of Adephaga are currently placed in a single superfamily, the Caraboidea.

SUPERFAMILY CARABOIDEA

Family Carabidae (88 genera, 350 species)

The members of this large family are generally known as ground beetles (Fig. 18.1), and this includes the five species of tiger beetles, formerly placed in a separate family, the Cicindelidae. The predatory larvae (Fig. 18.2) and adults are either burrowing or surface-dwelling, taking a wide variety of prey, sometimes living on shrubs or even trees. Although many are polyphagous there are also specialist predators on groups such as snails, and some larvae are even ectoparasites on other beetles. Many species are nocturnal and spend much of their time under stones, and many will be

Fig. 18.1 Typical Carabidae (Photo: Colin Rew)

Fig. 18.2 Carabidae larva (Photo: Roger Key)

Fig. 18.4 *Cylindera germanica* larva (Carabidae) (Photo: Roger Key)

Fig. 18.3 *Cicindela maritima* (Carabidae) (Photo: Roger Key)

Fig. 18.5 *Pterostichus niger* (Carabidae) (Photo: Roger Key)

found in coastal or riverside habitats such as shingle banks.

The tiger beetles, such as *Cicindela campestris*, the Green tiger beetle, are usually seen running around dry habitats in the sunshine; *C. maritima*, the Dune tiger beetle (Fig. 18.3) is a local species on coastal sand dunes. Tiger beetle larvae are active predators like the adults (Fig. 18.4). There are many common ground beetles, including *Carabus violaceus*, the Violet ground beetle, *Nebria brevicollis* and several species of *Pterostichus* (Fig. 18.5). *Brachinus crepitans*, the Bombardier beetle, is often mentioned in general books from its habit of emitting a noxious spray when disturbed, but it is a very local species and little is known of its life history. The genus *Notiophilus* has a distinctive appearance, with very large eyes (Fig. 18.6).

The following species are on the UKBAP list: *Agonum scitulum*, *Amara famelica* (Early sunshiner), *Amara fusca* (Wormwood moonshiner), *Anisodactylus*

Fig. 18.6 *Notiophilus* sp. (Carabidae) (Photo: Colin Rew)

nemorivagus (Heath short-spur), *Anisodactylus poeciloides* (Saltmarsh short-spur), *Bembidion humerale* (Thorne pin-palp), *Bembidion quadripustulatum* (Scarce four-dot pin-palp), *Bembidion testaceum* (Pale pin-palp), *Brachinus sclopeta* (Bombardier beetle), *Bracteon argenteolum* (Silt silver-spot), *Calosoma inquisitor* (Lesser searcher), *Carabus intricatus* (Blue ground beetle), *Carabus monilis* (Necklace ground beetle), *Chlaenius tristis* (Black night-runner), *Cicindela hybrida* (Northern dune tiger beetle), *Cicindela sylvatica* (Heath tiger beetle), *Cylindera germanica* (Cliff tiger beetle), *Harpalus froelichii* (Brush-thighed seed-eater), *Harpalus honestus* (St Bees seed-eater), *Harpalus melancholicus*, *Lebia cyanocephala* (Blue plunderer), *Ophonus laticollis* (Set-aside downy-back), *Ophonus melletii* (Mellet's downy-back), *Ophonus puncticollis*, *Ophonus stictus* (Oolite downy-back), *Panagaeus cruxmajor* (Crucifix ground beetle), *Philorhizus quadrisignatus*, *Philorhizus vectensis*, *Poecilus kugelanni* (Kugelann's green clock), *Pogonus luridipennis* (Yellow pogonus).

Identification: Lindroth's (1974) handbook was completely updated by Luff (2007), which includes colour photos of 147 species. A useful simplified guide is by Forsythe (2000) and an illustrated guide to the European fauna is given by Trautner and Geigenmüller (1987). The subfamily Carabinae is often divided into several tribes, and many genera are split into subgenera.

British subfamilies and genera:
Brachininae: *Brachinus*
Carabinae: *Abax, Acupalpus, Aepopsis, Aepus, Agonum, Amara, Anchomenus, Anisodactylus, Anthracus, Apristus, Asaphidion, Badister, Bembidion, Blemus, Blethisa, Bracteon, Bradycellus, Broscus, Calathus, Callistus, Calodromius, Calosoma, Carabus, Chlaeniellus, Chlaenius, Cillenus, Clivina, Cychrus, Cymindis, Demetrias, Diachromus, Dicheirotrichus, Dromius, Drypta, Dyschirius, Elaphrus, Eurynebria, Laemostenus, Lebia, Leistus, Licinus, Limodromus, Lionychus, Loricera, Masoreus, Microlestes, Miscodera, Nebria, Notiophilus, Ocys, Odacantha, Odontium, Olisthopus, Oodes, Ophonus, Oxypselaphus, Panagaeus, Paradromius, Paranchus, Patrobus, Pelophila, Perileptus, Philorhizus, Platyderus, Platynus, Poecilus, Pogonus, Polistichus, Porotachys, Pterostichus, Scybalicus, Sericoda, Sphodrus, Stenolophus, Stomis, Syntomus, Synuchus, Tachys, Tachyta, Tachyura, Thalassophilus, Trechicus, Trechoblemus, Trechus, Zabrus*
Cicindelinae: *Cicindela, Cylindera*
Omophroninae: *Omophron*

Family Dytiscidae (28 genera, 118 species)
The diving beetles are well known as attractive inhabitants of various freshwater habitats, with the larger species reaching lengths of around 40 mm, though the smaller ones may be only about 1 mm. The larvae are active predators, taking almost any other insects as well as tadpoles and small fish (Fig. 18.7) and the adults may be predators or scavengers. Some adults are large and strikingly coloured, like the iconic *Dytiscus marginalis*, the Great diving beetle (Fig. 18.8), though this a large family, and there are many difficulties in identifying some to species level. Other common genera include *Acilius, Colymbetes, Hydaticus* and *Hyphydrus* (Figs. 18.9 & 18.10).

Graphoderus bilineatus is listed on the Bern Convention and *G. zonatus* (Spangled diving beetle) is listed on the Wildlife and Countryside Act 1981.

Fig. 18.7 *Dytiscus marginalis* **larva and prey (Dytiscidae) (Photo: Roger Key)**

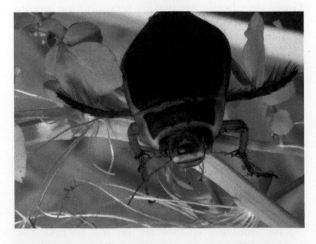

Fig. 18.8 *Dytiscus marginalis* **(Dytiscidae) (Photo: Roger Key)**

Fig. 18.9 *Hydaticus transversalis* (Dytiscidae) (Photo: Robin Williams)

Fig. 18.11 *Gyrinus* sp. (Gyrinidae) (Photo: Robin Williams)

Fig. 18.10 *Hyphydrus ovatus* (Dytiscidae) (Photo: Robin Williams)

The following species are on the UKBAP list: *Agabus brunneus* (Brown diving beetle), *Bidessus minutissimus* (Minutest diving beetle), *Bidessus unistriatus* (One-grooved diving beetle), *Graphoderus zonatus* (Spangled diving beetle), *Hydroporus necopinatus* subsp. *roni* (Ron's diving beetle), *Hydroporus rufifrons* (Oxbow diving beetle) and *Laccophilus poecilus* (Sussex diving beetle).

Identification: as well as the standard guides to water beetles, Nilsson and Holmen (1995) is useful, as is Nilsson (1982) for the larvae.

British subfamilies and genera:
  Agabinae: *Agabus, Ilybius, Platambus*
  Colymbetinae: *Colymbetes, Rhantus*
  Copelatinae: *Liopterus*

Dytiscinae: *Acilius, Cybister, Dytiscus, Graphoderus, Hydaticus*

Hydroporinae: *Bidessus, Deronectes, Graptodytes, Hydroglyphus, Hydroporus, Hydrovatus, Hygrotus, Hyphydrus, Laccornis, Nebrioporus, Oreodytes, Porhydrus, Scarodytes, Stictonectes, Stictotarsus, Suphrodytes*

Laccophilinae: *Laccophilus*

Family Gyrinidae (2 genera, 12 species)
Adult Gyrinidae are known as whirligig beetles from their habit of congregating in large numbers on the surface of ponds and swimming rapidly in circles (Fig. 18.11). The adults feed on small insects trapped on the water surface, but the larvae are deep-water or bottom feeders. *Gyrinus natator* is usually quoted in general books as a typical example, but it is a scarce species probably only found in Ireland within the British Isles; *G. substriatus* is much more common.

Identification: as well as the standard guides to water beetles, Holmen (1987) is also useful.

British genera: *Gyrinus, Orectochilus*

Family Haliplidae (3 genera, 19 species)
Because they swim using their legs alternately, the Haliplidae are known as the crawling water beetles in the USA, though this name is not commonly used in Britain. They are small beetles, 3–5 mm long, and usually yellowish in colour. Unusually the larvae in this family feed mainly on filamentous algae, but the adults seem to be more polyphagous. Several species of *Haliplus* are quite common and widespread.

Fig. 18.12 *Hygrobia hermanni* (Hygrobiidae) (Photo: Roger Key)

Identification: as well as the standard guides to water beetles, Vondel (1997) is also useful.

British genera: *Brychius, Haliplus, Peltodytes*

Family Hygrobiidae (1 genus, 1 species)
This family is also known as the Paelobiidae. The single British species of screech beetles, *Hygrobia hermanni* (Fig. 18.12), is a distinctive black and yellow beetle 8–10 mm long, with a similarly coloured larva. It seems to be quite common in muddy ponds, where the larva feeds on prey such as *Tubifex* worms. The adult emits a startling squeak when handled.

Identification: the standard guides to water beetles cover this species.

British genus: *Hygrobia*

Family Noteridae (1 genus, 2 species)
There are just two species of burrowing water beetles in Britain. They are rather distinctive reddish-brown beetles, around 4–5 mm long, and found in still water, often in dense aquatic vegetation, though little is known of their life histories. *Noterus clavicornis* is common and widespread, but *N. crassicornis* is much scarcer.

Identification: the standard guides to water beetles cover this family.

British genus: *Noterus*

SUBORDER MYXOPHAGA

SUPERFAMILY SPHAERIUSOIDEA

Family Sphaeriusidae (1 genus, 1 species)
This family was previopusly known as the Microsporidae. The only British species is *Sphaerius* *acaroides* (also known as *Microspora acaroides*), a tiny beetle less than 1 mm long that seems to prefer damp habitats. Although assumed to be uncommon it is easily overlooked because of its size, and to the casual observer it can resemble a mite.

British genus: *Sphaerius*

SUBORDER POLYPHAGA

Infraorder Bostrichiformia

SUPERFAMILY BOSTRICHOIDEA

As well as being defined by morphological characters, the beetles in this superfamily all share the ability to live on dry food with no need to drink. Most are associated with timber, and several have become pests, such as the furniture beetle or woodworm, *Anobium punctatum*. Some have become pests of stored products, so the group as a whole attracts much attention from economic entomologists. Identification of most of the common species can be made with standard works such as Joy (1932) with further reference to *Die Käfer Mitteleuropas*. The more difficult genera need specialized papers, which are listed by Hammond and Hine (1999).

Family Anobiidae (26 genera, 53 species)
Most members of this family, known generally as furniture beetles, are wood-boring insects and some of these are synanthropic pests, which can cause structural damage to timber in houses. A few have become cosmopolitan pests of stored products, and these include *Stegobium paniceum*, the Biscuit or Bread beetle and *Lasioderma serricorne*, the Cigarette or Tobacco beetle. The most common timber pest is *Anobium punctatum*, the Furniture beetle, whose larva is commonly known as woodworm (Fig. 18.13). The more notorious *Xestobium rufovillosum*, the Death-watch beetle (Fig. 18.14), is actually very localized and is usually to confined to ancient oak wood, whether in living trees or structural timbers.

This family also includes the spider beetles, which have sometimes been regarded as a separate family, the Ptinidae; they get their common name from the rounded body and long legs. Although some members of the Ptininae such as *Ptinus fur* and *Pt. tectus* (Fig. 18.15) feed on stored products, most of the native species in the group are found in nests or tree holes, where they feed on a variety of dry organic materials or fungi.

The Dorcatominae are a rather specialized subfamily; they are small rounded beetles around 2 mm in length and all are associated with fungi including puffballs and bracket fungi.

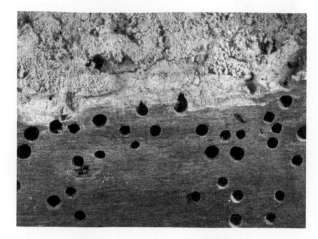

Fig. 18.13 *Anobium punctatum* woodworm exit holes (Anobiidae) (Photo: Roger Key)

Fig. 18.14 *Xestobium rufovillosum* (Anobiidae) (Photo: Roger Key)

Fig. 18.15 *Ptinus tectus* (Anobiidae) (Photo: Roger Key)

Identification: although Joy (1932) can be used for most of the common species, it is also useful to consult Freude et al. (1969) and Lohse and Lucht (1992); more specialized references are in Cooter and Barclay (2006).

British subfamilies and genera:
  Anobiinae: *Anobium, Gastrallus, Hadrobregmus, Priobium, Stegobium*
  Dorcatominae: *Anitys, Caenocara, Dorcatoma*
  Dryophilinae: *Dryophilus, Grynobius*
  Ernobiinae: *Ernobius, Ochina, Xestobium*
  Eucradinae: *Hedobia*
  Ptilininae: *Ptilinus*
  Ptininae: *Gibbium, Mezium, Niptus, Pseudeurostus, Ptinus, Sphaericus, Stethomezium, Tipnus, Trigonogenius*
  Xyletininae: *Lasioderma, Xyletinus*

Family Bostrichidae (3 genera, 3 species)
This family is sometimes known as the false powder-post beetles, from their similarity to the wood-boring Lyctidae; like several other families in this group the head is almost completely concealed from above by the hood-like pronotum. *Bostrichus capucinus* is the only species native to Britain, but there are no recent records from the mature woods which it prefers. The other two species are introduced: *Stephanopachys substriatus* is rarely recorded, but *Rhyzopertha dominica*, the Lesser grain borer, can be a serious pest in flour or grain stores.

British subfamilies and genera:
  Bostrichinae: *Bostrichus*
  Dinoderinae: *Rhyzopertha, Stephanopachys*

Family Dermestidae (13 genera, 40 species)
Although some members of this group are found outdoors, they are best known as pest species, both in the house and in commercial premises; these are known as hide beetles, carpet beetles or larder beetles. One of the most common genera is *Anthrenus*; various species are pests in museums, destroying unprotected zoological collections since they feed on chitin and keratin. *A. verbasci* is the Varied carpet beetle, whose larvae can be a serious pest on woollen materials, though the adults are often seen outside on flowers where they feed on nectar or pollen (Fig. 18.16). *Attagenus pellio* is another pest with similar habits, and *Dermestes lardarius*, the Larder or Bacon beetle, feeds on various animal products, though it is also common outdoors on carcasses, in birds' nests and so on. In commercial premises *Trogoderma granarium*, the Khapra beetle, can be problem in stored grain and similar materials. Most adults in this group have a

Fig. 18.16 *Anthrenus* sp. (Dermestidae) (Photo: Roger Key)

Fig. 18.17 *Ctesias serra* larva (Dermestidae) (Photo: Roger Key)

pattern of coloured scales or hairs on the body, and the larvae are extremely hairy, giving them the common name of woolly bears (Fig. 18.17). The larva of *Ctesias serra* has become known as the Cobweb beetle from its habit of stealing insects from spiders' webs; its long erectile hairs seem to make it immune to attacks from the spiders.

Identification: all the British species, as well the commonly imported ones, can be readily identified using Peacock's (1993) handbook, which also covers the larvae.

British subfamilies and genera:
  Anthreninae: *Anthrenus*
  Attageninae: *Attagenus*
  Dermestinae: *Dermestes*
  Megatominae: *Anthrenocerus, Ctesias, Globicornis, Megatoma, Orphinus, Reesa, Trogoderma*

Thorictinae: *Thorictodes*
Thylodriadinae: *Thylodrias*
Trinodinae: *Trinodes*

Family Lyctidae (1 genus, 2 species)
These are the powder-post beetles, which were sometimes included in the Bostrichidae in earlier classifications, though their heads are not covered by the pronotum. Their common name comes from the very fine powder produced by the wood-boring larvae. Of the two British species of *Lyctus*, *L. brunneus* is a cosmopolitan pest, widely introduced in timber products.

British genus: *Lyctus*

SUPERFAMILY DERODONTOIDEA

Family Derodontidae (1 genus, 1 species)
The only British species of Derodontidae is *Laricobius erichsoni*, about 2mm long, and first recorded in 1971. It is associated with conifers, and both adults and larvae feed on conifer woolly aphids (Adelgidae).

Identification: Peacock (1993).

British genus: *Laricobius*

Infraorder Cucujiformia

SUPERFAMILY CHRYSOMELOIDEA

The classification of the families within this group is still the subject of debate and some authors would place the Cerambycidae in their own subfamily; see the notes in Cooter and Barclay (2006).

Family Cerambycidae (49 genera, 65 species)
The longhorn beetles have always attracted the attention of collectors, with their large size, often very long antennae and striking colour patterns. Some adults feed on foliage, and others are frequently found on flowers. The larvae are phytophagous, either in wood or other plant tissues and some can become pests of forest trees; two species are internal feeders in the stems of various herbaceous plants. One commonly seen species is *Aromia moschata*, called the Musk beetle from the characteristic scent it emits when disturbed; it is found on willows. *Hylotrupes bajulus* is the House longhorn beetle, which can occasionally cause damage to structural softwood timbers though it is a very local species. *Rhagium mordax* (Fig. 18.18) is a widespread species often found on flowers; its larva (Fig. 18.19) feeds on dead deciduous wood. *Clytus arietus* is known as the Wasp beetle, not just because of its black and

Fig. 18.18 *Rhagium mordax* (Cerambycidae) (Photo: Roger Key)

Fig. 18.20 Metallic leaf-beetle (Chrysomelidae) (Photo: Colin Rew)

Fig. 18.19 *Rhagium* sp. larva (Cerambycidae) (Photo: Roger Key)

yellow stripes, but also because of its wasp-like behaviour, scurrying around and waving its antennae.

Adult cerambycids are short-lived and some species may well be under recorded; nonetheless several longhorns on the British list are likely to be extinct here and many of the remaining species are dependent on the survival of ancient woodland.

Oberea oculata (Eyed longhorn beetle) is on the UKBAP list.

Identification: most of the adult longhorn beetles can be identified using the standard works such as Joy (1932). Duffy's (1952) handbook is difficult to use and out of date, and it is necessary to refer to some recent continental works such as Bílý and Mehl (1989) and Bense (1995); see Hodge and Jones

(1995) and Cooter and Barclay (2006) for additional references. There are several popular guides to the more common species, such as Hickin (1987). Duffy's (1953) monograph on the timber beetle larvae is still useful, but more recent works are listed in Cooter and Barclay (2006).

British subfamilies and genera:

Cerambycinae: *Anaglyptus, Aromia, Callidium, Cerambyx, Clytus, Glaphyra, Gracilia, Hylotrupes, Molorchus, Nathrius, Obrium, Phymatodes, Plagionotus, Poecilium, Pyrrhidium, Trinophylum*

Lamiinae: *Acanthocinus, Agapanthia, Lamia, Leiopus, Mesosa, Oberea, Phytoecia, Pogonocherus, Saperda, Stenostola, Tetrops*

Lepturinae: *Alosterna, Anastrangalia, Anoplodera, Dinoptera, Grammoptera, Judolia, Leptura, Lepturobosca, Pachytodes, Paracorymbia, Pedostrangalia, Pseudovadonia, Rhagium, Rutpela, Stenocorus, Stenurella, Stictoleptura, Strangalia*

Prioninae: *Prionus*

Spondylidinae: *Asemum, Arhopalus, Tetropium*

Family Chrysomelidae (63 genera, 273 species)
The members of this large family are generally known as leaf beetles; many adults are metallic coloured (Fig. 18.20) and nearly all species feed on living plant tissue to the extent that some are regarded as agricultural or horticultural pests. On the other hand, a few species have been used in attempting biological control of weed plants. Each of the nine subfamilies has its own characteristics and habitat preferences. The Bruchinae are sometimes called pea or bean weevils, but seed beetles is

Fig. 18.21 *Bruchus rufimanus* (Chrysomelidae) (Photo: Roger Key)

Fig. 18.23 *Cryptocephalus coryli* (Chrysomelidae) (Photo: Roger Key)

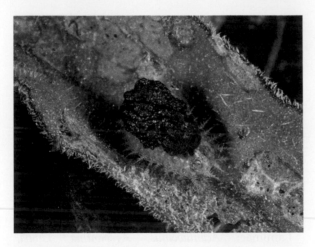

Fig. 18.22 *Cassida rubiginosa* (Chrysomelidae) (Photo: Roger Key)

a better term; some earlier works treat them as a distinct family. The adults are pollen-feeders found on a variety of flowers (Fig. 18.21) but the larvae live inside the seeds of Fabaceae and can be pests of stored pulses. Common species include *Bruchus pisorum*, the Pea beetle and *Acanthoscelides obtectus*, the Bean beetle.

The Cassidinae are the tortoise beetles, greenish with metallic stripes; they have a broad margin all around the body under which the head and appendages can be retracted like a tortoise. The unusual larvae feed on various herbaceous plants; they have spiny processes on the body with which they carry a pile of their excrement as camouflage (Fig. 18.22).

Adult Chrysomelinae are often brightly coloured; both larvae and adults feed in the open on their host plants. Some are very rare, like *Chrysolina cerealis*, known only from Snowdonia, and some can become minor pests, such as *C. americana*, on culinary herbs. *Chrysomela populi* is a red beetle that can be found in great numbers on willows and poplars. The colourfully named Bloody-nosed beetle, *Timarcha tenebricosa*, emits a red fluid from its mouth to deter predators.

The Criocerinae are rather elongate chrysomelids that often feed on grasses and related plants. Species of *Oulema* can be minor pests on cereal crops; *Crioceris asparagi* is the Asparagus beetle, which can be a pest in domestic and commercial asparagus beds; and *Lilioceris lilii* is a bright red beetle that can cause much damage to plants of *Lilium* and *Fritillaria*.

The larvae of Cryptocephalinae are unusual in having close associations with ants. Some live within ants' nests, feeding on dead plant matter, and others live under stones near the nests. The adults can be found on woody shrubs and small trees. *Cryptocephalus coryli* (Fig. 18.23), the Hazel pot beetle, is now a rare species. Like other members of this group the deep-red female, up to 8 mm long, coats her eggs in a 'pot' made from excrement, then drops the egg to the ground; the growing larva adds to this case throughout its life.

The Donaciinae are unique among the chrysomelids in being aquatic, and they are often covered in the standard texts on water beetles. The adults can be found on the flowers and leaves of various emergent plants, but the larvae are entirely aquatic, feeding on the roots and rhizomes of plants such as water-lilies. They breathe by inserting a pair of hollow spines on the eighth abdominal segment into the airways of plant stems. Several species of *Donacia* are brightly coloured and metallic.

The Eumolpinae contain only one rare British species, *Bromius obscurus*, which feeds on willow-herbs, often alongside rivers.

The Galerucinae include the flea beetles, named from their ability to jump. Several species, especially of the genus *Phyllotreta*, are pests on many cruciferous plants, particularly young brassicas. *Phyllotreta nemorum* is the Turnip flea beetle; *Psylliodes affinis* is known as the Potato flea beetle, but is usually found on nightshades.

The Lamprosomatinae are represented only by *Oomorphus concolor*; its case-bearing larva feeds on a variety of plants, including ivy, often in woodland areas.

Chrysolina cerealis (Rainbow leaf beetle) is listed on the Wildlife and Countryside Act 1981. The following species are on the UKBAP list: *Chrysolina graminis* (Tansy beetle), *Cryptocephalus coryli* (Hazel pot beetle, see Fig. 18.23), *Cryptocephalus decemmaculatus* (Ten-spotted pot beetle), *Cryptocephalus exiguus* (Pashford pot beetle), *Cryptocephalus nitidulus* (Shining pot beetle), *Cryptocephalus primarius* (Rock-rose pot beetle), *Cryptocephalus punctiger*, *Cryptocephalus sexpunctatus* (Six-spotted pot beetle), *Donacia aquatica* (Zircon reed beetle), *Donacia bicolora* (Two-tone reed beetle), *Psylliodes luridipennis* (Lundy cabbage flea beetle).

Identification: there is no single work that can be used for the identification of British chrysomelids, and it is often necessary to refer to European works such as Mohr (1966). However, there are several smaller works covering various subfamilies, tribes or genera, as well as some larvae; see the lists of references in Hammond and Hine (1999) and Cooter and Barclay (2006).

British subfamilies and genera:

Bruchinae: *Acanthoscelides, Bruchidius, Bruchus, Callosobruchus*

Cassidinae: *Cassida, Hypocassida, Pilemostoma*

Chrysomelinae: *Chrysolina, Chrysomela, Gastrophysa, Gonioctena, Hydrothassa, Phaedon, Phratora, Plagiodera, Prasocuris, Timarcha*

Criocerinae: *Crioceris, Lema, Oulema, Lilioceris*

Cryptocephalinae: *Clytra, Cryptocephalus, Labidostomis, Smaragdina*

Donaciinae: *Donacia, Macroplea, Plateumaris*

Eumolpinae: *Bromius*

Galerucinae: *Agelastica, Altica, Aphthona, Apteropeda, Batophila, Calomicrus, Chaetocnema, Crepidodera, Derocrepis, Diabrotica, Dibolia, Epitrix, Galeruca, Galerucella, Hermaeophaga, Hippuriphila, Lochmaea, Longitarsus, Luperomorpha, Luperus, Lythraria, Mantura, Mniophila, Neocrepidodera, Ochrosis, Phyllobrotica, Phyllotreta, Podagrica, Psylliodes, Pyrrhalta, Sermylassa, Sphaeroderma, Xanthogaleruca*

Lamprosomatinae: *Oomorphus*

Family Megalopodidae (1 genus, 3 species)
This family was formerly included within the Chrysomelidae. The three British species of *Zeugophora* are around 3mm long, and are found on poplars and willows where they can cause 'netting' of the leaves; the larvae are leaf-miners, producing black blotch-mines.

Identification: see the references in Cooter and Barclay (2006).

British genus: *Zeugophora*

Family Orsodacnidae (1 genus, 2 species)
This family was formerly regarded as a subgroup of the Chrysomelidae. The life histories of the two species of *Orsodacne* are little studied, though the larvae may be root-feeders. The adults are usually found on flowers (Fig. 18.24).

Identification: see the notes in Cooter and Barclay (2006).

British genus: *Orsodacne*

SUPERFAMILY CLEROIDEA

A high proportion of species in this subfamily are predators, both as larvae and adults, though some adults are also pollen-feeders, occasionally exclusively. Many are also associated with wood where some larvae are highly specific predators on other

Fig. 18.24 *Orsodacne cerasi* (Orsodacnidae) (Photo: Roger Key)

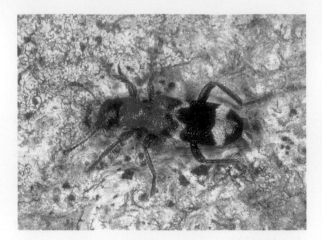

Fig. 18.25 *Thanasimus formicarius* (Cleridae) (Photo: Roger Key)

Fig. 18.26 *Thanasimus formicarius* larva (Cleridae) (Photo: Roger Key)

wood-boring larvae such as Anobiidae.Some are found under tree bark, others on carrion, and a few have become stored-product pests.

In many cases the adults of the British species can be reliably identified with the standard works such as Joy (1932) although a comparison with appropriate sections of *Die Käfer Mitteleuropas* is also recommended. Several species have been added to the British list recently, see Hodge and Jones (1995); additional references will be found in Hammond and Hine (1999) and Cooter and Barclay (2006).

Family Cleridae (10 genera, 14 species)

Although fourteen species are currently on the British list, there are no recent records of four of these, and two are accidental introductions. They are sometimes known as chequer beetles from the regular coloured patterns on at least some species, and they are noticeably hairy. Both larvae and adults are predatory, some feeding on anobiid beetles; *Necrobia* species are usually found on dry animal carcasses, and can be a pest on meat products. *Thanasimus formicarius* (Fig. 18.25) is commonly found in woodland; it is a mimic of a mutillid wasp, and its larva is a typical predator on other wood-boring beetle larvae (Fig. 18.26).

British subfamilies and genera:

  Clerinae: *Opilo, Thanasimus, Trichodes*
  Korynetinae: *Korynetes, Necrobia*
  Tarsosteninae: *Paratillus, Tarsostenus*
  Thaneroclinae: *Thaneroclerus*
  Tillinae: *Tilloidea, Tillus*

Family Dasytidae (4 genera, 9 species)

This family is sometimes regarded as a subgroup of the Malachiidae. Like the Cleridae the adults are

Fig. 18.27 *Psilothrix viridicoeruleus* (Dasytidae) (Photo: Roger Key)

usually hairy and are often seen on flowers; the larvae are predators on insects in decaying wood. Several species are coastal in distribution, and *Psilothrix viridicoeruleus* is often common on dunes in southern Britain (Fig. 18.27).

British subfamilies and genera:

  Dasytinae: *Dasytes, Dolichosoma, Psilothrix*
  Rhadalinae: *Aplocnemus*

Family Malachiidae (10 genera, 16 species)

The families Melyridae and Malachiidae are now considered as separate families, and all the British species are in the newly defined Malachiidae. Although the larvae in this group are predatory, the adults are often pollen-feeders and can be a common sight on flowers. Many species are brightly coloured, especially in the genus *Malachius* (Figs. 18.28

Fig. 18.28 *Malachius bipustulatus* (Malachiidae) (Photo: Roger Key)

Fig. 18.30 *Malachius bipustulatus* pupa (Malachiidae) (Photo: Roger Key)

Fig. 18.29 *Malachius aeneus* (Malachiidae) (Photo: Roger Key)

Fig. 18.31 *Thymalus limbatus* (Trogossitidae) (Photo: Roger Key)

& 18.29); the juvenile stages are often under bark (Fig. 18.30).

Hypebaeus flavipes (Moccas beetle) is listed on the Wildlife and Countryside Act 1981. *Malachius aeneus* (Scarlet malachite beetle) is on the UKBAP list.

British genera: *Anthocomus, Axinotarsus, Cerapheles, Clanoptilus, Cordylepherus, Ebaeus, Hypebaeus, Malachius, Sphinginus, Troglops*

Family Phloiophilidae (1 genus, 1 species)
The only British species is *Phloiophilus edwardsii*, associated with fungi on dead wood. The adult is around 3 mm long and is found locally on oak trees.

British genus: *Phloiophilus*

Family Trogossitidae (5 genera, 5 species)
Of the five species in this family *Lophocateres pusillus* is a commonly introduced cosmopolitan pest of stored products. The only species with a common

name is *Tenebroides mauritanicus*, the Cadelle, another pest on stored flour and grain. The most common British species is *Thymalus limbatus*, found under the bark of various trees (Fig. 18.31); most of the species associated with wood are assumed to be fungus feeders. In some classifications the Peltidae are treated as a distinct family; the family name is sometimes spelled Trogositidae.

British subfamilies and genera:
  Lophocaterinae: *Lophocateres*
  Peltinae: *Ostoma, Thymalus*
  Trogossitinae: *Nemozoma, Tenebroides*

SUPERFAMILY CUCUJOIDEA

Most members of this group are mycophagous as larvae and adults though inevitably there are many

Fig. 18.32 *Biphyllus lunatus* (Biphyllidae) (Photo: Roger Key)

Fig. 18.33 *Cerylon ferrugineum* (Cerylonidae) (Photo: Roger Key)

exceptions. As listed below, some are pollen-feeders, some are predatory and some have mouthparts specifically adapted for sucking fluids from fungal hyphae.

Family Alexiidae (1 genus, 1 species)
This family was formerly treated as a subfamily within the Endomychidae. The only British species is *Sphaerosoma pilosum*, a tiny hemispherical beetle around 1.5 mm long. It is found under moss and decaying vegetation, associated with fungi.

Identification: Joy (1932).

British genus: *Sphaerosoma*

Family Biphyllidae (2 genera, 2 species)
The two British species feed on fungi on woodland trees. *Biphyllus lunatus* (Fig. 18.32) is usually found on ash trees in the fruiting body of fungi.

Identification: the two species can be distinguished using standard works such as Joy (1932).

British genera: *Biphyllus, Diplocoelus*

Family Bothrideridae (3 genera, 5 species)
The members of this small family are associated with fungi, but are rarely encountered. Three species are either rare or extinct in Britain, being confined to rotten oaks in ancient woodland. The two species of *Anommatus* are found on subterranean decaying vegetable matter or underneath rotten wood.

Identification: four of the British species can be recognized using standard works such as Joy (1932), but see Cooter and Barclay (2006) for extra notes.

British subfamilies and genera:
Anommatinae: *Anommatus*
Teredinae: *Oxylaemus, Teredus*

Family Byturidae (1 genus, 2 species)
The more common of the two species in this family, *Byturus tomentosus* feeds on the fruit of *Rubus* species; it is known as the Raspberry beetle, as it can sometimes cause damage to commercial crops. The yellowish and hairy adults are often seen on the flowers where they eat pollen.

Identification: Vogt (1969) gives reliable characters to separate the two species.

British genus: *Byturus*

Family Cerylonidae (2 genera, 5 species)
Of the five British species in this small family, one has not been recorded since the early 19th century and *Murmidius segregatus* is an introduction on stored products. The three species in *Cerylon* (Fig. 18.33) are found on fungi under the bark and in rotting wood of ancient deciduous trees, often associated with Scolytidae. Some members of this family have styliform mouthparts that may be adapted for piercing and sucking liquid from fungal hyphae.

Identification: Joy (1932).

British genera: *Cerylon, Murmidius*

Family Coccinellidae (30 genera, 53 species)
The ladybirds are probably the best known group of beetles to the layman, and this familiarity is coupled with the idea that identification depends on examining the distinctive colour patterns. However, several species show colour polymorphism and in these cases accurate determination depends on morphological characters. Of the native species, *Adalia decempunctata* (10-spot ladybird) probably shows the greatest polymorphism, with numerous colour forms and patterns, which have

Fig. 18.34 *Subcoccinella vigintiquattuorpunctata* larva (Coccinellidae) (Photo: Roger Key)

Fig. 18.36 *Harmonia axyridis* colour form (Coccinellidae) (Photo: Roger Key)

Fig. 18.35 Cluster of *Coccinella septempunctatus*, 7-spot ladybirds (Coccinellidae) (Photo: Peter Barnard)

Fig. 18.37 *Harmonia axyridis* colour form (Coccinellidae) (Photo: Roger Key)

led to much taxonomic confusion in the past; despite its common name it frequently does not have 10 spots. A few species feed on mildews or directly on plants, but most are specialized predators on aphids and coccids; since these groups are often plant pests the ladybirds are then regarded as beneficial insects to gardeners and farmers. Ladybird larvae can have bright colours and distinctive patterns of setae (Figs. 18.34 & 18.38). It is common to see clusters of adults of various species (Fig. 18.35), sometimes as the results of mass swarming, and the overwintering species also form dense clusters when hibernating.

The most notorious species is *Harmonia axyridis*, the Harlequin ladybird, which arrived in Britain in

2004 and has now spread over most of the country. It is a native of Asia that is now almost cosmopolitan and is sometimes difficult to identify because of its extreme polymorphism (Figs. 18.36 & 18.37). Concern was raised that it might out-compete native species when food became scarce, and may even prey on other species; a national survey was therefore undertaken, and work still continues (www.harlequin-survey.org). Its larva has a very distinctive appearance (Fig. 18.38).

A useful work on the biology of the whole family is Majerus (1994) and, despite its limited-sounding title, the guide by Hawkins (2000) is also very useful. There is a good website covering all the British coccinellids (www.ladybird-survey.org).

Identification: the literature for identifying coccinellids can be divided into two distinct types. There are simplified guides to the most common

Fig. 18.38 *Harmonia axyridis* larva (Coccinellidae) and aphid prey (Photo: Peter Barnard)

Fig. 18.39 *Atomaria nitidula* (Cryptophagidae) (Photo: Roger Key)

species that use the (often unreliable) colour patterns such as Majerus et al. (2006), though this does include *Harmonia axyridis*. The specialized keys that use reliable morphological characters such as Pope (1953) are more difficult to use (and out of date), but a good compromise is the handbook by Majerus and Kearns (1989), which also includes a simplified key, though lacking *H. axyridis*.

British subfamilies and genera:

Chilocorinae: *Chilocorus, Exochomus, Platynaspis*
Coccidulinae: *Clitostethus, Coccidula, Cryptolaemus, Hyperaspis, Nephus, Rhyzobius, Rodolia, Scymnus, Stethorus*
Coccinellinae: *Adalia, Anatis, Anisosticta, Aphidecta, Calvia, Coccinella, Coccinula, Halyzia, Harmonia, Hippodamia, Myrrha, Myzia, Propylea, Psyllobora, Tytthaspis, Vibidia*
Epilachninae: *Henosepilachna, Subcoccinella*

Family Corylophidae (4 genera, 11 species)
Most species in this small family are only around 1 mm long; they are mycophagous and are usually found in rotten wood, decaying vegetation or under bark. Although one or two species are widespread in Britain, several are only found in the southern England.

Identification: this relies on specialist literature, listed in Cooter and Barclay (2006), such as Bowestead (1999).

British subfamilies and genera:

Corylophinae: *Corylophus, Sericoderus*
Orthoperinae: *Orthoperus*
Rypobiinae: *Rypobius*

Family Cryptophagidae (11 genera, 103 species)
This is a large family for which the life histories of many species are little studied. All are small and elongate species, usually 3–4 mm long, with distinctly clubbed antennae. Most species seem to be associated with rotting vegetation, but there are several exceptions: *Telmatophilus* is found in marsh plants; several Cryptophaginae are found in bracket fungi or under bark, and some are associated with birds' nests or the nests of social Hymenoptera. *Atomaria nitidula* (Fig. 18.39) is one of the common species found in compost heaps and similar situations.

Identification: although some progress can be made with standard works such as Joy (1932) there are some difficult genera; Lohse (1967) is useful, and several specialized references are given in Cooter and Barclay (2006).

British subfamilies and genera:

Atomariinae: *Atomaria, Ephistemus, Ootypus*
Hypocoprinae: *Hypocoprus*
Cryptophaginae: *Antherophagus, Caenoscelis, Cryptophagus, Henoticus, Micrambe, Paramecosoma*
Telmatophilinae: *Telmatophilus*

Family Cucujidae (1 genus, 2 species)
With the removal of the Silvanidae as a separate family, the Cucujidae now include just one genus, *Pediacus*, with two British species. Both are probably more common than records suggest, though *P. dermestoides* (Fig. 18.40) seems more widespread than *P. depressus*; both are found under the bark of decaying mature trees, associated with fungi.

British genus: *Pediacus*

Fig. 18.40 *Pediacus dermestoides* (Cucujidae) (Photo: Roger Key)

Fig. 18.42 *Triplax russica* (Erotylidae) (Photo: Roger Key)

Fig. 18.41 *Endomychus coccineus* (Endomychidae) (Photo: Roger Key)

Fig. 18.43 *Dacne bipustulata* (Erotylidae) (Photo: Roger Key)

Family Endomychidae (5 genera, 8 species)
All the members of this small family are mycophagous; five are small, around 2 mm long, and brownish in colour, but three species are larger, up to 6 mm, and more brightly coloured and patterned. The best known is probably *Endomychus coccineus* (Fig. 18.41), which is easily mistaken for a ladybird (Coccinellidae); its larvae are equally brightly coloured, and easily seen on the wood fungi on which they feed.

Identification: standard works such as Joy (1932) can be used, though the members of this family are widely scattered amongst various groups; Peez (1967) is useful for the genus *Holoparamecus*.

British genera: *Endomychus, Holoparamecus, Lycoperdina, Mycetaea, Symbiotes*

Family Erotylidae (3 genera, 7 species)
The British species in this small family are usually 3–6 mm long, and often have some orange coloration, though the exact patterning is variable and not reliable for identification. All species are mycophagous, and all are associated with tree fungi, sometimes under bark. Typical species are *Triplax russica* (Fig. 18.42) and *Dacne bipustulata* (Fig. 18.43).

Identification: most species can be determined with standard works such as Joy (1932); more information is given in Cooter and Barclay (2006).

British genera: *Dacne, Triplax, Tritoma*

Fig. 18.44 *Notolaemus unifasciatus* (Laemophloeidae) (Photo: Roger Key)

Fig. 18.45 *Stephostethus lardarius* (Latridiidae) (Photo: Roger Key)

Family Kateretidae (3 genera, 9 species)
This small family of pollen beetles was formerly placed as a subfamily of the Nitidulidae, often spelled as Cateretinae. The beetles are small, around 2–4 mm long, and both larvae and adults live in flowers where they feed on pollen or on flower-buds. The two species of *Brachypterus* are commonly found together on nettles; *B. urticae* is generally more abundant than *B. glaber*.

Identification: Kirk-Spriggs (1996).

British genera: *Brachypterolus, Brachypterus, Kateretes*

Family Laemophloeidae (4 genera, 11 species)
This family was formerly included within the Cucujidae. Most species are found under bark or otherwise associated with dead wood, such as *Notolaemus* (Fig. 18.44), although a few have become pests on stored products.

British genera: *Cryptolestes, Laemophloeus, Notolaemus, Placonotus*

Family Languriidae (1 genus, 1 species)
Sometimes known as lizard beetles, this family includes a few cosmopolitan pest species on stored products, but most species probably feed on decaying vegetation and fungi. *Cryptophilus integer* was added to the British list in 2007.

British genus: *Cryptophilus*

Family Latridiidae (13 genera, 56 species)
The family name was formerly spelled Lathridiidae. Members of this family are sometimes called mould beetles; they live on fungi in various situations such as compost heaps, under bark and leaf-litter. There are also some species that are only found indoors, on damp walls and inside unventilated cupboards; these are sometimes known as plaster beetles. Other species, such as *Stephostethus lardarius* (Fig. 18.45), are found on flowers.

Identification: the usual beetle guides are of little use in this family; Peez (1967) and Rücker (1992) should be consulted, and specialized references are given by Cooter and Barclay (2006).

British subfamilies and genera:
Corticariinae: *Corticaria, Corticarina, Cortinicara, Melanophthalma, Migneauxia*
Latridiinae: *Adistemia, Cartodere, Dienerella, Enicmus, Latridius, Lithostygnus, Stephostethus, Thes*

Family Monotomidae (3 genera, 23 species)
This family was formerly known as the Rhizophagidae, and the two subfamilies exhibit rather different life styles. Species of *Monotoma* are curiously slow-moving beetles, around 2 mm long, found in compost heaps and similar rotting vegetation, and two of the species are myrmecophilous. The Rhizophaginae are usually found under the bark of dead or moribund trees; some species are found underground and some are associated with fungi. A few species, such as *Rhizophagus dispar* (Fig. 18.46) are predators on other insect larvae.

Identification: Peacock (1977) covers nearly all the British species; see also Cooter and Barclay (2006).

Fig. 18.46 *Rhizophagus dispar* (Monotomidae) (Photo: Roger Key)

Fig. 18.48 *Pocadius ferrugineus* (Nitidulidae) (Photo: Roger Key)

Fig. 18.47 *Meligethes* larva (Nitidulidae) (Photo: Roger Key)

Fig. 18.49 *Pocadius ferrugineus* larvae (Nitidulidae) (Photo: Roger Key)

British subfamilies and genera:
Monotominae: *Monotoma*
Rhizophaginae: *Cyanostolus, Rhizophagus*

Family Nitidulidae (16 genera, 89 species)
This is the main family of pollen beetles, from which the Kateretidae were split. Most species are small, from 2–6 mm, and they show a wide range of life histories. Despite the common name of the family, it is usually only members of the genus *Meligethes* that are found swarming in large numbers over flowers in the summer; this is the largest genus in the family with 36 species. Although the larvae (Fig. 18.47) are highly specific to particular host plants, the adults are found on a wider range of flowers.

Within the Carpophilinae there are some cosmopolitan pest species that live on dried fruit, particularly when mould is present; larvae of *Epuraea* are found in the tunnels of boring beetles such as Scolytinae, but their biology is practically unknown. In the Nitidulinae both *Nitidula* and *Omosita* can be found on dry animal matter and bones, and some are found in birds' nests, especially of predatory species. *Amphotis marginata* is myrmecophilous, associated with *Lasius fuliginosus*, and some other genera such as *Pocadius* (Figs. 18.48 & 18.49) are found on fungi, with others on sap-runs.

Identification: Kirk-Spriggs (1996) covers the Meligethinae, and gives brief notes on the other subfamilies; see also Spornraft (1967, 1992) and Audisio (1993). Specialized references are given by Cooter and Barclay (2006).

British subfamilies and genera:

Carpophilinae: *Carpophilus, Epuraea, Urophorus*

Cryptarchinae: *Cryptarcha, Glischrochilus, Pityophagus*

Cybocephalinae: *Cybocephalus*

Meligethinae: *Meligethes, Pria*

Nitidulinae: *Amphotis, Cychramus, Nitidula, Omosita, Pocadius, Soronia, Thalycra*

Family Phalacridae (3 genera, 15 species)

The British members of this small family are less than 3 mm long, and have a globular appearance with characteristic microsculpturing on the dorsal side. Adults of this group are often seen on flowers, but the larvae are found on a variety of plant material. Some species are found on grasses and sedges infested with smuts and rusts, and there could be a general association with such fungi. *Stilbus testaceus* (Fig. 18.50) is one of the more common species, and this genus is associated with reeds or dried grass.

Identification: all the British species are readily identified with Thompson's (1958) handbook.

British genera: *Olibrus, Phalacrus, Stilbus*

Family Silvanidae (10 genera, 12 species)

This family was previously included in the Cucujidae. Although most species are found under bark, a few have become cosmopolitan pests on stored products, the best known being *Oryzaephilus surinamensis*, the Saw-toothed grain beetle, which has six 'teeth' on each side of the thorax. *Silvanus unidentatus* (Fig. 18.51) is a more typical flat bark beetle, around 2 mm long.

British subfamilies and genera:

Brontinae: *Cryptamorpha, Dendrophagus, Psammoecus, Uleiota*

Silvaninae: *Ahasverus, Cathartus, Nausibius, Oryzaephilus, Silvanoprus, Silvanus*

Family Sphindidae (2 genera, 2 species)

The members of this small family are associated with myxomycete slime-moulds on wood; *Sphindus dubius* (Fig. 18.52) is just over 2 mm long.

British genera: *Aspidiphorus, Sphindus*

Fig. 18.51 *Silvanus unidentatus* (Silvanidae) (Photo: Roger Key)

Fig. 18.50 *Stilbus testaceus* (Phalacridae) (Photo: Roger Key)

Fig. 18.52 *Sphindus dubius* (Sphindidae) (Photo: Roger Key)

SUPERFAMILY CURCULIONOIDEA

All the members of this group are known generally as weevils. The family-level classification has changed greatly in recent years, and will probably continue to do so for some time. The group is characterized by the elongate rostrum on the head, which bears the modified mouthparts. Although some other beetle families have similar modifications, the weevils are unusual in using the mouthparts to cut into plant tissue before ovipositing. Nearly all weevils are phytophagous, although a few specialize on fungi with others on decaying wood; a few are predatory. There is hardly a group of vascular plants, as well as ferns, that is not attacked by weevils, and their apodous larvae are internal feeders in all kinds of plant tissues; some cause gall formation. Some wood-feeding species have become forestry pests and others may be pests on fruit trees and field crops; the seed-feeding groups can be a problem in stored products.

The most important series of works for identifying British weevils are the handbooks by Morris (1990, 1997, 2002, 2008) and these should be consulted for all aspects of the biology of this group and higher classification, as well as species identification. A summary of the Curculionoidea is given in Cooter and Barclay (2006), and there is a useful overview by Morris (1991).

Family Anthribidae (8 genera, 9 species)
One of the families of orthocerous weevils (Morris, 1990). None of the British species in this small family is common; some species such as *Platystomos albinus* (Fig. 18.53) are associated with dead wood; some like *Platyrhinus resinosus* live on fungi, others are on living plants, and some even feed on scale insects (Coccidae). The Urodontinae were previously treated as a distinct family.

Identification: Morris (1990).

British subfamilies and genera:
    Anthribinae: *Anthribus, Dissoleucas, Enedreytes, Platyrhinus, Platystomos*
        Choraginae: *Araecerus, Choragus*
        Urodontinae: *Bruchela*

Family Apionidae (33 genera, 87 species)
One of the families of orthocerous weevils (Morris, 1990). Apart from a few species found on trees nearly all the species have specific relationships with various plant families, especially Fabaceae, Polygonaceae and Malvaceae. One unusual species, *Melanapion minimum*, the Sallow guest weevil, is an inquiline in sawfly galls on *Salix*. Common species include *Apion frumentarium* (Fig. 18.54) on *Rumex*; a much rarer species is *Exapion genistae* (Fig. 18.55) on *Genista*.

Exapion genistae (Petty whin weevil, Fig. 18.55) and *Melanapion minimum* (Sallow guest weevil) are on the UKBAP list.

Identification: Morris (1990). Many of the genera listed here were previously regarded as subgenera.

British genera: *Acentrotypus, Aizobius, Apion, Aspidapion, Betulapion, Catapion, Ceratapion, Cyanapion, Diplapion, Eutrichapion, Exapion, Helianthemapion, Hemitrichapion, Holotrichapion, Ischnopterapion, Ixapion, Kalcapion, Malvapion, Melanapion, Omphalapion, Oxystoma, Perapion, Pirapion, Protapion, Protopirapion, Pseudapion,*

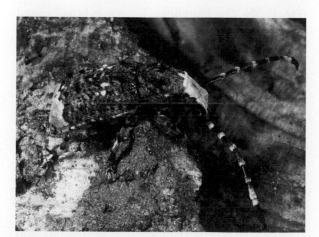

Fig. 18.53 *Platystomos albinus* (Anthribidae) (Photo: Roger Key)

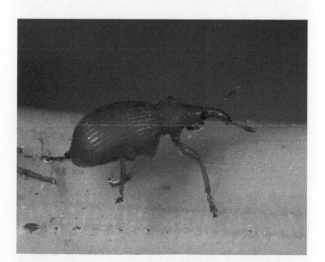

Fig. 18.54 *Apion frumentarium* (Apionidae) (Photo: Roger Key)

Fig. 18.55 *Exapion genistae* (Apionidae) (Photo: Roger Key)

Fig. 18.57 *Curculio nucum* (Curculionidae) (Photo: Roger Key)

Fig. 18.56 *Apoderus coryli* larva (Attelabidae) (Photo: Roger Key)

Fig. 18.58 *Curculio nucum* larva (Curculionidae) (Photo: Roger Key)

Pseudaplemonus, Pseudoprotapion, Rhopalapion, Squamapion, Stenopterapion, Synapion, Taeniapion

Family Attelabidae (2 genera, 2 species)
One of the families of orthocerous weevils (Morris, 1990); both British species are leaf-rollers, *Attelabus nitens* on oak, and *Apoderus coryli* (Fig. 18.56) on hazel.

Identification: Morris (1990).

British genera: *Apoderus, Attelabus*

Family Curculionidae (152 genera, 475 species)
These are sometimes known as the true weevils; as the largest family of weevils they naturally encompass a wide variety of lifestyles. Some are pests on various plants, such as *Anthonomus pomorum*, the Apple blossom weevil. A strikingly long rostrum is seen in *Curculio nucum* (Fig. 18.57), used to bore into hazelnuts, in which the larvae feed and develop

(Fig. 18.58). The Entiminae, known as broad-nosed weevils, do not use their broad rostrum to penetrate plant tissue either for feeding or oviposition; they include *Sitona lineatus*, the Pea weevil.

Some curculionids are economically important as wood-borers such *Hylobius abietis*, the Pine weevil. The Scolytinae, known as the the bark beetles, are sometimes treated as a distinct family, and they include the notorious *Scolytus scolytus*, the Elm bark beetle, which is the major vector of Dutch Elm disease (Fig. 18.59). Several larvae in this subfamily produce characteristic patterns of tunnels, including *Scolytus multistriatus* (Fig. 18.60) and *Hylesinus varius*, the Ash bark beetle (Fig. 18.61).

Bagous nodulosus (Flowering rush weevil) and *Orchestes testaceus* (Alder flea weevil) are on the UKBAP list.

Fig. 18.59 *Scolytus scolytus*, the Elm bark beetle (Curculionidae) (Photo: Roger Key)

Fig. 18.61 Galleries of *Hylesinus varius*, the Ash bark beetle (Curculionidae) (Photo: Peter Barnard)

Fig. 18.60 Galleries of *Scolytus multistriatus* (Curculionidae) (Photo: Roger Key)

Identification: Morris (1997, 2002, 2008).

British subfamilies and genera:

Bagoinae: *Bagous*

Baridinae: *Aulacobaris, Baris, Cosmobaris, Limnobaris, Melanobaris*

Ceutorhynchinae: *Amalorrhynchus, Amalus, Calosirus, Ceutorhynchus, Coeliodes, Coeliodinus, Datonychus, Drupenatus, Ethelcus, Eubrychius, Glocianus, Hadroplontus, Micrelus, Microplontus, Mogulones, Mononychus, Nedyus, Neophytobius, Parethelcus, Pelenomus, Phytobius, Poophagus, Rhinoncus, Rutidosoma, Sirocalodes, Stenocarus, Tapeinotus, Thamiocolus, Trichosirocalus, Zacladus*

Cossoninae: *Cossonus, Euophryum, Macrorhyncolus, Pentarthrum, Phloeophagus, Pselactus, Pseudophloeophagus, Rhopalomesites, Rhyncolus, Stereocorynes*

Cryptorhynchinae: *Acalles, Cryptorhynchus, Kyklioacalles*

Curculioninae: *Acalyptus, Anthonomus, Archarius, Brachonyx, Cionus, Cleopomiarus, Cleopus, Curculio, Dorytomus, Ellescus, Gymnetron, Isochnus, Mecinus, Miarus, Orchestes, Orthochaetes, Pachytychius, Pseudorchestes, Pseudostyphlus, Rhamphus, Rhinusa, Rhynchaenus, Sibinia, Smicronyx, Tachyerges, Tychius*

Cyclominae: *Gronops*

Entiminae: *Andrion, Attactagenus, Barynotus, Barypeithes, Brachyderes, Brachysomus, Caenopsis, Cathormiocerus, Charagmus, Coelositona, Graptus, Liophloeus, Neliocarus, Omiamima, Otiorhynchus, Pachyrhinus, Peritelus, Philopedon, Phyllobius, Polydrusus, Sciaphilus, Sitona, Strophosoma, Tanymecus, Trachyphloeus, Tropiphorus*

Hyperinae: *Hypera, Limobius*

Lixinae: *Bothynoderes, Cleonis, Coniocleonus, Larinus, Lixus, Rhinocyllus*

Mesoptiliinae: *Magdalis*

Molytinae: *Anchonidium, Anoplus, Hylobius, Leiosoma, Lepyrus, Liparus, Mitoplinthus, Pissodes, Syagrius, Trachodes*

Orobitidinae: *Orobitis*

Scolytinae: *Cryphalus, Crypturgus, Dendroctonus, Dryocoetes, Ernoporicus, Ernoporus, Hylastes, Hylastinus, Hylesinus, Hylurgops, Ips, Kissophagus, Lymantor, Orthotomicus, Phloeosinus, Phloeotribus, Pityogenes, Pityophthorus, Polygraphus, Pteleobius, Scolytus, Taphrorychus, Tomicus, Trypodendron, Trypophloeus, Xyleborinus, Xyleborus, Xylechinus, Xylocleptes*

Tanysphyrinae: *Tanysphyrus*

149

Fig. 18.62 *Dryophthorus corticalis* (Dryophthoridae) (Photo: Roger Key)

Fig. 18.63 *Notaris scirpi* (Erirhinidae) (Photo: Roger Key)

Family Dryophthoridae (2 genera, 4 species)
These were formerly treated as a subfamily of the Curculionidae, and were sometimes known as the Rhynchophorinae. *Dryophtherus corticalis* (Fig. 18.62) is a very local species, found in rotten oak wood in ancient forests. The three members of *Sitophilus* are all pests of stored grain, *S. granarius* the Grain or Granary weevil, being well known.

Identification: Morris (2002).

British genera: *Dryophthorus, Sitophilus*

Family Erirhinidae (6 genera, 12 species)
Previously treated as a subfamily of the Curculionidae, this family is unusual in being at least semi-aquatic. The larvae feed on waterside plants such as rushes and sedges, with at least one feeding on horsetails (*Equisetum*). Some are found on floating plants, though they have no adaptations for a truly aquatic existence. *Notaris scirpi* (Fig. 18.63) is an uncommon species usually associated with *Typha* and *Carex*; *Procas granulicollis* is found at damp, peaty sites.

Identification: Morris (2002).

British genera: *Grypus, Notaris, Procas, Stenopelmus, Thryogenes, Tournotaris*

Family Nanophyidae (2 genera, 2 species)
This family was previously treated as a subgroup of Apionidae; it is one of the families of orthocerous weevils (Morris, 1990) although the antennae are geniculate in this group. *Nanophyes marmoratus* is a common inhabitant of Purple loosestrife; *Dieckmanniellus gracilis* is apparently less common, being confined to Water purslane.

Identification: Morris (1990).

British genera: *Dieckmanniellus, Nanophyes*

Family Nemonychidae (1 genus, 1 species)
One of the families of orthocerous weevils (Morris, 1990); the only British species is *Cimberis attelaboides*, which seems to be confined to Scots pine.

Identification: Morris (1990).

British genus: *Cimberis*

Family Platypodidae (1 genus, 1 species)
Sometimes regarded as a subfamily of the Curculionidae; the only established British species seems to be *Platypus cylindrus*, the Pinhole borer, though there are several records of introduced species. The larvae bore into tree stumps such as oak.

British genus: *Platypus*

Family Raymondionymidae (1 genus, 1 species)
This group was previously treated as a subgroup of the Curculionidae; the only British species is the little-known *Ferreria marqueti*, whose larva is assumed to be a subterranean root-feeder that may be associated with conifers.

Identification: Morris (2002).

British genus: *Ferreria*

Family Rhynchitidae (7 genera, 18 species)
These were previously treated as a subfamily of the Attelabidae. They have unusual habits in that the adult female often creates a shelter for the developing larvae by rolling one or more leaves, sometimes after partially cutting through the petiole.

Fig. 18.64 *Byctiscus populi* (Rhynchitidae) (Photo: Roger Key)

Fig. 18.65 *Hylecoetus dermestoides* (Lymexylidae) (Photo: Roger Key)

Byctiscus populi (Poplar leaf-rolling weevil, Fig. 18.64) is on the UKBAP list.

Identification: Morris (1990).

British genera: *Byctiscus, Deporaus, Involvulus, Lasiorhynchites, Neocoenorrhinus, Rhynchites, Temnocerus*

SUPERFAMILY LYMEXYLOIDEA

Family Lymexylidae (2 genera, 2 species)
There are only two species in this family, and neither is commonly found. Their larvae are wood-borers in various deciduous trees, and the narrow-bodied adults are remarkable for the tufted maxillary palps in the males. The two British species are *Hylecoetus dermestoides* (Fig. 18.65) and *Lymexylon navale*.

Fig. 18.66 *Anthicus antherinus* (Anthicidae) (Photo: Roger Key)

British genera: *Hylecoetus, Lymexylon*

SUPERFAMILY TENEBRIONOIDEA (= HETEROMERA)

The larvae of tenebrionoid beetles exploit a wide range of habitats and feeding habits, though many seem to be associated with fungi on rotten wood. Some are scavengers, others mine plant stems, a few are predators and some are even parasitoids of bees and wasps. The adults are sometimes large and conspicuous but may be rarely seen as they are short lived.

Family Aderidae (3 genera, 3 species)
All three species in this family are uncommon. They are all small, around 2 mm long, and their larvae are found in rotting wood or vegetation, while the few adult records were obtained by beating tree foliage.

Identification: Buck (1954).

British genera: *Aderus, Euglenes, Vanonus*

Family Anthicidae (6 genera, 13 species)
These small beetles, up to 5 mm long, can have a rather ant-like appearance because of their long legs and narrow waisted thorax. Their larvae are usually found in decaying vegetation such as compost heaps, and some are found on salt-marshes. *Anthicus antherinus* (Fig. 18.66) is one of the more brightly coloured species.

Identification: Buck (1954), with additional references given by Hammond and Hine (1999).

British subfamilies and genera:
Anthicinae: *Anthicus, Cordicomus, Cyclodinus, Omonadus, Stricticomus*
Notoxinae: *Notoxus*

Family Ciidae (7 genera, 22 species)

The name of this family has been variously spelled Cisidae, Cissidae and Cioidae. Most ciids are brown or black beetles, 1–4 mm long, found in various tree fungi, including bracket fungi, and some fungal associations are quite specific. Many species are more common in the south of England, though a few are found only in Scotland.

Identification: Lohse (1967) and see the specialized references in Cooter and Barclay (2006).

British genera: *Cis, Ennearthron, Octotemnus, Orthocis, Rhopalodontus, Strigocis, Sulcacis*

Family Melandryidae (10 genera, 17 species)

Members of this family are sometimes known as false darkling beetles from their similarity to the Tenebrionidae. All are associated with fungi, and most are confined to ancient woodland. Nearly all are scarce species, with *Orchesia undulata* perhaps the most frequently found. Like the Mordellidae, most species have powerful hind legs and can jump and somersault if attacked. Many adults are inconspicuously coloured (Fig. 18.67) but some have more distinct markings and may be found on flowers (Fig. 18.68).

Melandrya barbata (Bearded false darkling beetle) is on the UKBAP list.

Identification: Buck (1954).

British subfamilies and genera:
Melandryinae: *Abdera, Anisoxya, Hypulus, Melandrya, Orchesia, Phloiotrya, Wanachia, Xylita*
Osphyinae: *Conopalpus, Osphya*

Family Meloidae (3 genera, 10 species)

The meloids are known generally as oil beetles or blister beetles from the vesicant oil secreted by many species. The most notorious member of the family is *Lytta vesicatoria*, the so-called Spanish fly, which is an occasional migrant from southern Europe; its secretion is cantharidin, supposedly an aphrodisiac. Most meloid species are rarely seen, either because there are no recent records or because they are occasional migrants. The species in *Meloe* are large black beetles with a swollen appearance especially in the females; they have short elytra and are flightless. The group demonstrates hypermetamorphosis, with more than one form of larva; the first instars are so-called triungulin larvae, which get their name from having three-clawed feet, and they are transported by solitary bees to their nests, where the beetle larvae eat the bees' eggs and food stores.

Meloe proscarabaeus (Black oil-beetle), *Meloe rugosus* (Rugged oil-beetle) and *Meloe violaceus* (Violet oil-beetle, Fig. 18.69) are on the UKBAP list.

Identification: Buck (1954), with extra references given by Cooter and Barclay (2006).

British subfamilies and genera:
Meloinae: *Lytta, Meloe*
Nemognathinae: *Sitaris*

Fig. 18.67 *Orchesia micans* (Melandryidae) (Photo: Roger Key)

Fig. 18.68 *Osphya bipunctata* (Melandryidae) (Photo: Roger Key)

Fig. 18.69 *Meloe violaceus* (Meloidae) (Photo: Roger Key)

Fig. 18.71 *Mycetophagus quadripustulatus* (Mycetophagidae) (Photo: Roger Key)

Fig. 18.70 *Mordellistena pumila* (Mordellidae) (Photo: Roger Key)

Family Mordellidae (5 genera, 17 species)
Members of this family all have a characteristic pointed tip to the abdomen (Fig. 18.70) but identification to species level is not easy. The adults are usually found on flowers that produce much pollen such as umbellifers, but the larval habitats are not well studied. Some are found inside the stems of herbaceous or woody plants. The adult beetles can jump and somersault if disturbed.

Identification: the standard literature is of little use in this family, and Buck's (1954) handbook is out of date; see Cooter and Barclay (2006) for specialized literature.

British genera: *Mordella, Mordellistena, Mordello-chroa, Tomoxia, Variimorda*

Family Mycetophagidae (6 genera, 13 species)
The members of this small family of mycophagous beetles can often be found in bracket fungi or sometimes under bark. The largest genus is *Mycetophagus*, which also includes the largest species, up to 8 mm long; species such as *M. quadripustulatus* (Fig. 18.71) are frequently found under rotting logs.

Although the standard works such as Joy (1932) will identify most species, there are some specialized references listed in Cooter and Barclay (2006).

British genera: *Eulagius, Litargus, Mycetophagus, Pseudotriphyllus, Triphyllus, Typhaea*

Family Mycteridae (1 genus, 1 species)
The only British species is *Mycterus curculioides*, which has been collected from flowers such as umbellifers, though there are no recent records. It was formerly placed in the Pythidae.

Identification: Buck (1954).

British genus: *Mycterus*

Family Oedemeridae (4 genera, 10 species)
Most members of this small family have a rather distinctive appearance; they are elongate with a soft cuticle, and the elytra often taper to a point or at least are separated by a gap apically. Most species are around 8–10 mm in length, with long antennae; the larvae are found in rotting wood or plant stems, and the adults feed on pollen. *Oedemera nobilis* is commonly found in grassland; as in several other

Fig. 18.72 *Oedemera nobilis* male (Oedemeridae) (Photo: Roger Key)

Fig. 18.74 *Pyrochroa serraticornis* larva (Pyrochroidae) (Photo: Roger Key)

Fig. 18.73 *Pyrochroa serraticornis* (Pyrochroidae) (Photo: Roger Key)

Fig. 18.75 *Metoecus paradoxus* (Ripiphoridae) (Photo: Roger Key)

species the male hind femora are greatly enlarged (Fig. 18.72).

Identification: Buck's (1954) handbook is now out of date, and the more recent references listed in Hammond and Hine (1999) should be consulted.

British genera: *Chrysanthia, Ischnomera, Nacerdes, Oedemera*

Family Pyrochroidae (2 genera, 3 species)
The cardinal beetles get their name from their bright red coloration. *Pyrochroa serraticornis* (Fig. 18.73) is distinguished by its red head and is widespread throughout Britain; *P. coccinea* has a black head and is more restricted to the south. The adults are short lived and not often seen, but the equally distinctive larvae are often easy to find under tree bark (Fig. 18.74).

Identification: Buck (1954).

British genera: *Pyrochroa, Schizotus*

Family Pythidae (1 genus, 1 species)
The only British species is *Pytho depressus*, which is known only from Scotland where its larvae can be found under the bark of Scots pine trees. This family was previously part of the Salpingidae.

Identification: Buck (1954).

British genus: *Pytho*

Family Ripiphoridae (1 genus, 1 species)
The only British species is *Metoecus paradoxus*, and its larvae prey on the larvae of social wasps such as *Vespula vulgaris*. This so-called brood parasitoid is probably fairly common, though there are few definitive records. The pointed elytra and pectinate antennae are characteristic (Fig. 18.75).

Identification: Buck (1954).

British genus: *Metoecus*

Family Salpingidae (6 genera, 11 species)
Some members of this small family have the head produced into a short rostrum which gives them a superficial similarity to the curculionoid weevils. The larvae all live in dead wood, usually found under the bark, and often associated with other wood-boring beetles. *Aglenus brunneus* was previously placed in the Colydiidae.

Identification: Buck (1954).

British subfamilies and genera:
 Agleninae: *Aglenus*
 Salpinginae: *Lissodema*, *Rabocerus*, *Salpingus*, *Sphaeriestes*, *Vincenzellus*

Family Scraptiidae (2 genera, 14 species)
This family has sometimes been treated as a subfamily of the Mordellidae and the larvae of most species are assumed to live in dead wood, though some may be in living plant stems. All species are small, around 2–4 mm long. The three species of *Scraptia* are generally associated with mature woodland; some of the species of *Anaspis* are frequently be seen in large numbers on flowers (Fig. 18.76).

Identification: Buck's (1954) handbook is now superseded by Levey (2009).

British genera: *Anaspis*, *Scraptia*

Family Tenebrionidae (33 genera, 47 species)
Worldwide this is a very large family of beetles, but there are only a moderate number of species in Britain. Most are dark brown or black, hence the common name of darkling beetles, but the family shows a great range of morphological forms as well as in size. There is also wide variation in habits and habitats though the family can be divided into two fairly distinct groups, the mainly native species that live in natural habitats, and the synanthropic species, some of which are cosmopolitan stored-product pests. Tenebrionid larvae have a rather uniform shape, evenly cylindrical with a tough shiny cuticle (Fig. 18.77).

 The outdoor species include *Nalassus laevioctostriatus*, whose larvae are found in soil in a wide range of habitats; some species are mainly coastal in distribution, including *Phylan gibbus* and *Opatrum sabulosum* (Fig. 18.78). Many of the species associated with human habitations are declining with

Fig. 18.77 *Tenebrio molitor* larvae, mealworms (Tenebrionidae) (Photo: Peter Barnard)

Fig. 18.76 *Anaspis maculata* (Scraptiidae) (Photo: Roger Key)

Fig. 18.78 *Opatrum sabulosum* (Tenebrionidae) (Photo: Roger Key)

Fig. 18.79 *Tenebrio molitor* pupae (Tenebrionidae) (Photo: Peter Barnard)

Fig. 18.80 *Tenebrio molitor* adults at various stages of maturity (Tenebrionidae) (Photo: Peter Barnard)

better hygiene and a reduction in the amount of horse-drawn transport, an example being *Blaps mucronata*, the Churchyard beetle, which at 25 mm was a strikingly large and shiny black inhabitant of cellars and barns; it is still widespread though less common than previously. Some species are still potentially serious pests of stored products such as flour and grain, and these include *Tribolium confusum*, the Confused flour beetle. Perhaps the most familiar is *Tenebrio molitor*, the Mealworm beetle (Figs. 18.79 & 18.80), which is widely bred on a commercial scale as food for birds and reptiles.

The Alleculinae and Lagriinae are sometimes treated as distinct families.

Identification: Brendell (1975) and Buck (1954) for the Alleculinae and Lagriinae.

British subfamilies and genera:
Alleculinae: *Cteniopus, Gonodera, Isomira, Mycetochara, Omophlus, Prionychus, Pseudocistela*
Diaperinae: *Alphitophagus, Corticeus, Crypticus, Diaperis, Gnatocerus, Myrmechixenus, Pentaphyllus, Phaleria, Platydema, Scaphidema*
Lagriinae: *Lagria*
Tenebrioninae: *Alphitobius, Blaps, Bolitophagus, Eledona, Helops, Latheticus, Melanimon, Nalassus, Opatrum, Palorus, Phylan, Tenebrio, Tribolium, Uloma, Xanthomus*

Family Tetratomidae (2 genera, 4 species)
All the members of this small family are associated with fungi. One species, *Hallomenus binotatus* was formerly placed in the Melandryidae. *Tetratoma fun-*

Fig. 18.81 *Tetratoma fungorum* (Tetratomidae) (Photo: Roger Key)

gorum (Fig. 18.81) is a distinctively coloured beetle, often found on bracket fungi.

Identification: Buck (1954).

British genera: *Hallomenus, Tetratoma*

Family Zopheridae (9 genera, 12 species)
Many of the species in this family were formerly placed in the Colydiidae, which is now regarded as a subfamily only. Most species are associated with fungi on trees, but some are known to be predators on other beetles such as Platypodidae. The most

Fig. 18.82 *Bitoma crenata* (Zopheridae) (Photo: Roger Key)

Fig. 18.84 *Agrilus biguttatus* (Buprestidae) (Photo: Roger Key)

Fig. 18.83 *Orthocerus clavicornis* (Zopheridae) (Photo: Roger Key)

common species is probably the distinctively patterned *Bitoma crenata* (Fig. 18.82); one remarkable species is *Orthocerus clavicornis*, which is found on sand dunes and has distinctive antennae (Fig. 18.83).

Identification: the standard works such as Joy (1932) are adequate, but there have been many taxonomic changes and the additional references listed by Hammond and Hine (1999) should be consulted.

British subfamilies and genera:

Colydiinae: *Aulonium, Bitoma, Cicones, Colydium, Endophlaeus, Langelandia, Orthocerus, Synchita*

Pycnomerinae: *Pycnomerus*

Infraorder Elateriformia

SUPERFAMILY BUPRESTOIDEA

The only family in the Buprestoidea is the Buprestidae, known generally as jewel beetles. This is a large family worldwide, and many of its members are brightly coloured, often metallic, and popular among collectors. It is predominantly a tropical family, so the small number of British species is not suprising, though around 200 are known from Europe.

Family Buprestidae (5 genera, 14 species)

The jewel beetles are generally confined to the south of England; most are local and some are known from very few records. All the larvae are phytophagous, feeding internally and often under tree bark. Adults are generally day-flying and are often seen on flowers. The two British species of *Aphanisticus* live in the stems of rushes and sedges, and the larvae of *Trachys* are leaf-miners on a variety of plant species; these are the smaller species in the family. The larvae of the larger species, those of *Agrilus, Anthaxia* and *Melanophila*, are found under tree bark. The species most likely to be seen are *Agrilus biguttatus* (formerly known as *A. pannonicus*, Fig. 18.84) and *A. laticornis*. *Trachys minuta* (Fig. 18.85) is usually associated with willow and hazel.

The only commonly imported species is the large North American *Buprestis auricollis*; its larva can live in coniferous timber for over 30 years, and the adult emerges long after the timber has been prepared and used in construction.

Identification: Levey (1977), with the addition of *Agrilus sulcicollis*; see Cooter and Barclay (2006) for

Fig. 18.85 *Trachys minuta* (Buprestidae) (Photo: Roger Key)

Fig. 18.86 *Byrrhus fasciatus* (Byrrhidae) (Photo: Roger Key)

extra references. The northern European work by Bílý (1982) is also useful.

British subfamilies and genera:
Agrilinae: *Agrilus, Aphanisticus, Trachys*
Buprestinae: *Anthaxia, Melanophila*

SUPERFAMILY BYRRHOIDEA

The families in this superfamily are sometimes divided into separate groups, with the terrestrial Byrrhidae in their own superfamily, and the remaining, largely aquatic, families in the Dryopoidea. The aquatic families are generally covered by the standard works on water beetles (see suborder Adephaga).

Family Byrrhidae (7 genera, 13 species)
Members of this small family are generally known as pill beetles from their small ovoid form (Fig. 18.86); when disturbed they contract the legs and fold the head down, making them even more globular (Fig. 18.87). Both adults and larvae are slow moving, living on the surface or in leaf-litter, where they feed mainly on mosses and liverworts, though some species are found on herbaceous roots. Some have specialized habitats on moorland and heaths, and even coastal shingles. The commonest British species is probably *Simplocaria semistriata*.

Curimopsis nigrita (Mire pill-beetle) is listed on the the Wildlife and Countryside Act 1981 and is on the UKBAP list.

Identification: the European key by Paulus (1979) covers the British species; for other specialized papers see Cooter and Barclay (2006).

Fig. 18.87 *Byrrhus fasciatus* underside (Byrrhidae) (Photo: Roger Key)

British subfamilies and genera:
Byrrhinae: *Byrrhus, Cytilus, Morychus, Porcinolus, Simplocaria*
Syncalaptinae: *Chaetophora, Curimopsis*

Family Dryopidae (2 genera, 9 species)
Although adult dryopids are entirely aquatic, and are covered in the standard texts on water beetles, the larvae are at best semi-aquatic. Only *Pomatinus substriatus* lives in deep water; the larvae of *Dryops* are usually found in wet mud or sometimes damp wood, and may be some distance from any open water.

Identification: the species of *Dryops* are difficult to identify and it is advisable to consult the works of Jäch (1992) or Olmi (1976).

British genera: *Dryops, Pomatinus*

Family Elmidae (8 genera, 12 species)
Known as riffle beetles, the elmids are closely related to the Dryopidae, although their larvae are all entirely aquatic; the family is sometimes known as the Elminthidae. Both larvae and adults feed on algae or vegetable matter. The adults have a covering of hydrofuge hairs that traps an air layer; this functions as a plastron so the adults never need to come to the surface to breathe.

Identification: Holland's (1972) handbook covered all the British species except *Oulimnius major*; all species are covered by Friday (1988).

British genera: *Elmis, Esolus, Limnius, Macronychus, Normandia, Oulimnius, Riolus, Stenelmis*

Family Heteroceridae (2 genera, 9 species)
Although treated as honorary water beetles, all stages of heterocerids feed on algae and decaying vegetable matter in the mud by the side of water-bodies (Fig. 18.88), where the adults often form tunnels. Several species are associated with brackish water in salt-marshes.

Identification: Clarke (1973).

British genera: *Augyles, Heterocerus*

Family Limnichidae (1 genus, 1 species)
The only British species in this family is *Limnichus pygmaeus*; the few records are from England and Wales but at less than 2 mm long this beetle may be overlooked; it is generally found at the side of fast-running water, often in the splash zone.

British genus: *Limnichus*

Family Psephenidae (1 genus, 1 species)
The larvae of this group are known as water pennies from their flattened shape; they feed on algae in fast-running water. The short-lived adults of the only British species, *Eubria palustris* (Fig. 18.89), are less than 2 mm long and are terrestrial, living by the side of the water-courses.

British genus: *Eubria*

Family Ptilodactylidae (1 genus, 1 species)
The larvae and adults of this family are usually associated with leaf-litter and rotting vegetation. The sole British species is *Ptilodactyla exotica*, an introduced species that can only breed in glasshouses. See Cooter and Barclay (2006) for information.

British genus: *Ptilodactyla*

SUPERFAMILY DASCILLOIDEA

Family Dascillidae (1 genus, 1 species)
The larvae of the only British species, *Dascillus cervinus*, live in the soil and feed on roots. The adults are rather hairy (Fig. 18.90) and are often seen on flowers, though the species is only locally common.

British genus: *Dascillus*

SUPERFAMILY ELATEROIDEA

Within this group the families Cantharidae, Drilidae, Lycidae and Lampyridae are sometimes placed in a separate superfamily, the Cantharoidea; leaving the

Fig. 18.88 *Heterocerus* sp. (Heteroceridae) (Photo: Roger Key)

Fig. 18.89 *Eubria palustris* (Psephenidae) (Photo: Roger Key)

Fig. 18.90 *Dascillus cervinus* (Dascillidae) (Photo: Roger Key)

Fig. 18.91 *Cantharis* larva (Cantharidae) (Photo: Roger Key)

click beetles (Elateridae, Eucnemidae and Throscidae) in a more narrowly defined Elateroidea.

Family Cantharidae (7 genera, 41 species)

The soldier beetles are a very familiar group, as some species are very commonly seen on flower heads; they are noted for having very soft elytra, and often have bright coloration. The blue or black species are sometimes loosely known as sailor beetles, in contrast to the red or yellow soldier beetles, from the similarity to military uniforms. The two main subfamilies have a rather different appearance; the Cantharinae includes larger (7–15 mm), more brightly coloured species, and the Malthininae are smaller (3–6 mm) and dark, though often with yellow-tipped elytra. Both adults and larvae are general predators on other insects though they sometimes feed on plants as well. Cantharine larvae (Fig. 18.91) and pupae (Fig. 18.92) are usually found in the soil or leaf-litter, whereas malthinine larvae are usually under bark or in rotting timber. Several species of *Cantharis* are commonly seen (Fig. 18.93); *Rhagonycha fulva* (Fig. 18.94) seems to spend all its time mating on umbellifer flowers, but this is because mating is a protracted process. *Silis ruficollis*, the only member of the subfamily Silinae, is a local species found in damp habitats; it resembles a small cantharine.

Identification: the standard books such as Joy (1932) work most of the time, but it is useful to consult European works such as Dahlgren (1979) and Wittmer (1979). For more specialized papers see Cooter and Barclay (2006).

British subfamilies and genera:

Cantharinae: *Ancistronycha, Cantharis, Podabrus, Rhagonycha*

Fig. 18.92 *Cantharis cryptica* pupa (Cantharidae) (Photo: Roger Key)

Fig. 18.93 *Cantharis pallida* (Cantharidae) (Photo: Roger Key)

Fig. 18.94 *Rhagonycha fulva* (Cantharidae) (Photo: Roger Key)

Fig. 18.95 *Ampedus cardinalis* larva (Elateridae) (Photo: Roger Key)

Malthininae: *Malthinus, Malthodes*
Silinae: *Silis*

Family Drilidae (1 genus, 1 species)
The larvae of this small family are predators on snails. The adults do not feed but show a remarkable sexual dimorphism, even more pronounced than that of the lampyrid glow-worms. Males of the only British species, *Drilus flavescens*, are normal-looking, fully winged beetles with strongly pectinate antennae; the rarely seen females are wingless and larviform with a small head and short antennae and they are several times the size of the males.

British genus: *Drilus*

Family Elateridae (38 genera, 73 species)
The large family of click beetles is probably the most familiar in the superfamily. They all have rear-facing points on the back of the pronotum, and have a mechanism on the underside of the thorax that can be suddenly released, springing the insect several centimetres into the air with an audible click. The beetle can often land, cat-like, back on its feet and can right itself if deliberately placed on its back.

Elaterid larvae are known as wireworms, some being particularly tough-skinned and elongate (Fig. 18.95). Many live in the soil and feed on roots; some species of *Agriotes* can be pests on crops or grass pastures. A few species are found only in shingle habitats or dunes, and some are known to be predators. Around half of the British species are found in dead wood, though many of these require established trees in mature woodland and the scarcity of this habitat is making some of the elaterids rare. Some of these wood-associated species live in rot-

Fig. 18.96 *Agrypnus murinus* (Elateridae) (Photo: Roger Key)

holes in living trees, others on dead wood only, and some under bark.

Some of the more common species are in the genera *Agrypnus* (Fig. 18.96), *Cardiophorus* (Fig. 18.97), *Ctenicera*, *Agriotes* and *Athous*. Some groups such as *Ampedus* are notoriously difficult to identify to species.

Limoniscus violaceus (Violet click beetle) is listed on the Wildlife and Countryside Act 1981 and on the Habitats Directive. The following species are on the UKBAP list: *Ampedus rufipennis* (Red-horned cardinal click beetle), *Anostirus castaneus* (Chestnut click beetle), *Lacon querceus* (Oak click beetle), *Limoniscus violaceus* (Violet click beetle), *Megapenthes lugens*, *Melanotus punctolineatus* (Sandwich click beetle) and *Synaptus filiformis* (Hairy click beetle).

Fig. 18.97 *Cardiophorus vestigialis* (Elateridae) (Photo: Roger Key)

Fig. 18.98 *Melasis buprestoides* (Eucnemidae) (Photo: Roger Key)

Identification: some success can be had with the standard works such as Joy (1932), but more recent European works such as Platia (1994) and Laibner (2000) are important; specialized references are listed by Hammond and Hine (1999) and Cooter and Barclay (2006).

British subfamilies and genera:
Agrypninae: *Agrypnus*, *Lacon*
Cardiophorinae: *Cardiophorus*, *Dicronychus*
Denticollinae: *Actenicerus*, *Anostirus*, *Aplotarsus*, *Athous*, *Calambus*, *Cidnopus*, *Ctenicera*, *Denticollis*, *Diacanthous*, *Hemicrepidius*, *Kibunea*, *Limoniscus*, *Paraphotistus*, *Prosternon*, *Selatosomus*, *Stenagostus*
Elaterinae: *Adrastus*, *Agriotes*, *Ampedus*, *Brachygonus*, *Dalopius*, *Elater*, *Ischnodes*, *Megapenthes*, *Panspaeus*, *Procraerus*, *Sericus*, *Synaptus*
Hypnoidinae: *Hypnoidus*
Melanotinae: *Melanotus*
Negastriinae: *Fleutiauxellus*, *Negastrius*, *Oedostethus*, *Zorochros*

Family Eucnemidae (5 genera, 6 species)
The members of this small family are sometimes called false click beetles; they are little studied, and three of the British species were only recorded in the last 50 years or so. They are all woodland species, where the larvae live in dead wood. Some species seem to be associated with particular trees such as beech or oak; others seem to use a wider range of tree species. Although some larvae produce characteristic tunnels in the wood, it is not clear what their food actually is, and it may involve certain fungi or slime-moulds. *Melasis buprestoides* is the most common species and at 9–10mm in length is also the largest (Fig. 18.98).

Fig. 18.99 *Lampyris noctiluca*, glow-worm larva (Lampyridae) (Photo: Roger Key)

Identification: the standard works such as Joy (1932) are of little use because only half the British species were known at the time. Muona's (1993) review of the whole family is very useful, and specialized papers are listed by Cooter and Barclay (2006).

British subfamilies and genera:
Eucneminae: *Eucnemis*
Melasinae: *Epiphanis*, *Hylis*, *Melasis*, *Microrhagus*

Family Lampyridae (3 genera, 3 species)
Of the three species of glow-worms recorded in Britain, only *Lampyris noctiluca* is likely to be seen, and even that species is becoming rather locally distributed. Glow-worm larvae (Fig. 18.99) are predators on snails, but the adults do not feed. The

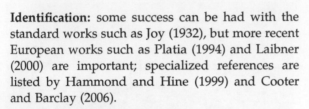

Fig. 18.100 *Lampyris noctiluca* male (Lampyridae) (Photo: Roger Key)

Fig. 18.102 *Platycis minutus* (Lycidae) (Photo: Roger Key)

Fig. 18.101 *Lampyris noctiluca* female glowing (Lampyridae) (Photo: Roger Key)

male is a normal looking beetle (Fig. 18.100) that can produce only a faint glow, but the wingless, larviform female is well known for its ability to produce a strong greenish glow at the end of the abdomen (Fig. 18.101) with which it attracts the attention of the male.

Phosphaenus hemipterus is a European species that has occasionally been recorded in south-east England, and *Lamprorhiza splendidula* is known from only a few 19th-century records.

Identification: see Cooter and Barclay (2006) for specialized references.

British genera: *Lampyris, Lamprohiza, Phosphaenus*

Family Lycidae (3 genera, 4 species)
The members of this small family are all strikingly crimson coloured; their larvae live in well-decayed wood in ancient woodland which means that, like

several Elateridae, they are scarce. Two species are most frequent in the Scottish Highlands, the other two being more common in south-east England. All are moderate-sized beetles, up to about 13 mm in length, and even the misleadingly named *Platycis minutus* (Fig. 18.102) can be up 10 mm long.

Platycis cosnardi (Cosnard's net-winged beetle) is on the UKBAP list.

Identification: Cooter and Barclay (2006) give the necessary references.

British genera: *Dictyoptera, Platycis, Pyropterus*

Family Throscidae (2 genera, 5 species)
Although they are small and short-bodied, the Throscidae are closely related to the Elateridae and they can jump in a similar way, though this behaviour is not often seen. None of the British species exceeds 3.5 mm in length; when disturbed they tuck their appendages under their body, rather like the Byrrhidae, which makes them even less conspicuous. The biology of most species is not known but the larva of at least one species, *Trixagus dermestoides*, is known to live underground, feeding on mycorrhizal fungi on tree roots.

Identification: apart from the more recent addition of *Aulonothroscus brevicollis*, the identification of Throscidae is relatively straightforward using the standard works such as Joy 1932); extra references are given by Cooter and Barclay (2006).

British genera: *Aulonothroscus, Trixagus*

SUPERFAMILY SCIRTOIDEA

Although generally regarded as a good superfamily, the three families in this group, the Scirtidae,

Eucinetidae and Clambidae are different in many ways, both in morphology and life histories.

Family Clambidae (2 genera, 10 species)

This is a small family, and all its members are less than 2 mm long, globular in shape and capable of rolling up into balls. Most species are associated with rotting vegetation, where both adults and larvae are probably mycophagous; some species seem to be associated with myxomycete slime-moulds.

Identification: Johnson's (1966) handbook covers all except some recent additions; see Cooter and Barclay (2006).

British genera: *Calyptomerus*, *Clambus*

Family Eucinetidae (1 genus, 1 species)

Many Eucinetidae are associated with the fruiting bodies of myxomycete slime-moulds in wooded areas and adults are sometimes seen on flowers. The only British species, *Eucinetus meridionalis*, was first recorded in 1968.

Identification: see references in Hammond and Hine (1999).

British genus: *Eucinetus*

Family Scirtidae (7 genera, 20 species)

This family was formerly known as the Helodidae. The larvae of this group are entirely aquatic filter-feeders and look rather like isopod wood-lice; some species are found in tree rot-holes where they help to break down leaf debris. The adults are soft bodied, often reddish brown, and with long antennae (Fig. 18.103); they can be found on vegetation near the larval habitats, sometimes on flowers. Species in the genus *Scirtes* have large hind femora and can jump, rather like flea-beetles in the Chrysomelidae.

Identification: because of difficulties with some genera in this family it is necessary to consult specialized works; see Hammond and Hine (1999) and Cooter and Barclay (2006) for references.

British genera: *Cyphon*, *Elodes* (formerly *Helodes*), *Hydrocyphon*, *Microcara*, *Odeles*, *Prionocyphon*, *Scirtes*

Infraorder Scarabaeiformia

SUPERFAMILY SCARABAEOIDEA

This superfamily includes the stag beetles, dung beetles and related groups, which were formerly placed in just four families; this number has now been inflated to ten. However, the group remains well defined, recognized by the asymmetrical lamellate club of the antennae, which gave them the older name of Lamellicornia.

Britton's (1956) handbook to this group was updated by Jessop's (1986) work; for further specialized notes see Hammond and Hine (1999) and Cooter and Barclay (2006).

Family Aegialiidae (1 genus, 3 species)

This group was formerly treated as a subfamily of the Scarabaeidae.

None of the three British species of *Aegialia* is particularly common; *A. arenaria* is a coastal species found on sand dunes (Fig. 18.104).

Identification: Jessop (1986); see under Scarabeoidea.

British genus: *Aegialia*

Fig. 18.103 *Odeles marginatus* (Scirtidae) (Photo: Roger Key)

Fig. 18.104 *Aegialia arenaria* (Aegialidae) (Photo: Roger Key)

Fig. 18.105 *Aphodius rufipes* (Aphodiidae) (Photo: Roger Key)

Fig. 18.107 *Cetonia aurata* larva (Cetoniidae) (Photo: Roger Key)

Fig. 18.106 Larva of *Aphodius* (Aphodiidae) (Photo: Roger Key)

Family Aphodiidae (34 genera, 55 species)
This family, the largest in the Scarabaeoidea, was formerly regarded as a subfamily of the Scarabaeidae. With 42 British species, the genus *Aphodius* (Fig. 18.105) is by far the largest and a few species, such as *A. rufipes*, are attracted to light, being frequent visitors to the lepidopterist's light trap. Nearly all are dung feeders (Fig. 18.106), but unlike some other families they do not burrow underneath the dung or bury it, although they sometimes use existing burrows made by other groups. Often a single cow-pat is host to several different species of *Aphodius*, though not necessarily simultaneously because of a succession of species over time.

Identification: Jessop (1986), although three species have since been added to the British list; see Cooter

and Barclay (2006). Some of the genera listed below are regarded as subgenera by some authors.

British subfamilies and genera:
Aphodiinae: *Acrossus, Agoliinus, Agrilinus, Ammoecius, Aphodius, Bodilus, Calamosternus, Chilothorax, Colobopterus, Esymus, Euheptaulacus, Euorodalus, Eupleurus, Heptaulacus, Labarrus, Limarus, Liothorax, Melinopterus, Nimbus, Otophorus Oxyomus Phalacronothus, Plagiogonus, Planolinus, Sigorus, Teuchestes, Volinus*
Eupariinae: *Saprosites*
Psammodinae: *Brindalus, Diastictus, Pleurophorus, Psammodius, Rhyssemus, Tesarius*

Family Bolboceratidae (1 genus, 1 species)
The only British species, *Bolboceras armiger*, was formerly placed in the genus *Odontaeus*, also spelled *Odonteus*, in the family Geotrupidae. It is apparently associated with hypogeal fungi like truffles and is uncommon, found only in dry areas of south-east England.

Identification: Jessop (1986); see under Scarabeoidea.

British genus: *Bolboceras*

Family Cetoniidae (4 genera, 6 species)
The Cetoniidae is sometimes considered as a subfamily of Scarabaeidae.

The larvae of rose chafers are usually found in the soil or decaying wood, where they feed on decomposing plant material (Fig. 18.107). The brightly coloured adults are usually day-flying; they can be seen on flowers and feed on pollen, nectar, sap or fruits.

Fig. 18.108 *Cetonia aurata*, the Rose chafer (Cetoniidae) (Photo: Roger Key)

Fig. 18.109 Underside of *Geotrupes stercorarius* (Geotrupidae) with phoretic mites (Photo: Colin Rew)

Cetonia aurata, the Rose chafer (Fig. 18.108) is by far the most common species, and it was formerly considered a pest on cultivated rose bushes because it nibbles the flowers. Although the markings can be highly variable, the species of *Cetonia* are basically metallic green in colour, in contrast to species of *Trichius*, which are orange and black, giving them the common name of bee beetles.

Gnorimus nobilis (Noble chafer) and *Gnorimus variabilis* (Variable chafer) are on the UKBAP list.

Identification: Jessop (1986); see under Scarabeoidea.

British subfamilies and genera:
Cetoniinae: *Cetonia, Protaetia*
Trichiinae: *Gnorimus, Trichius*

Family Geotrupidae (4 genera, 7 species)
The dor beetles are all dung feeders, which burrow under the dung and then drag it into their burrows as food for the developing larvae.

Geotrupes stercorarius is the most common dor beetle, which seems to prefer horse dung, and is usually infested with mites (Fig. 18.109), giving it the alternative name of Lousy watchman. It can stridulate by rubbing the hind coxae together. The males of *Typhaeus typhoeus*, the Minotaur beetle, have three characteristic thoracic horns that are used in fighting (Fig. 18.110).

Identification: Jessop (1986); see under Scarabeoidea.

British genera: *Anoplotrupes, Geotrupes, Trypocopris, Typhaeus*

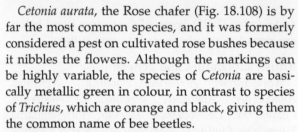

Fig. 18.110 *Geotrupes stercorarius* and *Typhaeus typhoeus* (Geotrupidae) (Photo: Roger Key)

Family Lucanidae (4 genera, 4 species)
The stag beetles are a well-known group; their larvae can live for several years in decaying wood. Although four species are on the British list, *Platycerus caraboides* has not been recorded since the early 19th century, and is probably extinct here. *Sinodendron cylindricum* is probably the most widespread species although still very local; it burrows into dead wood of a wide range of deciduous trees. The Lesser stag beetle, *Dorcus parallelipipedus* (Figs. 18.111 & 18.112) is generally more common in southern England on several deciduous trees. The most iconic species is *Lucanus cervus*, the Stag beetle, which is mainly found in south-east England; the larvae are unusual in living underground, feeding on the decaying roots of dead tree stumps.

Fig. 18.111 *Dorcus parallelipipedus*, the Lesser stag beetle (Lucanidae) (Photo: Roger Key)

Fig. 18.113 *Serica brunnea* larva (Melolonthidae) (Photo: Roger Key)

Fig. 18.112 *Dorcus parallelipipedus* larva (Lucanidae) (Photo: Roger Key)

Fig. 18.114 *Melolontha melolontha*, the Cockchafer (Melolonthidae) (Photo: Roger Key)

Lucanus cervus (Stag beetle) is listed on the Wildlife and Countryside Act 1981 and on the Habitats Directive; it is also on the UKBAP list.

Identification: Jessop (1986); see under Scarabeoidea.

British subfamilies and genera:
Dorcinae: *Dorcus*
Lucaninae: *Lucanus*, *Platycerus*
Syndesinae: *Sinodendron*

Family Melolonthidae (5 genera, 7 species)

The Melolonthinae were formerly regarded as a subfamily of the Scarabaeidae. This is one of the main groups of chafer beetles in Britain; the larvae are all subterranean root-feeders (Fig. 18.113) and, where they occur in large numbers, can be a serious pest on cultivated plants. As in the Lucanidae the

larvae can take two or more years to develop; the adults feed on leaves. One of the largest and best known species is *Melolontha melolontha*, the Common cockchafer (Fig. 18.114), which can reach 30 mm in length and is a common sight in May and June in southern Britain, where the adults are attracted to lights and often enter houses. *Amphimallon solstitialis* is the Summer chafer; it has similar habits to the cockchafer but is more coastal in distribution. Among the smaller species is *Serica brunnea*, up to 10 mm long.

Identification: Jessop (1986); see under Scarabeoidea.

British subfamilies and genera:
Melolonthinae: *Amphimallon*, *Melolontha*, *Polyphylla*
Sericinae: *Omaloplia*, *Serica*

167

Fig. 18.115 *Phyllopertha horticolor*, the Garden chafer (Rutelidae) (Photo: Roger Key)

Fig. 18.116 Large population of *Onthophagus* (Scarabaeidae) (Photo: Roger Key)

Family Rutelidae (3 genera, 3 species)
This family was previously treated as a subfamily of the Scarabaeidae and, like the Melolonthidae, they are known generally as chafer beetles. As in the latter family the larvae feed on plant roots and other organic matter underground, and the adults are leaf-feeders. One species has a common name, *Phyllopertha horticola*, the Garden chafer (Fig. 18.115) and this is the only one to reach pest proportions. The larvae feed on grass roots and can cause damage to pastures and golf courses, though the species seems to be less common than previously in many parts of the country.

Identification: Jessop (1986); see under Scarabeoidea.

British subfamilies and genera:
  Hopliinae: *Hoplia*
  Rutelinae: *Anomala, Phyllopertha*

Family Scarabaeidae (2 genera, 9 species)
Worldwide this family of dung beetles includes the well-known scarab beetles that roll balls of dung. The British species do not roll dung balls but they dig vertical burrows underneath the dung in which they lay their eggs. The burrows vary in form depending on the species, but all are provisioned with dung on which the developing larvae feed. Many species are now quite rare or very local, but some *Onthophagus* species can be seen in large numbers (Fig. 18.116).

A brief world overview of dung beetles is given by Ridsdill-Smith and Simmons (2009).

Identification: Jessop (1986); see under Scarabeoidea.

British genera: *Copris, Onthophagus*

Family Trogidae (1 genus, 3 species)
The Trogidae are rather different from the rest of the Scarabaeoidea in that both adults and larvae feed on dry animal remains, such as bones, dried carcasses and sometimes owl pellets. They can often be found in the nests of predatory birds such as owls, though only one species, *Trox scaber*, seems at all common in Britain; it is around 6–8mm long and is cosmopolitan in distribution.

Identification: Jessop (1986); see under Scarabeoidea.

British genus: *Trox*

Infraorder Staphyliniformia

SUPERFAMILY HYDROPHILOIDEA

In some classifications this group is split into two: the Histeroidea including the terrestrial Histeridae and Sphaeritidae, and the 'aquatic' Hydrophiloidea, but this is not often followed, if only because the Hydrophilidae themselves have a clearly terrestrial group of dung-feeding beetles. All the families have many morphological characters in common, and they all have predaceous larvae.

Family Georissidae (1 genus, 1 species)
The only British species is *Georissus crenulatus*, a very local species found in mud by the side of rivers; it is less than 2mm long.

British genus: *Georissus*

Family Helophoridae (1 genus, 20 species)
This is not an easy group to identify, but several species of *Helophorus* are quite common around the

Fig. 18.117 *Helophorus brevipalpis* (Helophoridae) (Photo: Roger Key)

Fig. 18.118 *Hister* sp. (Histeridae) (Photo: Roger Key)

margins of temporary pools. They have a characteristic grooved pattern on the prothorax (Fig. 18.117).

Helophorus laticollis (New Forest mud beetle) is on the UKBAP list.

Identification: Angus (1992) is generally more useful for separating the critical species than the standard guides to water beetles.

British genus: *Helophorus*

Family Histeridae (20 genera, 51 species)
This family is sometimes included with the Sphaeritidae in a separate superfamily, the Histeroidea. They are found in a wide range of habitats; some are associated with fungi, seaweed, rotten wood, bark, dung and carrion; others are found in particular habitats such as rabbit burrows, and some have specific associations with other insects such as ants. Many are regarded as rare or local species but a few, such as *Margarinotus cadaverinus* and *M. carbonarius*, *Onthophilus striatus*, *Acritus nigricornis* and *Saprinus semistriatus*, are quite common (Fig. 18.118).

Identification: Halstead's (1963) handbook covers all except a couple of species recently added to the British list (Cooter & Barclay, 2006).

British subfamilies and genera:
Abraeinae: *Abraeus, Acritus, Aeletes, Halacritus, Plegaderus, Teretrius*
Dendrophilinae: *Carcinops, Dendrophilus, Kissister, Paromalus*
Hetaeriinae: *Haeterius*
Histerinae: *Atholus, Hister, Margarinotus*
Onthophilinae: *Onthophilus*

Saprininae: *Gnathoncus, Hypocaccus, Myrmetes, Saprinus*
Tribalinae: *Epierus*

Family Hydrochidae (1 genus, 7 species)
Members of this family are rather elongate water beetles, with distinctive sculpturing over the body, usually found around the margins of ponds, though their lifecycles have not been intensively studied.

Hydrochus nitidicollis (Gravel water beetle) is on the UKBAP list.

Identification: the standard guides to water beetles cover this family.

British genus: *Hydrochus*

Family Hydrophilidae (18 genera, 70 species)
The members of this large family are sometimes known as the water scavenger beetles, and they include the Silver water beetle; they range in size from 1 mm to nearly 50 mm. There are two distinct subfamilies: the Hydrophilinae, which are aquatic in all stages, the adults having fringes of hairs on their legs for swimming, and the Sphaeridiinae. The subfamily Sphaeridiinae is mostly associated with dung or decaying vegetation, though a few are aquatic with one or two species living on seaweed. The aquatic species breathe by means of an air bubble held on the underside of their bodies, giving them a silvery appearance, in contrast to groups like Dytiscidae, which trap air under the elytra. Their larvae are carnivorous (Fig. 18.119), but adults tend to be scavengers or herbivorous; although they can swim rather slowly many species crawl around on submerged vegetation.

Fig. 18.119 *Hydrochara caraboides* larva (Hydrophilidae) (Photo: Roger Key)

Fig. 18.121 *Sphaeridium scarabaeoides* (Hydrophilidae) (Photo: Roger Key)

Fig. 18.120 *Hydrophilus piceus* (Hydrophilidae) (Photo: Robin Williams)

Hydrophilus piceus, the Great silver beetle (Fig. 18.120) can reach almost 50 mm in length, and is one of the largest beetles in Britain. It was formerly very common, but loss of suitable habitats is making it more scarce in some areas. It has a sharp spine on the underside of the thorax which can pierce the skin of an unwary collector.

Among the Sphaeridiinae, the species of *Sphaeridium* (Fig. 18.121) live only on dung, whereas in the genus *Cercyon* some live on dung, others on decaying vegetation, sometimes at the margins of ponds.

Hydrochara caraboides (Lesser silver water beetle) and *Paracymus aeneus* (Bembridge beetle) are listed on the Wildlife and Countryside Act 1981.

Identification: most of the aquatic and semi-aquatic species can be identified with Friday's (1988) key. Hansen's (1987) work gives more detail, and covers the terrestrial Sphaeridiinae. Also useful is Lohse and Vogt (1971).

British subfamilies and genera:

Hydrophilinae: *Anacaena, Berosus, Chaetarthria, Cymbiodyta, Enochrus, Helochares, Hydrobius, Hydrochara, Hydrophilus, Laccobius, Limnoxenus, Paracymus*

Sphaeridiinae: *Cercyon, Coelostoma, Cryptopleurum, Dactylosternum, Megasternum, Sphaeridium*

Family Spercheidae (1 genus, 1 species)
Members of this family of water beetles were previously placed in the Hydrophilidae. The only British species is *Spercheus emarginatus*, and it may no longer occur here as there are no recent records.

British genus: *Spercheus*

Family Sphaeritidae (1 genus, 1 species)
This family is sometimes included with the Histeridae in a separate superfamily, the Histeroidea. The only British species is *Sphaerites glabratus*, a rare northern species usually associated with fungi.

Identification: Halstead (1963).

British genus: *Sphaerites*

SUPERFAMILY STAPHYLINOIDEA

The families in this group show a wide range of morphology and, to some extent, life histories; however, there are several biological common factors. In general the larvae and adults share similar habitats, and most species need moisture to survive, unlike other families that can exploit dried foodstuffs. With a few exceptions the staphylinoid beetles do not feed on vascular plants and many are associated with fungi; in some families predators are common.

Family Hydraenidae (5 genera, 30 species)
The Hydraenidae are very small beetles, 1–3 mm long, with distinctive sculpturing. Some species live in mud at the margins of ponds and small streams, though some species of *Ochthebius* are coastal, living in brackish water. Their larvae are not particularly adapted to aquatic life, and many are semi-terrestrial. All stages apparently feed on algae.

Ochthebius poweri (Rockface beetle) is on the UKBAP list.

Identification: the standard guides to water beetles cover this family; Lohe and Vogt (1971) is also useful.

British subfamilies and genera:
  Hydraeninae: *Hydraena, Limnebius*
  Ochthebiinae: *Aulacochthebius, Enicocerus, Ochthebius*

Family Leiodidae (21 genera, 93 species)
Most of the species in this large family seem to be associated with fungi in a variety of habitats, including subterranean fungi, bark, leaf-litter, and rotting vegetation; many can be collected by sieving these substrates, using pitfall traps or flight-interception traps. There is still much work to be done on the lifecycles in this group, and the larvae of several common species are completely unknown. Some of the better known species include *Leiodes calcarata, Anisotoma humeralis* (Fig. 18.122) and *Colon brunneum*.

Identification: at present, identification is only possible using a series of specialized keys; see Hammond and Hine (1999) and Cooter and Barclay (2006) for details.

British subfamilies and genera:
  Cholevinae: *Apocatops, Catopidius, Catops, Choleva, Nemadus, Nargus, Parabathyscia, Ptomaphagus, Sciodrepoides*
  Coloninae: *Colon*
  Leiodinae: *Agaricophagus, Agathidium, Aglyptinus, Amphicyllis, Anisotoma, Colenis, Hydnobius, Leiodes, Liocyrtusa, Sogda, Triarthron*
  Platypsyllinae: *Leptinus*

Family Ptiliidae (18 genera, 75 species)
The tiny feather-wing beetles are scarcely known to many entomologists, yet they form a moderately large family. No species is much more than 1 mm in length (Fig. 18.123), and most species feed on fungal spores and hyphae in habitats such as rotten wood, leaf-litter or dung as well as in the larger fungi. Because of their small size and specialized habitats, this is not a well-known group, and most species are collected by sieving suitable substrates. Identification depends on microscopic characters, so permanent slide-mounting is often required. Some genera such as *Acrotrichis*, with 26 British species, are particularly difficult to determine.

Identification: there are no recent works on the British fauna; standard books like Joy (1932) are too out of date for this family. It is necessary to consult works such as Besuchet and Sundt (1971) with additional notes in Lohse and Lucht (1989); see also Hammond and Hine (1999) for extra references.

Fig. 18.122 *Anisotoma humeralis* (Leiodidae) (Photo: Roger Key)

Fig. 18.123 *Ptenidium gressneri* (Ptiliidae) (Photo: Roger Key)

British subfamilies and genera:

Acrotrichinae: *Acrotrichis, Actinopteryx, Baeocrara, Nephanes, Smicrus*

Ptiliinae: *Actidium, Euryptilium, Micridium, Microptilium, Millidium, Nossidium, Oligella, Ptenidium, Pteryx, Ptiliola, Ptiliolum, Ptilium, Ptinella*

Family Scydmaenidae (9 genera, 32 species)

The members of this family only reach about 2 mm in length, and both larvae and adults are believed to feed on small invertebrates such as mites in habitats such as leaf-litter, rotten wood, moss and under bark. One of the most common British species is *Cephennium gallicum*, found in compost heaps and similar places. Difficult groups to identify include the genera *Eutheia, Neuraphes* and *Scydmoraphes*.

Identification: although the main genera and more common species can be identified with standard books such as Joy (1932), it is necessary to use works such as Franz and Besuchet (1971) and other specialized references; see Hammond and Hine (1999) and Cooter and Barclay (2006) for details.

British genera: *Cephennium, Euconnus, Eutheia, Euthiconus, Microscydmus, Nevraphes, Scydmaenus, Scydmoraphes, Stenichnus*

Family Silphidae (7 genera, 21 species)

Most of the burying beetles, sometimes known as carrion beetles, are quite large species, between 10–25 mm in length. Not all are associated with carrion; some are found in fungi and others feed on living vegetation. A few are specialized predators, with some *Silpha* species feeding on snails, after breaking through their shells.

Members of the genus *Nicrophorus* are known as the Sexton beetles, or Burying beetles; the largest is *N. humator*, at about 25 mm in length and completely black except for orange antennal clubs, and the most common is *N. vespilio*, which, like most others in the genus, has orange bands across the elytra (Fig. 18.124).

Identification: is reasonably straightforward using standard works such as Joy (1932); Freude's (1971) key is definitive. See also Hammond and Hine (1999) for other references.

British subfamilies and genera:

Silphinae: *Aclypea, Dendroxena, Necrodes, Oiceoptoma, Silpha, Thanatophilus*

Nicrophorinae: *Nicrophorus*

Family Staphylinidae (264 genera, c. 1000 species)

The rove beetles (known colloquially as 'staphs' among coleopterists) constitute the largest family of

Fig. 18.124 *Nicrophorus* sp. (Silphidae) (Photo: Colin Rew)

Fig. 18.125 *Tasgius ater* (Staphylinidae) (Photo: Roger Key)

beetles in Britain, and they pose many identification problems. The Pselaphinae were previously treated as a distinct family, as were the Scaphidiinae, though not all authors accept the latter as merely a subfamily of the Staphylinidae. Most members of the family are readily recognized by the short elytra (Fig. 18.125), though this character is not universal, nor is it unique to the staphylinids. Nearly all species are active predators, though some feed on fungi, algae or rotting vegetation.

Some of the larger species, as in the genus *Paederus* for example, have distinctive colour patterns and can be quite easy to identify, but the majority of staphylinids are small and inconspicuous. The largest British species is the fairly common *Ocypus olens* Devil's coach-horse, which often alarms people by its size and habit of raising its abdomen up like a scorpion.

Fig. 18.126 *Scaphidium quadrimaculatum* (Staphylinidae) (Photo: Roger Key)

Fig. 18.127 *Claviger testaceus* (Staphylinidae) with *Lasius flavus* (Photo: Roger Key)

Scaphidium quadrimaculatum (Fig. 18.126) is a strikingly marked species often found under bark. The two British members of the genus *Claviger* are mymecophilous, and *C. testaceus* is found commonly in the nests of the ant genus *Lasius* (Fig. 18.127).

Meotica anglica (Shingle rove beetle), *Stenus longitarsis*, *Stenus palposus* (Lough Neagh camphor beetle) and *Thinobius newberyi* (Newbery's rove beetle) are on the UKBAP list.

Identification: is a specialized matter for most Staphylinidae, and the literature in Hammond and Hine (1999) and Cooter and Barclay (2006) should be consulted. Tottenham's (1954) key is still useful for some subfamilies such as most of the Steninae. Pearce's (1957) handbook to the Pselaphinae is still useful though not complete and is usefully supplemented by Besuchet (1974). The recent handbook by Lott (2009) provides keys to the species of three subfamilies and includes several colour photos.

British subfamilies and genera:

Aleocharinae: *Acrotona, Actocharis, Adota, Agaricochara, Alaobia, Aleochara, Alevonota, Alianta, Aloconota, Amarochara, Amidobia, Amischa, Anomognathus, Anopleta, Arena, Atheta, Autalia, Badura, Bessobia, Bohemiellina, Bolitochara, Borboropora, Boreophilia, Brachida, Brachyusa, Brundinia, Cadaverota, Callicerus, Calodera, Ceritaxa, Chaetida, Coprothassa, Cordalia, Cousya, Crataraea, Cypha, Cyphea, Dacrila, Dadobia, Dalotia, Dasygnypeta, Datomicra, Deinopsis, Dexiogyia, Diglotta, Dilacra, Dimetrota, Dinaraea, Dinarda, Dochmonota, Drusilla, Enalodroma, Encephalus, Euryusa, Falagria, Falagrioma, Geostiba, Gnypeta, Gymnusa, Gyrophaena, Halobrecta, Haploglossa, Heterota, Holobus, Homalota, Homoeusa, Hydrosmecta, Hygronoma, Hygropora, Ilyobates, Ischnoglossa, Ischnopoda, Leptusa, Liogluta, Lomechusa, Lomechusoides, Lyprocorrhe, Meotica, Microdota, Mniusa, Mocyta, Mycetota, Myllaena, Myrmecocephalus, Myrmecopora, Myrmoecia, Nehemitropia, Neohilara, Notothecta, Ocalea, Ocyusa, Oligota, Oreostiba, Ousipalia, Oxypoda, Pachnida, Pachyatheta, Parameotica, Paranopleta, Pella, Philhygra, Phloeopora, Phytosus, Placusa, Plataraea, Pseudomicrodota, Pseudopasilia, Pycnota, Rhagocneme, Rhopalocerina, Schistoglossa, Silusa, Stichoglossa, Tachyusa, Tachyusida, Tetralaucopora, Thamiaraea, Thecturota, Thiasophila, Thinobaena, Thinonoma, Tinotus, Traumoecia, Trichiusa, Xenota, Zyras*

Euaesthetinae: *Euaesthetus*

Habrocerinae: *Habrocerus*

Micropeplinae: *Arrhenopeplus, Micropeplus*

Omaliinae: *Acidota, Acrolocha, Acrulia, Anthobium, Anthophagus, Coryphium, Deliphrum, Dropephylla, Eucnecosum, Eudectus, Eusphalerum, Geodromicus, Hadrognathus, Hapalaraea, Hypopycna, Lesteva, Micralymma, Olophrum, Omalium, Orochares, Paraphloeostiba, Philorinum, Phloeonomus, Phloeostiba, Phyllodrepa, Phyllodrepoidea, Xylodromus, Xylostiba*

Oxyporinae: *Oxyporus*

Oxytelinae: *Anotylus, Aploderus, Bledius, Carpelimus, Coprophilus, Deleaster, Manda, Ochthephilus, Oxytelus, Planeustomus, Platystethus, Syntomium, Teropalpus, Thinobius, Thinodromus*

Paederinae: *Achenium, Astenus, Hypomedon, Lathrobium, Lithocharis, Lobrathium, Medon, Ochthephilum, Paederidus, Paederus, Pseudomedon, Rugilus, Scopaeus, Sunius*

Phloeocharinae: *Phloeocharis*

Piestinae: *Siagonium*

Proteininae: *Megarthrus, Metopsia, Proteinus*

Pselaphinae: *Amauronyx, Batrisodes, Bibloplectus, Bibloporus, Brachygluta, Bryaxis, Bythinus, Claviger, Euplectus, Fagniezia, Plectophloeus, Pselaphaulax, Pselaphus, Reichenbachia, Rybaxis, Trichonyx, Trimium, Tychobythinus, Tychus*

Pseudopsinae: *Pseudopsis*

Scaphidiinae: *Scaphidium, Scaphisoma, Scaphium*

Staphylininae: *Acylophorus, Astrapaeus, Atrecus, Bisnius, Cafius, Creophilus, Dinothenarus, Emus, Erichsonius, Euryporus, Gabrius, Gabronthus, Gauropterus, Gyrohypnus, Heterothops, Hypnogyra, Leptacinus, Megalinus, Neobisnius, Nudobius, Ocypus, Ontholestes, Othius, Phacophallus, Philonthus, Platydracus, Quedius, Rabigus, Remus, Staphylinus, Tasgius, Velleius, Xantholinus*

Steninae: *Dianous, Stenus*

Tachyporinae: *Bolitobius, Bryophacis, Bryoporus, Cilea, Ischnosoma, Lamprinodes, Lordithon, Mycetoporus, Parabolitobius, Sepedophilus, Tachinus, Tachyporus*

Trichophyinae: *Trichophya*

REFERENCES

ANGUS, R.B. 1992. Insecta: Coleoptera: Hydrophilidae: Helophorinae. *Süsswasserfauna Mitteleuropas* 20: 144 pp.

AUDISIO, P. 1993. Coleoptera Nitidulidae – Kateretidae. *Fauna d'Italia* 32: xvi + 971 pp.

BALFOUR-BROWNE, F. 1940. *British water beetles*. Vol. 1. Ray Society, London.

BALFOUR-BROWNE, F. 1950. *British water beetles*. Vol. 2. Ray Society, London [Facsimile reprint 1964].

BALFOUR-BROWNE, F. 1958. *British water beetles*. Vol. 3. Ray Society, London [All 3 vols. on CD from Pisces Conservation].

BENSE, U. 1995. Longhorn beetles: illustrated key to the Cerambycidae and Vesperidae of Europe. Josef Margraf, Germany.

BESUCHET, C. 1974. Pselaphidae. *Die Käfer Mitteleuropas*. Vol. 5: 305–62. Goecke & Evers, Krefeld.

BESUCHET, C. & SUNDT, E. 1971. Familie: Ptiliidae. *Die Käfer Mitteleuropas*. Vol. 3: 311–42. Goecke & Evers, Krefeld.

BÍLÝ, S. 1982. The Buprestidae (Coleoptera) of Fennoscandia and Denmark. *Fauna Entomologica Scandinavica* 10: 109 pp.

BÍLÝ, S. & MEHL, O. 1989. Longhorn beetles (Coleoptera, Cerambycidae) of Fennoscandia and Denmark. *Fauna Entomologica Scandinavica* 22: 203 pp.

BOOTH, R.G., COX, M.L. & MADGE, R.B. 1990. *Coleoptera*. IIE guides to insects of importance to man. 3. CAB International, Wallingford.

BOWESTEAD, S. 1999. A revision of the Corylophidae (Coleoptera) of the West Palaearctic region. *Instrumenta Biodiversitatis* 3: 203 pp. Muséum d'histoire naturelle, Geneva.

BRENDELL, M.J.D. 1975. Tenebrionidae. *Handbooks for the identification of British insects* 5(10): 22 pp.

BRITTON, E.B. 1956. Scarabaeoidea (Lucanidae, Trogidae, Geotrupidae, Scarabaeidae). *Handbooks for the identification of British insects* 5(2): 29 pp.

BUCK, F.D. 1954. Lagriidae, Alleculidae, Tetratomidae, Melandryidae, Salpingidae, Pythidae, Mycteridae, Oedemeridae, Mordellidae, Scraptiidae, Pyrochroidae, Rhipiphoridae, Anthicidae, Aderidae and Meloidae. *Handbooks for the identification of British insects* 5(9): 30 pp.

CLARKE, R.O.S. 1973. Heteroceridae. *Handbooks for the identification of British insects* 5(2c): 15 pp.

COOTER, J. & BARCLAY, M. 2006. *A coleopterist's handbook* (4th edn.). Amateur Entomologist's Society, London.

CROWSON, R.A. 1956. Coleoptera: introduction and key to families. *Handbooks for the identification of British insects* 4(1): 59 pp.

CROWSON, R.A. 1981. *The biology of the Coleoptera*. Academic Press, London.

DAHLGREN, G. 1979. Cantharidae (except Malthinini). *Die Käfer Mitteleuropas*. Vol. 6: 18–39. Goecke & Evers, Krefeld.

DENTON, J. 2007. *Water bugs and water beetles of Surrey*. Surrey Wildlife Trust, Pirbright.

DUFFY, E.A.J. 1952. Cerambycidae. *Handbooks for the identification of British insects* 5(12): 18 pp.

DUFFY, E.A.J. 1953. *A monograph of the immature stages of British and imported timber beetles (Cerambycidae)*. British Museum (Natural History), London.

FORSYTHE, T. 2000. *Ground beetles* (2nd edn.). Naturalists' Handbooks no. 8. Richmond Publishing, Slough.

FOWLER, W.W. 1886–91. *The Coleoptera of the British islands*, 5 vols (in large paper edn.). Reeve, London.

FOWLER, W.W. & DONISTHORPE, H.St J.K. 1913. *The Coleoptera of the British islands*, vol. 6 (Suppl.). Reeve, London.

FRANZ, H. & BESUCHET, C. 1971. Scydmaenidae. *Die Käfer Mitteleuropas*. Vol. 3: 271–303. Goecke & Evers, Krefeld.

FREUDE, H. 1971. Silphidae. *Die Käfer Mitteleuropas*. Vol. 3: 190–201. Goecke & Evers, Krefeld.

FREUDE, H., HARDE, K.W. & LOHSE, G.A. (eds.) 1969. Lyctidae, Bostrychidae, Anobiidae, Ptinidae. *Die Käfer Mitteleuropas*. Vol. 8: 7–74. Goecke & Evers, Krefeld.

FRIDAY, L.E. 1988. *Key to the adults of British water beetles*. Field Studies Council (AIDGAP) no. 188. 152 pp.

HALSTEAD, D.G.H. 1963. Coleoptera Histeroidea. Sphaeritidae and Histeridae. *Handbooks for the identification of British insects* 4(10): 16 pp.

HAMMOND, P.M. & HINE, S.J. 1999. Coleoptera: the beetles. In: BARNARD, P.C. (ed.) *Identifying British insects and arachnids: an annotated bibliography of key works*. Cambridge University Press, Cambridge, pp. 80–138.

HANSEN, M. 1987. The Hydrophiloidea (Coleoptera) of Fennoscandia and Denmark. *Fauna Entomologica Scandinavica* 18: 254 pp.

HARDE, K.W. & HAMMOND, P.M. 1984. *A field guide in colour to beetles*. Octopus Books, London [English edn. edited and with additional introductory material by

P.M. Hammond. Original German title: *Die Kosmos – Käferführer*. Kosmos Verlag, Stuttgart, 1981].

HAWKINS, R. 2000. *Ladybirds of Surrey*. Surrey Wildlife Trust, Woking.

HICKIN, N.E. 1963. *The insect factor in wood decay*. Hutchinson, London.

HICKIN, N.E. 1987. *Longhorn beetles of the British Isles*. Shire Publications, Princes Risborough.

HINTON, H.E. 1945. *A monograph of the beetles associated with stored products*. Vol. I. British Museum (Natural History), London [reprinted 1963, Johnson Reprint].

HODGE, P.J. & JONES, R.A. 1995. *New British beetles: species not in Joy's practical handbook*. British Entomological and Natural History Society, Hurst.

HOLLAND, D.G. 1972. A key to the larvae, pupae and adults of the British species of Elminthidae. *Scientific Publications of the Freshwater Biological Association* 26: 58 pp.

HOLMEN, M. 1987. The aquatic Adephaga (Coleoptera) of Fennoscandia and Denmark. 1. Gyrinidae, Haliplidae, Hygrobiidae and Noteridae. *Fauna Entomologica Scandinavica* 20: 168 pp.

HYMAN, P.S. & PARSONS, M.S. 1992. Review of the scarce and threatened Coleoptera of Great Britain. 1. *UK Nature Conservation* 3: 448 pp. Joint Nature Conservation Committee, Peterborough.

HYMAN, P.S. & PARSONS, M.S. 1994. Review of the scarce and threatened Coleoptera of Great Britain. 2. *UK Nature Conservation* 12: 248 pp. Joint Nature Conservation Committee, Peterborough.

JÄCH, M. 1992. Dryopidae. *Die Käfer Mitteleuropas*. Vol. 13: 67–9. Goecke & Evers, Krefeld.

JESSOP, L. 1986. Dung beetles and chafers. Coleoptera: Scarabaeoidea (New edn.). *Handbooks for the identification of British insects* 5(11): 53 pp.

JOHNSON, C. 1966. Clambidae. *Handbooks for the identification of British insects* 4(6a): 13 pp.

JOY, N.H. 1932. *A practical handbook of British beetles*, 2 vols. Witherby, London [Reprinted 1976 and 1997 by Classey, Farringdon, also on CD from Pisces Conservation].

KIRK-SPRIGGS, A.H. 1996. Pollen beetles. Coleoptera: Kateretidae and Nitidulidae: Meligethinae. *Handbooks for the identification of British insects* 5(6a): 157 pp.

KLAUSNITZER, B. 1978. Ordnung Coleoptera (Larven). *Bestimmungsbücher zur Bodenfauna Europas*. Junk, The Hague.

KLAUSNITZER, B. 1991, 1994, 1996. *Die Käfer Mitteleuropas, Larven*. Vols. 1, 2 & 3. Goecke & Evers, Krefeld.

LAIBNER, S. 2000. *Elateridae of the Czech and Slovak Republics*. Kabourek, Zlín.

LAWRENCE, J., HASTINGS, A., DALLWITZ, M. & PAINE, T. 1993. *Beetle larvae of the world*. CSIRO, Australia.

LEVEY, B. 1977. Buprestidae. *Handbooks for the identification of British insects* 5(5a): 8 pp.

LEVEY, B. 2009 British Scraptiidae. *Handbooks for the identification of British insects* 5(18): 32 pp.

LINDROTH, C.H. 1974. Coleoptera: Carabidae. *Handbooks for the identification of British insects* 4(2): 148 pp.

LINSSEN, E.F. 1959. *Beetles of the British Isles*. 2 vols. Frederick Warne, London, The Wayside and Woodland Series.

LÖBL, I. & SMETANA, A. (eds.) 2003. *Catalogue of Palaearctic Coleoptera*. Vol. 1: Archostemata–Myxophaga–Adephaga. Apollo Books, Stenstrup, 819 pp.

LÖBL, I. & SMETANA, A. (eds.) 2004. *Catalogue of Palaearctic Coleoptera*. Vol. 2: Hydrophiloidea–Staphylinoidea. Apollo Books, Stenstrup, 942 pp.

LÖBL, I. & SMETANA, A. (eds.) 2006. *Catalogue of Palaearctic Coleoptera*. Vol. 3: Scarabaeoidea, Scirtoidea, Dascilloidea, Buprestoidea and Byrrhoidea. Apollo Books, Stenstrup, 690 pp.

LÖBL, I. & SMETANA, A. (eds.) 2007. *Catalogue of Palaearctic Coleoptera*. Vol. 4: Elateroidea, Derodontoidea, Bostrichoidea, Lymexyloidea, Cleroidea and Cucujoidea. Apollo Books, Stenstrup, 935 pp.

LÖBL, I. & SMETANA, A. (eds.) 2008. *Catalogue of Palaearctic Coleoptera*. Vol. 5: Tenebrionoidea. Apollo Books, Stenstrup, 670 pp.

LÖBL, I. & SMETANA, A. (eds.) 2010. *Catalogue of Palaearctic Coleoptera*. Vol. 6: Chrysomeloidea. Apollo Books, Stenstrup, 924 pp.

LOHSE, G.A. 1967. Cryptophagidae. *Die Käfer Mitteleuropas*. Vol. 7: 110–57. Goecke & Evers, Krefeld.

LOHSE, G.A. & LUCHT, W.H. 1989. Ptiliidae. *Die Käfer Mitteleuropas*. Vol. 12: 118–20. Goecke & Evers, Krefeld.

LOHSE, G.A. & LUCHT, W.H. 1992. Anobiidae. *Die Käfer Mitteleuropas*. Vol. 13: 176–9. Goecke & Evers, Krefeld.

LOHSE, G.A. & VOGT, H. 1971. Hydraenidae, Spercheidae, Hydrophilidae. *Die Käfer Mitteleuropas*. Vol. 3: 95–156. Goecke & Evers, Krefeld.

LOTT, D.A. 2009. The Staphylinidae (rove beetles) of Britain and Ireland. Part 5: Scaphidiinae, Piestinae, Oxytelinae. *Handbooks for the identification of British insects* 12(5): 100 pp.

LUFF, M.L. 2007. The Carabidae (ground beetles) of Britain and Ireland. *Handbooks for the identification of British insects* 4(2): 247 pp.

MAJERUS, M.E.N. 1994. *Ladybirds*. Collins New Naturalist, London.

MAJERUS, M.E.N. & KEARNS, P. 1989. *Ladybirds*. Naturalists Handbooks no. 10. Richmond Publishing, Slough.

MAJERUS, M., ROY, H., BROWN, P. & WARE, R. 2006. *Guide to ladybirds of the British Isles*. Field Studies Council, Preston Montford, OP 102.

McHUGH, J.V. & LIEBHERR, J.K. 2009. Coleoptera (beetles, weevils, fireflies). In: RESH, V.H. & CARDÉ, R.T. (eds.) *Encyclopedia of insects* (2nd edn.). Academic Press/Elsevier, San Diego & London, pp. 183–201.

MOHR, K.H. 1966. Chrysomelidae. *Die Käfer Mitteleuropas*. Vol. 9: 95–280. Goecke & Evers, Krefeld.

MORRIS, M.G. 1990. Orthocerous weevils. Coleoptera; Curculionidae (Nemonychidae, Anthribidae, Urodontidae, Attelabidae and Apionidae). *Handbooks for the identification of British insects* 5(16): 108 pp.

MORRIS, M.G. 1991. *Weevils*. Naturalists Handbooks no. 16. Richmond Publishing, Slough.

MORRIS, M.G. 1997. Broad–nosed weevils. Coleoptera: Curculionidae (Entiminae). *Handbooks for the identification of British insects* 5(17a): 106 pp.

MORRIS, M.G. 2002. True weevils (Part 1). Coleoptera: Curculionidae (Subfamilies Raymondionyminae to Smicronychinae). *Handbooks for the identification of British insects* 5(17b): 149 pp.

MORRIS, M.G. 2008. True weevils (Part 2). Coleoptera: Curculionidae, Ceutorhynchinae. *Handbooks for the identification of British insects* 5(17c): 129 pp.

MUONA, J. 1993. Review of the phylogeny, classification and biology of the family Eucnemidae (Coleoptera). *Entomologica Scandinavica* (Suppl.) 44: 133 pp.

NILSSON, A.N. 1982. A key to the larvae of the Fennoscandian Dytiscidae (Coleoptera). *Fauna Norrlandica* 2: 45 pp.

NILSSON, A.N. & HOLMEN, N. 1995. The aquatic Adephaga (Coleoptera) of Fennoscandia and Denmark. II. Dytiscidae. *Fauna Entomologica Scandinavica* 32: 195 pp. Brill, Leiden.

OLMI, M. 1976. Coleoptera Dryopidae, Elminthidae. *Fauna d'Italia* 12: 280 pp.

PAULUS, H.F. 1979. Byrrhidae. *Die Käfer Mitteleuropas.* Vol. 6: 328–50. Goecke & Evers, Krefeld.

PEACOCK, E.R. 1977. Rhizophagidae. *Handbooks for the identification of British insects* 5(5a): 19 pp.

PEACOCK, E.R. 1993. Adults and larvae of hide, larder and carpet beetles and their relatives (Coleoptera: Dermestidae) and of derodontid beetles (Coleoptera: Derodontidae). *Handbooks for the identification of British insects* 5(3): 144 pp.

PEARCE, E.J. 1957. Pselaphidae. *Handbooks for the identification of British insects* 4(9): 32 pp.

PEEZ, A. von 1967. Lathridiidae. *Die Käfer Mitteleuropas.* Vol. 7: 168–90. Goecke & Evers, Krefeld.

PLATIA, G. 1994. Coleoptera Elateridae. *Fauna d'Italia* 33: 429 pp.

POPE, R.D. 1953. Coccinellidae and Sphindidae. *Handbooks for the identification of British insects* 5(7): 12 pp.

RIDSDILL-SMITH, J. & SIMMONS, L.W. 2009. Dung beetles. In: RESH, V.H. & CARDÉ, R.T. (eds.) *Encyclopedia of insects* (2nd edn.). Academic Press/Elsevier, San Diego & London, pp. 304–7.

RÜCKER, W.H. 1992. Latridiidae. *Die Käfer Mitteleuropas.* Vol. 13: 139–60. Goecke & Evers, Krefeld.

SPORNRAFT, K. 1967. Nitidulidae. *Die Käfer Mitteleuropas.* Vol. 7: 20–79. Goecke & Evers, Krefeld.

SPORNRAFT, K. 1992. Nitidulidae. *Die Käfer Mitteleuropas.* Vol. 13: 91–110. Goecke & Evers, Krefeld.

TACHET, H., RICHOUX, P., BOURNAUD, M., & USSEGLIO-POLATERA, P. 2000. *Invertébrés d'eau douce: systématique, biologie, écologie.* CNRS Editions, Paris.

THOMPSON, R.T. 1958. Phalacridae. *Handbooks for the identification of British insects* 5(5b): 17 pp.

TOTTENHAM, C.E. 1954. Staphylinidae (Piestinae to Euaesthetinae). *Handbooks for the identification of British insects* 4(8a): 79 pp.

TRAUTNER, J. & GEIGENMÜLLER, K. 1987. Tiger beetles, ground beetles: illustrated key to the Cicindelidae and Carabidae of Europe. J. Margraf, Aichtal, Germany.

UNWIN, D.M. 1984. *A key to the families of British beetles (and Strepsiptera).* Field Studies Council (AIDGAP) no. 166: 48 pp. [reprinted with minor alterations, 1988].

VOGT, H. 1969. Byturidae. *Die Käfer Mitteleuropas.* Vol. 7: 19–20. Goecke & Evers, Krefeld.

VONDEL, B.J. van 1997. Insecta: Coleoptera: Haliplidae. *Süsswasserfauna von Mitteleuropa* 20(2): 95 pp.

WITTMER, W. 1979. Cantharidae, Malthinini. *Die Käfer Mitteleuropas.* Vol. 6: 40–51. Goecke & Evers, Krefeld.

WEBSITES

http://www.coleopterist.org.uk

Although intended as the website for the the journal *The Coleopterist*, this site has much information on the British species, including the definitive checklist, as well as notes on identification, a gallery of photos, and information on the British recording schemes.

http://wtaxa.csic.es

An on-line catalogue of weevil (Curculionoidea) names of the world.

http://www.johnwalters.co.uk

Contains a small series of identification guides to certain groups of beetles.

http://markgtelfer.co.uk/beetles

Has many useful notes on identification, corrections to published keys and so on.

www.ladybird-survey.org

Site for information about the British ladybirds (Coccinellidae).

www.harlequin-survey.org

Site about the invasive Harlequin ladybirds (*Harmonia axyridis*).

http://www.brc.ac.uk/DBIF/homepage.aspx

A general site providing records of the foodplants of British insects, which is useful for the phytophagous beetles.

http://www.thewcg.org.uk

The Watford Coleoptera Group has a useful website with many photos, information on identification and so on.

http://www.leafmines.co.uk/index.htm

A project to provide information on all the British leaf-mining groups.

19 Order Diptera: the true flies

c. 7000 species in 103 families

The exact relationships of the Diptera with other insect orders are still open to question; undoubtedly they belong in the loosely defined group of Panorpids, which includes the Lepidoptera, Mecoptera, Siphonaptera and Trichoptera. They are sometimes grouped with the Mecoptera and Siphonaptera in the Antliophora, but some recent studies have questioned the monophyly of the scorpionflies and there are suggestions that the Diptera may be closely allied with a narrower subgroup of the Mecoptera. There are even proposals that the Strepsiptera, traditionally placed with the Coleoptera, could be the sister-group of the Diptera; many of these issues are discussed in Grimaldi and Engel (2005).

The latest list of British Diptera includes 7032 species (Chandler, 2010), which may be just ahead of the Hymenoptera, though this number will certainly have changed by the time the current book is published. Just like the coleopterists and hymenopterists, those who study Diptera will argue that the flies are biologically the most diverse of any insect order; as with those other groups the case is strong. Diptera can be predators on other arthropods, including many insect groups; they can also be mycophagous, phytophagous, leaf-miners, dung-feeders and even snail-killers; some are blood-feeders on humans and other animals and can be important vectors of diseases; others are ectoparasites or endoparasites on a wide variety of hosts.

In general the Diptera are easy to recognize as a group; with only a single pair of functional wings and the hind wings modified into halteres, used as balancing organs during flight. Many families are extremely familiar to the layman, as shown by the great number of common names at both species and family level, such as hoverflies, mosquitoes, midges, house-flies, bluebottles and crane-flies. Clearly there are some large and brightly coloured groups that have always attracted the amateur entomologist, but many species are small, black and difficult to identify, and these groups are more often tackled by the specialist. Even so, there are many families of British Diptera for which there are no complete identification guides, especially in the acalyptrates. Even some economically important groups such as the Cecidomyiidae (gall-midges) are in urgent need of revision.

The standard book for providing a comprehensive overview of the whole group was Colyer and Hammond (1968); this pioneering work is still useful although inevitably the classification is out of date and much more information is available for some groups. An essential guide for any student of British Diptera is Chandler's (2010) *Dipterist's Handbook*, which includes sections by many expert authors on all aspects of the biology, study and classification. There are several keys to families available, including those in Colyer and Hammond (1968), Oldroyd (1970) and Unwin (1981); all have slightly different approaches and all are outdated to some extent. The book on the European families by Oosterbroek (2006) is extremely useful, both as a family key and as a source of information on each group. The identification at family level is sometimes covered by reasonably modern handbooks but in many cases it is necessary to consult a wide range of scattered publications in specialist journals. Many of these were listed by Wyatt and

The Royal Entomological Society Book of British Insects, First Edition. Peter C. Barnard.
© 2011 Royal Entomological Society. Published 2011 by Blackwell Publishing Ltd.

Chainey (1999), and the latest lists are in Chandler (2010). There is a standard series of monographic works on the European Diptera called *Die Fliegen der paläarktischen Region*, published in Stuttgart and originally edited by the German entomologist Erwin Lindner (1888–1988). The series still continues, but many of the early volumes are now very outdated; the more useful sections are listed under their individual authors. Another important European work is the catalogue by Soós and Papp (1984–93). The checklist of British Diptera by Kloet and Hincks (1976) was replaced by that of Chandler (1998), and frequent updates to this have been published; see Chandler (2010) for details.

For the study of the immature stages, Smith's (1989) handbook is invaluable, as are several sections in Chandler (2010). Oldroyd's (1964) book on the biology of flies is old but still very readable and useful, and the medically important groups are covered by Lane and Crosskey (1997).

There are around 19,000 species in 132 families in Europe, and 152,000 species in 160 families worldwide. A brief world overview of the Diptera is given by Merritt et al. (2009).

The higher classification used here still retains some traditional groupings such as Nematocera even though it is clear that these are paraphyletic groups. Some authors use terms like 'Lower Diptera' as informal groups until the phylogeny is clarified. Even at family level there are many changes being proposed, particularly in the acalyptrates.

HIGHER CLASSIFICATION OF BRITISH DIPTERA

Suborder Brachycera
  Infraorder Asilomorpha
    Superfamily Asiloidea
      Family Asilidae (17 genera, 29 species)
      Family Bombyliidae (4 genera, 9 species)
      Family Scenopinidae (1 genus, 2 species)
      Family Therevidae (6 genera, 14 species)
    Superfamily Empidoidea
      Family Atelestidae (1 genus, 2 species)
      Family Dolichopodidae (40 genera, 293 species)
      Family Empididae (18 genera, 212 species)
      Family Hybotidae (20 genera, 178 species)
      Family Microphoridae (1 genus, 3 species)
    Superfamily Nemestrinoidea
      Family Acroceridae (2 genera, 3 species)
  Infraorder Muscomorpha (Aschiza)
    Superfamily Lonchopteroidea
      Family Lonchopteridae (1 genus, 7 species)
    Superfamily Platypezoidea
      Family Opetiidae (1 genus, 1 species)
      Family Phoridae (23 genera, 329 species)
      Family Platypezidae (10 genera, 33 species)
    Superfamily Syrphoidea
      Family Pipunculidae (11 genera, 93 species)
      Family Syrphidae (70 genera, 276 species)
  Infraorder Muscomorpha (Schizophora Acalyptratae)
    Superfamily Carnoidea
      Family Braulidae (1 genus, 2 species)
      Family Canacidae (2 genera, 2 species)
      Family Carnidae (2 genera, 13 species)
      Family Chloropidae (39 genera, 177 species)
      Family Milichiidae (6 genera, 19 species)
      Family Tethinidae (3 genera, 10 species)
    Superfamily Conopoidea
      Family Conopidae (7 genera, 23 species)

    Superfamily Diopsoidea
      Family Megamerinidae (1 genus, 1 species)
      Family Psilidae (5 genera, 26 species)
      Family Strongylophthalmyiidae (1 genus, 1 species)
      Family Tanypezidae (1 genus, 1 species)
    Superfamily Ephydroidea
      Family Camillidae (1 genus, 5 species)
      Family Campichoetidae (1 genus, 2 species)
      Family Diastatidae (1 genus, 6 species)
      Family Drosophilidae (8 genera, 62 species)
      Family Ephydridae (40 genera, 147 species)
    Superfamily Lauxanioidea
      Family Chamaemyiidae (7 genera, 32 species)
      Family Lauxaniidae (13 genera, 54 species)
    Superfamily Nerioidea
      Family Micropezidae (5 genera, 10 species)
      Family Pseudopomyzidae (1 genus, 1 species)
    Superfamily Opomyzoidea
      Family Acartophthalmidae (1 genus, 2 species)
      Family Agromyzidae (19 genera, 392 species)
      Family Anthomyzidae (7 genera, 20 species)
      Family Asteiidae (3 genera, 8 species)
      Family Aulacigastridae (1 genus, 1 species)
      Family Clusiidae (4 genera, 10 species)
      Family Odiniidae (1 genus, 9 species)
      Family Opomyzidae (2 genera, 17 species)
      Family Periscelididae (1 genus, 3 species)
      Family Stenomicridae (1 genus, 2 species)
    Superfamily Sciomyzoidea
      Family Coelopidae (2 genera, 2 species)
      Family Dryomyzidae (4 genera, 5 species)
      Family Phaeomyiidae (1 genus, 2 species)
      Family Sciomyzidae (23 genera, 68 species)
      Family Sepsidae (6 genera, 29 species)

Superfamily Sphaeroceroidea
  Family Chyromyidae (3 genera, 11 species)
  Family Heleomyzidae (17 genera, 63 species)
  Family Sphaeroceridae (36 genera, 137 species)
Superfamily Tephritoidea
  Family Lonchaeidae (5 genera, 46 species)
  Family Pallopteridae (2 genera, 13 species)
  Family Piophilidae (12 genera, 14 species)
  Family Platystomatidae (2 genera, 2 species)
  Family Tephritidae (34 genera, 76 species)
  Family Ulidiidae (11 genera, 20 species)
Infraorder Muscomorpha (Schizophora Calyptratae)
  Superfamily Hippoboscoidea
    Family Hippoboscidae (10 genera, 14 species)
    Family Nycteribiidae (3 genera, 3 species)
  Superfamily Muscoidea
    Family Anthomyiidae (29 genera, 242 species)
    Family Fanniidae (2 genera, 60 species)
    Family Muscidae (40 genera, 282 species)
    Family Scathophagidae (23 genera, 54 species)
  Superfamily Oestroidea
    Family Calliphoridae (14 genera, 38 species)
    Family Oestridae (5 genera, 11 species)
    Family Rhinophoridae (6 genera, 8 species)
    Family Sarcophagidae (15 genera, 60 species)
    Family Tachinidae (141 genera, 261 species)
Infraorder Tabanomorpha
  Superfamily Stratiomyoidea
    Family Stratiomyidae (16 genera, 48 species)
    Family Xylomyidae (2 genera, 3 species)
  Superfamily Tabanoidea
    Family Athericidae (3 genera, 3 species)
    Family Rhagionidae (3 genera, 12 species)
    Family Spaniidae (2 genera, 3 species)
    Family Tabanidae (5 genera, 30 species)
Infraorder Xylophagomorpha
  Superfamily Xylophagoidea
    Family Xylophagidae (1 genus, 3 species)

Suborder Nematocera
  Infraorder Bibionomorpha
    Superfamily Bibionoidea
      Family Bibionidae (2 genera, 18 species)
    Superfamily Sciaroidea
      Family Bolitophilidae (1 genus, 17 species)
      Family Cecidomyiidae (140 genera, 652 species)
      Family Diadocidiidae (1 genus, 3 species)
      Family Ditomyiidae (2 genera, 3 species)
      Family Keroplatidae (15 genera, 52 species)
      Family Mycetophilidae (61 genera, 471 species)
      Family Sciaridae (19 genera, 266 species)
  Infraorder Culicomorpha
    Superfamily Chironomoidea
      Family Ceratopogonidae (20 genera, 170 species)
      Family Chironomidae (139 genera, 608 species)
      Family Simuliidae (3 genera, 35 species)
      Family Thaumaleidae (1 genus, 3 species)
    Superfamily Culicoidea
      Family Chaoboridae (2 genera, 6 species)
      Family Culicidae (6 genera, 34 species)
      Family Dixidae (2 genera, 15 species)
  Infraorder Psychodomorpha
    Superfamily Anisopodoidea
      Family Anisopodidae (1 genus, 4 species)
      Family Mycetobiidae (1 genus, 3 species)
    Superfamily Psychodoidea
      Family Psychodidae (20 genera, 99 species)
    Superfamily Scatopsoidea
      Family Scatopsidae (16 genera, 46 species)
    Superfamily Trichoceroidea
      Family Trichoceridae (2 genera, 10 species)
  Infraorder Ptychopteromorpha
    Superfamily Ptychopteroidea
      Family Ptychopteridae (1 genus, 7 species)
  Infraorder Tipulomorpha
    Superfamily Tipuloidea
      Family Cylindrotomidae (4 genera, 4 species)
      Family Limoniidae (49 genera, 214 species)
      Family Pediciidae (4 genera, 19 species)
      Family Tipulidae (8 genera, 87 species)

SPECIES OF CONSERVATION CONCERN

No species of Diptera have any legal protection, but the following 35 species are on the UKBAP list: *Botanophila fonsecai* (Anthomyiidae); *Asilus crabroniformis* (Asilidae); *Bombylius minor* and *Thyridanthrax fenestratus* (Bombyliidae); *Lipara similis* (Chloropidae); *Clusiodes geomyzinus* (Clusiidae); *Campsicnemus magius, Dolichopus laticola* and *Dolichopus nigripes* (Dolichopodidae); *Amiota variegate* (Drosophilidae); *Empis limata* and *Rhamphomyia hirtula* (Empididae); *Asindulum nigrum* (Keroplatidae); *Gnophomyia elsneri, Idiocera sexguttata, Lipsothrix ecu-*cullata, *Lipsothrix errans, Lipsothrix nervosa, Lipsothrix nigristigma* and *Rhabdomastix japonica* (Limoniidae); *Lonchaea ragnari* (Lonchaeidae); *Phaonia jaroschewskii* (Muscidae); *Neoempheria lineola* (Mycetophilidae); *Dorylomorpha clavifemora* (Pipunculidae); *Salticella fasciata* (Sciomyzidae); *Odontomyia hydroleon* (Stratiomyidae); *Blera fallax, Callicera spinolae, Chrysotoxum octomaculatum, Doros profuges, Eristalis cryptarum, Hammerschmidtia ferruginea* and *Myolepta potens* (Syrphidae); *Cliorismia rustica* (Therevidae); *Dorycera graminum* (Ulidiidae).

The following species were removed from the UKBAP list in 2007 for various reasons: *Bombylius discolor* (Bombyliidae); *Rhabdomastix laeta* (Limoniidae); *Spiriverpa lunulata* (Therevidae); *Tipula serrulifera* (Tipulidae).

Important works on the species of conservation concern were by Falk (1991), Falk and Chandler (2005) and Falk and Crossley (2005); the section on conservation in Chandler (2010) should also be consulted.

The Families of British Diptera

SUBORDER BRACHYCERA

If one accepts the informal grouping of the lower Diptera as the Nematocera then the other suborder, the Brachycera, are considered the more advanced group. They were previously divided into two subgroups with the Cyclorrhapha being a distinct suborder, but the current uncertainties about the exact relationships of all the higher groupings mean that different classifications will be found in almost every major work. Chandler (2010) summarizes some of the latest thoughts on this subject. At the time of works such as Colyer and Hammond (1968) the Nematocera and Brachycera *sensu stricto* were combined as the Orthorrhapha, in which the adult emerged from the pupa via a longitudinal slit; this was in contrast to the higher Cyclorrhapha in which an end-cap of the puparium separated via a circular split. It is now clear that these characters have no phylogenetic significance. Several families in the Brachycera exhibit characters that would be considered as primitive, such as multisegmented antennae, complete abdominal segmentation and complex wing venation, features shared with the Nematocera; the 'higher' groups show reduction in antennal and abdominal segmentation and a simplification of venation. Brachyceran larvae tend to have a reduced head capsule, with just a pair of sclerotized mouthhooks; pupation occurs within the last larval skin, known as the puparium.

Infraorder Asilomorpha

SUPERFAMILY ASILOIDEA

The well-illustrated book by Stubbs and Drake (2001) supersedes Oldroyd's (1969) handbook for the families in this group.

Family Asilidae (17 genera, 29 species)
Generally known as robber-flies, these are predators as both larvae and adults. The larvae live in the soil or in rotting wood where they feed on other insect

Fig. 19.1 *Asilus crabroniformis* (Asilidae) with prey (Photo: Robin Williams)

larvae. The fast-flying adults can be seen in sunshine sitting on vantage points, waiting for for other insects that they catch on the wing (Fig. 19.1).

Dasypogon diadema has not been seen since the 19th century and was probably an introduction.

Asilus crabroniformis (Hornet robberfly) is on the UKBAP list.

Identification: Stubbs and Drake (2001) cover the British species.

British subfamilies and genera:
Asilinae: *Asilus, Dysmachus, Eutolmus, Machimus, Neoitamus, Neomochtherus, Pamponerus, Philonicus, Rhadiurgus, Tolmerus*
Dasypogoninae: *Dasypogon, Leptarthrus*
Laphriinae: *Choerades, Laphria*
Leptogasterinae: *Leptogaster*
Stenopogoninae: *Dioctria, Lasiopogon*

Family Bombyliidae (4 genera, 9 species)
The bee-flies get their name from their resemblance to aculeate Hymenoptera, especially bumblebees, as they generally have broad hairy bodies (Fig. 19.2). Many species have a long proboscis with which they drink nectar while hovering in front of flowers (Fig. 19.3). Adults without such a proboscis settle on the flowers to feed, and they often take pollen. Bombyliid larvae are parasitoids or predators on the juvenile stages of other insects or on spider eggs. Several species are associated with social bees; the female lays her eggs near the nest entrance and the newly hatched larvae enter the nest, feeding on the cell contents as well as on the bee larvae. The best-known species is *Bombylius major*, which can be seen on the wing as early as the end of March if the weather is warm enough. All members of this family are temperature dependent, which explains why most are found in

Fig. 19.2 *Bombylius canescens* (Bombyliidae) (Photo: Roger Key)

Fig. 19.4 *Cliorismia rustica* larva (Therevidae) (Photo: Roger Key)

Fig. 19.3 *Bombylius discolor* (Bombyliidae) (Photo: Roger Key)

the southern half of Britain; several species are considered as rare.

Although still on the British list, *Villa venusta* is probably extinct here. *Bombylius minor* (Heath bee-fly) and *Thyridanthrax fenestratus* (Mottled bee-fly) are on the UKBAP list.

Identification: Stubbs and Drake (2001) cover all the British species.

British subfamilies and genera:
Bombyliinae: *Bombylius*
Exoprosinae: *Thyridanthrax, Villa*
Phthiriinae: *Phthiria*

Family Scenopinidae (1 genus, 2 species)
These are usually known as window-flies as the commonest place to find them is inside old build-

ings, often walking slowly around the windows. The larvae live in almost any kind of dry organic matter such as soil, wood-borings and domestic dust, where they prey on mites and insects, and some species are described as predators on the larvae of carpet-beetles (Dermestidae). One of the two British species, *Scenopinus niger*, lives outside where it is associated with rotting wood though it seems to be scarce. *S. fenestralis* is typically synanthropic and more widespread, though mainly in the southern half of Britain.

Identification: Stubbs and Drake (2001) cover both British species.

British genus: *Scenopinus*

Family Therevidae (6 genera, 14 species)
These are generally known as stiletto-flies because of their pointed abdomen, though this character is shared with many other families, and superficially they resemble Asilidae and Rhagionidae. The biology of this group is little studied but the known larvae are predatory on small invertebrates such as insect larvae, and are often found in dry sandy soils (Fig. 19.4). The males of many species have silvery hairs (Fig. 19.5), and they form short-lived mating swarms; adults are not seen on flowers and have occasionally been seen to take in fluids, including water. One of the most common species is *Thereva nobilitata*.

Cliorismia rustica (Southern silver stiletto-fly) is on the UKBAP list.

Identification: Stubbs and Drake (2001) cover all the British species.

Fig. 19.5 *Spiriverpa lunulata* (Therevidae) (Photo: Roger Key)

Fig. 19.6 *Dolichopus festivus* (Dolichopodidae) (Photo: Roger Key)

British genera: *Acrosathe, Cliorismia, Dialineura, Pandivirilia, Spiriverpa, Thereva*

SUPERFAMILY EMPIDOIDEA

Family Atelestidae (1 genus, 2 species)
Most members of this family were previously in the Empididae, but the only British genus was sometimes placed in the Platypezidae. Like the Empididae they are known as dance-flies; they are very small black flies, rarely more than 3 mm long, with large eyes. The biology of the group is unknown and the larvae are undescribed; only four species are known from Europe.

Identification: the British species were covered in Collin's (1961) review of the Empididae.

British genus: *Atelestus*

Family Dolichopodidae (40 genera, 293 species)
This family was formerly known as Dolichopidae; some authors include the Microphoridae in this group. They are sometimes called long-legged flies, but are colloquially known as 'dollies' among Dipterists. Many species are small, from 2–3 mm long, and not easy to identify; many have a greenish metallic colour and all have long legs. Dolichopodid larvae are found in damp soil, rotting wood and sometimes in the mud at the side of freshwater; all are predatory on small invertebrates, with the exception of the phytophagous species of *Thrypticus* that mine monocotyledous plants. The adults are also assumed to be predatory, though there are fewer observations of their behaviour. Some of the more common species are to be found in genera such as *Sciapus, Poecilobothrus* and *Dolichopus* (Fig. 19.6).

Campsicnemus magius (Fancy-legged fly), *Dolichopus laticola* (Broads long-legged fly) and *Dolichopus nigripes* (Bure long-legged fly) are on the UKBAP list.

Identification: Fonseca's (1978) handbook covers nearly all the British species; see also Stackelberg and Negrobov (1930–79) for a European overview, though incomplete. Chandler (2010) gives many additional references.

British subfamilies and genera:
Achalcinae: *Achalcus*
Diaphorinae: *Argyra, Chrysotus, Diaphorus, Melanostolus*
Dolichopodinae: *Dolichopus, Hercostomus, Muscidideicus, Ortochile, Poecilobothrus, Sybistroma, Tachytrechus*
Hydrophorinae: *Aphrosylus, Hydrophorus, Liancalus, Machaerium, Orthoceratium, Scellus, Schoenophilus, Thinophilus*
Medeterinae: *Cyrturella, Medetera, Systenus, Thrypticus*
Neurigoninae: *Neurigona*
Rhaphiinae: *Nematoproctus, Rhaphium*
Sciapodinae: *Sciapus*
Sympycninae: *Acropsilus, Anepsiomyia, Campsicnemus, Chrysotimus, Lamprochromus, Micromorphus, Micropygus, Sympycnus, Syntormon, Telmaturgus, Teuchophorus, Xanthochlorus*

Family Empididae (18 genera, 212 species)
This group has also been known as the Empidae, and their common name is dance-flies. The family

Fig. 19.7 Empidid (Photo: Colin Rew)

Fig. 19.8 Empidids mating (Photo: Colin Rew)

has been split into several smaller groups in recent years, and this process is likely to continue as there is still doubt over the monophyly of the present constitution. Some authors would separate the Trichopezinae as a separate family, the Brachystomatidae. The members of this large family have a superficial resemblance to the Asilidae, but are separated by differences in wing venation. The larvae are all predatory and are found in the soil, rotting wood or organic matter; a few are aquatic, but all feed on small invertebrates, often other Diptera larvae. Most adults are also fierce predators, with piercing mouthparts and grasping legs with which they catch flying prey (Fig. 19.7). After a mating swarm the male may give an insect as a 'nuptial' present to the female; while she eats it the male mates with her, grasping her with his middle and hind legs while hanging precariously on a leaf using only his front legs (Fig. 19.8).

Empis limata (English assassin fly) and *Rhamphomyia hirtula* (Mountain dance-fly) are on the UKBAP list.

Identification: Collin's (1961) book is still useful though inevitably out of date; see also Chvála (1994, 2005) for recent reviews of the European species and Chandler (2010) for many additional references.

British subfamilies and genera:
   Clinocerinae: *Clinocera, Dolichocephala, Kowarzia, Wiedemannia*
   Empidinae: *Empis, Hilara, Rhamphomyia*
   Hemerodromiinae: *Chelifera, Chelipoda, Dryodromya, Hemerodromia, Phyllodromia*
   Oreogetoninae: *Hormopeza, Iteaphila, Ragas*
   Trichopezinae: *Gloma, Heleodromia, Trichopeza*

Family Hybotidae (20 genera, 178 species)
This large group was originally included within the Empididae, and they have inherited the common name of dance-flies. Most are small dark species and, like the Empididae, they are active predators on other flying insects. The larvae are also predatory, associated with damp habitats such as soil, rotting wood and organic detritus.

Identification: Collin's (1961) book is still useful although now out of date; Chvála (1975, 1983) covered most of the British species, but see Chandler (2010) for many additional references.

British subfamilies and genera:
   Hybotinae: *Hybos, Syndyas, Syneches*
   Ocydromiinae: *Anthalia, Bicellaria, Euthyneura, Leptopeza, Ocydromia, Oedalea, Oropezella, Trichina, Trichinomyia*
   Tachydrominae: *Chersodromia, Crossopalpus, Drapetis, Platypalpus, Stilpon, Symballophthalmus, Tachydromia, Tachypeza*

Family Microphoridae (1 genus, 3 species)
Not all authors accept the separation of this family from the Empididae and some now place these species in the Dolichopodidae. All are very small species, less than 3 mm long, and are predators as both larvae and adults, although their biology is little studied.

Identification: the British species were covered by Collin (1961); see also Chvála (1983) for the northern European species.

British genus: *Microphor*

SUPERFAMILY NEMESTRINOIDEA

Family Acroceridae (2 genera, 3 species)

This family was formerly known as the Cyrtidae or the Ogcodidae. They have recently been given the common name of hunchback-flies on account of their strongly convex thorax in side view. Although mainly dark in colour, they usually have some pale-coloured markings on the abdomen. Adults form mating swarms but they have short flight periods so are not often seen. The larvae are notable for being internal parasitoids of spiders; the first instar is a highly mobile planidium, which penetrates the host spider through an intersegmental membrane on the leg and then develops inside. The host, which is usually a hunting spider such as a lycosid or salticid, remains alive until the acrocerid larva is ready to pupate.

Acrocera orbiculus is the most common species; it is sometimes placed in the genus *Paracrocera*.

Identification: Stubbs and Drake (2001) cover all the British species.

British genera: *Acrocera, Ogcodes*

Infraorder Muscomorpha (Aschiza)

The Muscomorpha correspond to the old suborder Cyclorrhapha, and are usually considered as including the most advanced families of Diptera. There are two subdivisions: the Schizophora, in which the emergence of the adult is aided by an inflatable sac on the head, known as the ptilinum, which helps to split the puparium and which contracts to a ptilinal suture after emergence; and the Aschiza, which have no ptilinum, and hence no ptilinal suture. The latter is clearly not monophyletic but is retained as an informal group for convenience.

SUPERFAMILY LONCHOPTEROIDEA

Family Lonchopteridae (1 genus, 7 species)

This family was sometimes known as the Musidoridae. All species are small brownish or yellowish flies with pointed wings (Fig. 19.9). They are unusual in showing sexual dimorphism in the wing ventation; in the male the anal vein meets the hind margin of the wing, but in the female it curves forward to meet the cubital vein. Most species are associated with humid habitats, ranging from marshes to damp woodland, where the larvae feed on rotting vegetation. *Lonchoptera furcata* is a cosmo-

Fig. 19.9 *Lonchoptera lutea* (Lonchopteridae) (Photo: Roger Key)

politan species, which seems to be parthenogenetic in at least part of its range; males are rarely seen though they may be under collected.

Identification: Smith's (1969b) handbook covers all the British species.

British genus: *Lonchoptera*

SUPERFAMILY PLATYPEZOIDEA

Family Opetiidae (1 genus, 1 species)

This group was formerly placed within the Platypezidae and both families have the common name of flat-footed flies. The sole British species is *Opetia nigra*, which is also the only species known in Europe. It is a small, dark fly, often found in wooded areas and the larva is known to feed in rotting wood. Males are more commonly seen than females, and they forming mating swarms.

Identification: Chandler (2001) reviewed this family together with the Platypezidae.

British genus: *Opetia*

Family Phoridae (23 genera, 329 species)

The members of the large family of scuttle-flies are all small with a humpbacked appearance; the adult flies have a characteristically fast way of scuttling around on leaves. Some species are less than 1 mm long and the males may form swarms before mating (Fig. 19.10). The larval biologies in this family are extremely varied and complex and for many species are as yet unknown. Some species are predators or parasitoids, some are true parasites and some are saprophagous, but it is not always possible to be

Fig. 19.10 *Phora atra*, mating pair (Phoridae) (Photo: Roger Key)

Fig. 19.11 *Spiniphora maculata* puparia in snail (Phoridae) (Photo: Roger Key)

sure without careful rearing because of the various habitats in which phorid larvae can be found. Some species are definitely predators on slug eggs, snails, insects such as aphids, and spider eggs; others parasitize millipedes or various stages of other insects. Where species have been reared from habitats such as fungi or dung they may have been preying on other insects in that substrate; on the other hand species apparently reared from dead insects may be genuine saprophages on moribund or dead hosts. *Spiniphora maculata* (Fig. 19.11) is an example of a species that is found in dead snails where it is probably saprophagous rather than a parasitoid.

Megaselia is by far the largest genus, with around 220 British species; some species such as *M. rufipes* are frequently found in houses. *Conicera tibialis* is the ominously named coffin-fly; its larvae have been found in long-dead human bodies though it is also found on carrion.

Identification: Fortunately the members of this traditionally 'difficult' group can now be identified using Disney (1983), with Disney (1989) for the genus *Megaselia*. For a general biological account of the family see Disney (1994). Recent additions to the British fauna are listed by Chandler (2010).

British genera: *Aenigmatias, Anevrina, Beckerina, Borophaga, Chaetopleurophora, Chonocephalus, Conicera, Diplonevra, Dohrniphora, Gymnophora, Gymnoptera, Hypocera, Megaselia, Metopina, Obscuriphora, Phalacrotophora, Phora, Plectanocnema, Pseudacteon, Puliciphora, Spiniphora, Triphleba, Woodiphora*

Family Platypezidae (10 genera, 33 species)

This family has also been known as the Clythiidae. They are known as flat-footed flies along with the Opetiidae, which were previously included here. Most species are small but often showing sexual dimorphism in coloration, with the males being mainly dark while the females often have brighter colours or patterns. Like the Phoridae they can be seen running around on vegetation, often in damp situations; the minute species of *Microsania* are attracted to the smoke of bonfires. Platypezid larvae are flattened with fringe-like processes along the body; they live in various habitats such as rotting vegetation or on roots in the soil.

Identification: The British species were covered by Chandler (2001); see also Chandler (2010) for additional references.

British subfamilies and genera:
Callomyiinae: *Agathomyia, Callomyia*
Microsaniinae: *Microsania*
Platypezinae: *Bolopus, Lindneromyia, Paraplatypeza, Platypeza, Polyporivora, Protoclythia, Seri*

SUPERFAMILY SYRPHOIDEA

Family Pipunculidae (11 genera, 93 species)

This family was previously known as the Dorilaidae or Dorylaidae; they are sometimes known as big-headed flies because of their relatively huge heads, which are made up almost entirely of the eyes (Fig. 19.12) though the flies are generally small. They are closely related to the Syrphidae and are just as good

Fig. 19.12 Pipunculid showing the enormous eyes (Photo: Roger Key)

Fig. 19.13 Hoverfly (Syrphidae) (Photo: Peter Barnard)

at hovering as the true hoverflies, but they are generally smaller and often stay lower to the ground so are less conspicuous. Some species can even hover inside the entomologist's glass collecting-tube. The larvae are remarkable for being internal parasitoids of hemipteran leafhoppers and planthoppers, and the large adult eyes are probably adapted for spotting these often cryptically coloured hosts. Eggs are laid internally in the bug using the female fly's long, pointed ovipositor, and the developing larva eats the entire host body contents, after which it enters the soil to pupate.

Dorylomorpha clavifemora (Clubbed big-headed fly) is on the UKBAP list.

Identification: Coe's (1966) handbook is still useful but several species have been added to the British list; for additional references see Chandler (2010).

British subfamilies and genera:
    Chalarinae: *Chalarus, Jassidophaga, Verrallia*
    Nephrocerinae: *Nephrocerus*
    Pipunculinae: *Cephalops, Cephalosphaera, Dorylomorpha, Eudorylas, Microcephalops, Pipunculus, Tomosvaryella*

Family Syrphidae (70 genera, 276 species)
The hoverflies are probably the best known family of Diptera, partly because of their conspicuous habit of hovering around flowers and also because of their bright colours (Fig. 19.13). The wasp-like coloration of some common species frequently misleads the layman into thinking that they are harmful, but they are generally considered beneficial garden species. Many species regularly visit flowers to feed on nectar and pollen, and they can be an important pollinating group. Even experi-

Fig. 19.14 Hoverfly larva (Syrphidae) (Photo: Roger Key)

enced entomologists sometimes assume that all syrphid larvae are predators on aphids and similar plant pests; in fact they exploit a very wide range of habitats and food sources. The predatory species are often brightly coloured with distinct patterns (Fig. 19.14). Some of the commonly seen species such as *Episyrphus balteatus* and *Syrphus ribesii* are indeed predatory on aphids and related insects; others are mycophagous, while some feed on sap-runs or in decaying wood or vegetation. A few such as *Cheilosia* species feed on living plants, while the larvae of *Merodon equestris* (Fig. 19.15) can be a pest in the bulbs of onions or ornamental plants such as daffodils. Some, like *Eristalis* species, are entirely aquatic; the larvae live in stagnant water such as ditches, filter-feeding on organic particles and they

Fig. 19.15 *Merodon equestris* (Syrphidae) (Photo: Roger Key)

Fig. 19.17 *Episyrphus balteatus* congregation and migration (Syrphidae) (Photo: Peter Barnard)

Fig. 19.16 Hoverfly venation (Syrphidae) (Photo: Colin Rew)

Fig. 19.18 *Rhingia campestris* (Syrphidae) (Photo: Robin Williams)

breathe by means of a long extendable tube that gives them the name of rat-tailed maggots. Perhaps the strangest larvae are those of *Microdon*; they live in ants' nests, feeding on discarded food particles and they so closely resemble small slugs that they have mistakenly been described as molluscs by earlier biologists.

Apart from their usually conspicuous colours and behaviour, hoverflies can be recognized by at least one easily observed charcter: vein M1 does not reach the wing margin but curves forward to meet the radial sector, resulting in a more or less continuous vein running pararallel to the inconspicuous unthickened hind margin (Fig.

19.16). Some common species such as *Episyrphus balteatus* can migrate in huge numbers to find new food sources, and often congregate in groups on flower-heads (Fig. 19.17). An unusual-looking but common species is *Rhingia campestris* (Fig. 19.18) with a very conspicuous 'snout' on the front of the head.

Blera fallax, Callicera spinolae (Golden hoverfly), *Chrysotoxum octomaculatum* (Broken-banded wasp-hoverfly), *Doros profuges* (Phantom hoverfly), *Eristalis cryptarum* (Bog hoverfly), *Hammerschmidtia ferruginea* (Aspen hoverfly) and *Myolepta potens* (Western wood-vase hoverfly) are on the UKBAP list.

Identification: Stubbs and Falk (2002) cover nearly all the British species; additions are listed by Chandler (2010). A useful brief introduction to the group is by Gilbert (1986) and an important guide to the larvae is Rotheray (1993). Veen (2004) also covers the British species, and has a different range of illustrations. Despite (or perhaps because of) the fact that this is a well-studied family there is currently no agreement about subdivisions within the family. There is much useful information about this family on the Hoverfly Recording Scheme website (http://www.hoverfly.org.uk).

British genera: *Anasimyia, Arctophila, Baccha, Blera, Brachyopa, Brachypalpoides, Brachypalpus, Caliprobola, Callicera, Chalcosyrphus, Chamaesyrphus, Cheilosia, Chrysogaster, Chrysotoxum, Criorhina, Dasysyrphus, Didea, Doros, Epistrophe, Episyrphus, Eriozona, Eristalinus, Eristalis, Eumerus, Eupeodes, Ferdinandea, Hammerschmidtia, Helophilus, Heringia, Lejogaster, Lejops, Leucozona, Mallota, Melangyna, Melanogaster, Melanostoma, Meligramma, Meliscaeva, Merodon, Microdon, Myathropa, Myolepta, Neoascia, Orthonevra, Paragus, Parasyrphus, Parhelophilus, Pelecocera, Pipiza, Pipizella, Platycheirus, Pocota, Portevinia, Psilota, Pyrophaena, Rhingia, Riponnensia, Scaeva, Sericomyia, Sphaerophoria, Sphegina, Syritta, Syrphus, Trichopsomyia, Triglyphus, Tropidia, Volucella, Xanthandrus, Xanthogramma, Xylota*

Infraorder Muscomorpha (Schizophora Acalyptratae)

If the Schizophora are accepted as a monophyletic group then their current subdivision leads to another pair of convenience groups, essentially based on the presence or absence of a small group of characters: the Calyptratae have well-developed calypters or squamae behind the bases of the wings, they have a complete transverse suture on the thorax, and a cleft in the antennal pedicel; the Acalyptratae lack these features.

Most are small flies, often cryptically coloured, although some families have strongly patterned wings. Many are seen as taxonomically difficult groups, which have not attracted the attention of amateur entomologists, so there is still a great deal of work to be done on many of these families, many of which show a huge variety of life-styles

From a practical point of view one could say that the Acalyptratae share another plesiomorphic character, a lack of identification guides. Many of these families are not covered by recent handbooks or any other faunal works, and it is often necessary to consult specialized references in a variety of scientific journals; see Wyatt and Chainey (1999), Oosterbroek (2006) and Chandler (2010) for lists of these references.

SUPERFAMILY CARNOIDEA

Family Braulidae (1 genus, 2 species)
The relationships of this family are unclear, and some previous authors placed them in the superfamily Ephydroidea; they are commonly known rather misleadingly as bee-lice. There are just three species in Europe; all are small flattened insects with no wings or halteres and with highly reduced eyes; the legs are long and the tarsal claws form combs of spines. All these adaptations enable them to cling like a louse or a mite to the hairs of their hosts, which are honey-bees. The larvae develop in the combs of the hive and feed on honey and pollen, but there is little evidence that these commensals cause any great damage to commercial bee-keeping; they may be less common than before because of control measues against *Varroa* mites.

Identification: see Chandler (2010).

British genus: *Braula*

Family Canacidae (2 genera, 2 species)
This family has also been known as the Canaceidae; some authors include the Tethinidae within this family. There are just four species in Europe, of which two are known from Britain. They are all small, greyish-brown flies that live on the sea-shore, where their larvae feed in the intertidal zone on algae and organic detritus.

Identification: see Chandler (2010).

British genera: *Canace, Xanthocanace*

Family Carnidae (2 genera, 13 species)
This family was formerly spelled Carniidae, and they were sometimes treated as a subfamily of the Milichiidae. They are all small, shiny black flies, rarely more than 2 mm long. The larvae of *Meoneura* feed in decaying organic matter, including dung, and the adults are found near such habitats. *Carnus* species live in birds' nests, where the larvae feed on general detritus in the nest. Newly emerged adults have normal wings, which they shed once a suitable nest for oviposition has been found; they then attach themselves to young birds and apparently feed on their blood using their proboscis.

Identification: Hennig's (1937) European review is rather out of date; see Chandler (2010) for additional references.

Fig. 19.19 *Meromyza pratorum* (Chloropidae) (Photo: Roger Key)

British genera: *Carnus, Meoneura*

Family Chloropidae (39 genera, 177 species)
The members of this large family are sometimes known as grass-flies, and a few species of economic significance have received individual names. They are small flies, usually either entirely black or with yellow or green markings (Fig. 19.19). The larvae of some species are found in decaying vegetation or rotting wood and similar habitats, but many feed on living plants. Some of these cause the formation of galls, such as species of *Lipara* on reeds (*Arundo*); these cigar-shaped galls (Fig. 19.20) often provide homes to other flies, including the chloropid *Cryptonevra flavitarsis*, which seems to be a regular inquiline.

The most notorious species is *Oscinella frit*, the Frit-fly, whose larvae can cause damage to cereal crops; the adult fly is black and just over 1 mm long. *Thaumatomyia notata*, the Small cluster-fly, frequently enters houses in large numbers in the autumn in order to over-winter; preferred buildings can be invaded every year. One unusual species is *Camarota curvipennis*, in which the wings curve back over the body almost like beetle elytra (Fig. 19.21).

Lipara similis (Cigarillo gall-fly) is on the UKBAP list.

Identification: there are no recent works covering the British fauna; see Chandler (2010) for specialized references. A useful review of the classification is by Andersson (1977).

British subfamilies and genera:
Chloropinae: *Camarota, Cetema, Chlorops, Cryptonevra, Diplotoxa, Epichlorops, Eurina, Eutropha,*

Fig. 19.20 *Lipara lucens* gall (Chloropidae) (Photo: Roger Key)

Fig. 19.21 *Camarota curvipennis* (Chloropidae) (Photo: Roger Key)

Lasiosina, Melanum, Meromyza, Neohaplegis, Platycephala, Pseudopachychaeta, Thaumatomyia

Oscinellinae: *Aphanotrigonum, Calamoncosis, Conioscinella, Dicraeus, Elachiptera, Eribolus, Gampsocera, Gaurax, Hapleginella, Incertella, Lasiambia, Lipara, Melanochaeta, Microcercis, Oscinella, Oscinimorpha, Oscinisoma, Polyodaspis, Rhopalopterum, Siphonella, Siphunculina, Speccafrons, Trachysiphonella, Tricimba*

Family Milichiidae (6 genera, 19 species)
The members of this family are sometimes known as jackal-flies, because some species have been observed waiting around near predators such as Asilidae, wasps and spiders to steal a surreptitious meal. Both larvae and adults feed on liquids and the larvae are found in dung, sap-runs and similar habitats. Some adults are known to feed on nectar, as well as imbibing fluids from carcasses or dung.

Identification: Hennig's (1937) European review is rather out of date; see Chandler (2010) for additional references.

British genera: *Desmometopa, Leptometopa, Madiza, Milichia, Neophyllomyza, Phyllomyza*

Family Tethinidae (3 genera, 10 species)
Some authors would place this group within the family Canacidae. Like the members of the latter family they are small yellowish or grey flies, only 2–3 mm long; they have a similar life-style, being found on the sea-shore, saltmarshes or other saline habitats, even in some polluted waters.

Identification: Chandler (2010) gives the necessary specialized references (under the family Canacidae).

British subfamilies and genera:
  Pelomyiinae: *Pelomyia, Pelomyiella*
  Tethininae: *Tethina*

SUPERFAMILY CONOPOIDEA

Family Conopidae (7 genera, 23 species)
Although the members of this family have no common name, many are brightly coloured species, often resembling solitary wasps. Their larvae are internal parasitoids of bees and wasps, the female conopid laying a single egg on the host, which she grasps with special appendages on her abdomen. The developing larva feeds on the haemolymph until the host dies, when the larva pupates and remains in the host body over the winter. The adult flies are usually seen on flowers, especially in the sunshine. Most members of the Conopinae have striking black and yellow markings (Fig. 19.22) whereas many of the Myopinae are smaller and more dull in colour, although some also have brighter markings (Fig. 19.23).

Identification: Smith's (1969a) handbook covers all the British species, but see Chandler (2010) for additional references.

British subfamilies and genera:
  Conopinae: *Conops, Leopoldius, Physocephala*
  Myopinae: *Myopa, Sicus, Thecophora, Zodion*

Fig. 19.22 *Conops quadrifasciatus* (Conopidae) (Photo: Roger Key)

Fig. 19.23 *Sicus ferrugineus* (Conopidae) (Photo: Peter Barnard)

SUPERFAMILY DIOPSOIDEA

Family Megamerinidae (1 genus, 1 species)
The only British species in this family is also the only one in Europe, *Megamerina dolium* (Fig. 19.24); it resembles an ichneumonid wasp and can be seen running around on vegetation in mature woodland. The larvae prey on other insects beneath the bark of dead or dying trees.

Identification: see Chandler (2010).

British genus: *Megamerina*

Family Psilidae (5 genera, 26 species)
Although this family has no common name it includes some species of *Psila* that are well known as agricultural or horticultural pests such as

Fig. 19.24 *Megamerina dolium* (Megamerinidae) (Photo: Roger Key)

Fig. 19.25 *Loxocera albiseta* (Psilidae) (Photo: Roger Key)

Ps. rosae, the Carrot fly. This lays its eggs in the soil near its host plant, and the larvae burrow into the roots; the direct feeding damage causes flaking of the carrot, and the leaves wilt and change colour. Species of *Loxocera* (Fig. 19.25) are very narrow-bodied flies with red and black coloration and they often resemble ichneumonid wasps; all have phytophagous larvae.

Identification: Goot and Veen (1996) cover the species of north-west Europe, but see Chandler (2010) for recent references.

British subfamilies and genera:
Loxocerinae: *Loxocera*
Psilinae: *Chamaepsila, Chyliza, Psila, Psilosoma*

Family Strongylophthalmyiidae (1 genus, 1 species)
This small family is sometimes included within the Tanypezidae. The sole British species is *Strongylopthalmyia ustulata*, and there is only one other species in Europe. This small black fly was not discovered in Britain until the 1980s and little is known about its biology. Its larvae are usually found under the bark of dead or moribund trees, and the adults are found near such hosts in damp woodland.

Identification: see Chandler (2010) for specialist references.

British genus: *Strongylopthalmyia*

Family Tanypezidae (1 genus, 1 species)
The sole British species in this family, *Tanypeza longimana*, is also the only one known in Europe. It is not often seen and its biology is not well studied, though the larvae are known to feed on rotting vegetation in damp woodland, often near standing water.

Identification: see Chandler (2010) for specialist references.

British genus: *Tanypeza*

SUPERFAMILY EPHYDROIDEA

Family Camillidae (1 genus, 5 species)
All the members of this family are small, dark flies, around 2–3 mm long. There are just five species recorded from Britain, with only three more known from Europe, all in the genus *Camilla*. Their biology is little studied, but the larvae apparently feed on organic matter; some species seem to be associated with rabbit burrows where they may feed on dung. The adults are usually seen visiting flowers.

Identification: see Chandler (2010) for specialized references.

British genus: *Camilla*

Family Campichoetidae (1 genus, 2 species)
This group was previously regarded as a subfamily of the Diastatidae. Members of the two families look rather similar though there are reliable small morphological differences; the life histories are similar, as far as is known. Adult campichoetids are often found by sweeping low vegetation in damp situations and most have some dark wing markings, which may be used in courtship displays, as in many other acalyptrate families. The larvae are assumed to feed in decaying vegetation.

191

Identification: see Chandler (2010) for specialized references.

British genus: *Campichoeta*

Family Diastatidae (1 genus, 6 species)

Much of the information about the Campichoetidae also applies to the Diastatidae. The adults are found in similar damp places such as marshy woodland and they also have dark wing markings which are different in most species. The larvae are associated with decaying vegetation, though very little is known about the life histories of most species.

Identification: see Chandler (2010) for specialized references.

British genus: *Diastata*

Family Drosophilidae (8 genera, 62 species)

Of the small group of families in the Ephydroidea, the Drosophilidae are perhaps the best known to non-specialists, if only for their nuisance value when they fly around ripe fruit, compost bins or glasses of wine and beer. Generally known as fruit-flies or vinegar-flies, most of their larvae feed on the micro-organisms in rotting fruits, fungi or other vegetable matter; others are found on sap-runs, and some feed in dung or carrion. One remarkable species, *Cacoxenus indagator*, is a parasitoid on solitary bees such as *Osmia*, while *Acletoxenus formosus* is a predator on Aleyrodidae (whiteflies). The adult drosophilids that are commonly seen in domestic situations are small yellow or brown flies, often striped, rather rounded in shape, and with red eyes. Although not strong or fast fliers, their persistence in hovering around attractive odours makes them a source of irritation. Species of *Drosophila*, especially *D. melanogaster*, have become of legendary significance in genetic studies, because they are easy to breed in laboratory conditions on artificial substrates, they have a rapid lifecycle and so-called 'giant' polytene chromosomes that are easy to observe; see O'Grady (2009) for a brief review.

Amiota variegata (Variegated fruit-fly) is on the UKBAP list.

Identification: Bächli et al. (2004) covered the north European species, but see Chandler (2010) for additional specialized references. Shorrocks (1972) gives a useful overview of the genus *Drosophila*.

British subfamilies and genera:
Drosophilinae: *Chymomyza, Drosophila, Scaptomyza*
Steganinae: *Acletoxenus, Amiota, Cacoxenus, Leucophenga, Stegana*

Fig. 19.26 *Ochthera manicata* (Ephydridae) (Photo: Roger Key)

Family Ephydridae (40 genera, 147 species)

The members of this large family are usually found in damp or aquatic habitats but, despite their common name of shore-flies, many species are found at inland sites by ponds, streams and marshes.

The small adult flies are often reluctant to take to the wing and can be seen running around on mud and similar habitats, often in large numbers. Several species are predatory as adults, some with raptorial front legs (Fig. 19.26). Some larvae are genuinely aquatic, living amongst filamentous algae and a few species are tolerant of organic pollution. Several are found in saltmarshes, and many of the aquatic species are phytophagous, even mining into the stems and leaves of aquatic or emergent plants. The larvae of *Notiphila* species tap into the air-supplies of plants, like some beetle larvae. Among the most commonly seen shore-flies are species of *Psilopa*, very small but brightly metallic, and often seen in large numbers around ponds and streams.

Identification: Becker's (1926) review is very out of date; see Chandler (2010) for additional references. Dahl (1959) is useful for some species as is Olafsson (1991); see also Zatwarnicki (1997) for the north European species.

British subfamilies and genera:
Discomyzinae: *Clanoneurum, Discomyza, Psilopa, Trimerina*
Ephydrinae: *Coenia, Ephydra, Eutaenionotum, Haloscatella, Lamproscatella, Limnellia, Paracoenia, Parydra, Philotelma, Scatella, Scatophila, Setacera, Teichomyza*
Gymnomyzinae: *Allotrichoma, Athyroglossa, Diclasiopa, Discocerina, Ditrichophora, Glenanthe, Gymno-*

clasiopa, Hecamede, Hecamedoides, Mosillus, Ochthera, Polytrichophora

Hydrelliinae: *Atissa, Hydrellia, Notiphila, Schema*

Ilytheinae: *Axysta, Hyadina, Ilythea, Nostima, Parydroptera, Pelina, Philygria*

SUPERFAMILY LAUXANIOIDEA

Family Chamaemyiidae (7 genera, 32 species)
This family has also been known as the Ochthiphilidae, and its members are sometimes called silver-flies. They are small species, often only 2–3 mm long, with a silvery grey appearance. The larvae of all species are specialized predators on the Hemipteran families of aphids, scale-insects, adelgids and psyllids; in this they resemble the activities of hoverfly larvae though they are much smaller. Adult chamaemyiids can often be swept from low vegetation and grassland where the larval prey is found. At least one species, *Chamaemyia flavipalpis*, is commonly found on the coast, often in sand-dune areas.

Identification: see Chandler (2010) for the necessary specialized references.

British subfamilies and genera:
Chamaemyiinae: *Acrometopia, Chamaemyia, Parochthiphila*
Leucopinae: *Leucopis, Leucopomyia, Lipoleucopis, Neoleucopis*

Family Lauxaniidae (13 genera, 54 species)
This family was formerly known as the Sapromyzidae, perhaps appropriately as many species are saprophagous as larvae. Adult lauxaniids are small flies, but sometimes brightly coloured or metallic. The larvae are generally found in decaying vegetation, where they probably feed on associated microorganisms.

Identification: see Chandler (2010) for the necessary specialized references.

British subfamilies and genera:
Homoneurinae: *Homoneura*
Lauxaniinae: *Aulogastromyia, Calliopum, Cnemacantha, Lauxania, Lyciella, Minettia, Peplomyza, Poecilolycia, Sapromyza, Sapromyzosoma, Tricholauxania, Trigonometopus*

SUPERFAMILY NERIOIDEA

Family Micropezidae (5 genera, 10 species)
This family was previously known as the Tylidae, they have the common name of stilt-legged flies or

Fig. 19.27 *Micropeza lateralis* (Micropezidae) (Photo: Roger Key)

just stilt-flies, from the characteristic way in which they stand, raised up on their long legs (Fig. 19.27). They are all very slender-bodied flies with small heads and are often found in damp places where they prey on other insects such as aphids or small Diptera such as Chironomidae. The larvae fall into three main groups: most are saprophagous, found in rotting vegetation; some are confined to decaying wood in ancient forests; others are phytophagous, feeding on the root-nodules of legumes.

Identification: Goot and Veen (1996) cover the species of north-west Europe; see Chandler (2010) for additional references.

British subfamilies and genera:
Calobatinae: *Calobata, Cnodacophora, Neria*
Micropezinae: *Micropeza*
Taeniapterinae: *Rainieria*

Family Pseudopomyzidae (1 genus, 1 species)
The sole British species, *Pseudopomyza atrimana*, is also the only the only species found in Europe; it was first found in Britain in the 1980s. It is a very small, dark fly, around 2 mm long, found in woodland, but its biology is little known, despite being quite widespread in continental Europe. It seems to be associated with rotting wood.

Identification: see Chandler (2010).

British genus: *Pseudopomyza*

SUPERFAMILY OPOMYZOIDEA

Family Acartophthalmidae (1 genus, 2 species)
This group was formerly included within the Clusiidae. The two British species are small dark

grey flies, around 2 mm long, and there is just one other species in Europe. The adults are often seen congregating on fungi, and the larvae are associated with rotting vegetation and wood, and also carrion.

Identification: Czerny's (1928) review covered this family within the Clusiidae, but the work is out of date; see Chandler (2010) for additional notes.

British genus: *Acartophthalmus*

Family Agromyzidae (19 genera, 392 species)

All the species in this large family have phytophagous larvae; most are leaf-miners, nearly always on angiosperms but with a few on pteridophytes, and some cause the formation of galls. They are generally very small, rather inconspicuous flies, and the adults are not often seen apart from some common species that can be swept in large numbers from grassland and low herbage. A more common way to study this group is to collect the mined leaves from the host plants and to rear the adults out. Most species are more-or-less monophagous, restricted to just a few species in one genus of plants, and this restriction aids the identification of species. Female agromyzids are known to pierce plants with their ovipositor and then to sample the sap by tasting, presumably to check the identification of the correct host plant before oviposition.

Inevitably in such a large phytophagous group, several species have become pests on various agriculturally and horticulturally important plants, from cereal crops to asparagus and willow trees. A well-known and very common species is *Phytomyza ilicis*, which forms blotch-mines on holly-tree leaves (Fig. 19.28); another common pest is *Chromatomyia syngenesiae*, the Chrysanthemum leaf-miner.

Fig. 19.28 *Phytomyza ilicis* mine (Agromyzidae) (Photo: Roger Key)

Identification: Spencer's (1972) handbook covered most of the British species, supplemented by Spencer (1976), but see also Chandler (2010) for the numerous additional references.

British subfamilies and genera:
Agromyzinae: *Agromyza, Hexomyza, Melanagromyza, Ophiomyia*

Phytomyzinae: *Amauromyza, Aulagromyza, Calycomyza, Cerodontha, Chromatomyia, Galiomyza, Gymnophytomyza, Liriomyza, Metopomyza, Napomyza, Nemorimyza, Phytobia, Phytoliriomyza, Phytomyza, Pseudonapomyza*

Family Anthomyzidae (7 genera, 20 species)

These small flies are mostly less than 3 mm long with rather narrow wings; they are often found in damp habitats including true wetlands, with some species known from sand dunes, though not exclusively. Most of the larvae are phytophagous, particularly on monocotyledons such as grasses and several marsh-loving plants like sedges and rushes; *Anthomyza gracilis* is one of the more common species. Larvae are often found in the leaf sheaths, while a few species are true miners, and some are known as inquilines in galls caused by other families such as Chloropidae. *Fungomyza albimana* is unusual in feeding on decaying fungi.

Identification: see Chandler (2010) for the necessary specialized references.

British genera: *Anagnota, Anthomyza, Cercagnota, Fungomyza, Paranthomyza, Stiphrosoma, Typhamyza*

Family Asteiidae (3 genera, 8 species)

The members of this family are small species of dark flies, sometimes with yellow markings, and usually under 3 mm in length. The adults can be found on flowers, sometimes on sap runs, and some congregate in large numbers in the autumn. The larvae are usually found in decaying plant material; at least one species is found on coastal marshes, and some are associated with rotting vetetation in marram grass on sand dunes. *Astiosoma rufifrons* has been found to be attracted to smouldering wood ash following bonfires; *Leiomyza* species are found on decaying fungi, like *Fungomyza* in the Anthomyzidae.

Identification: see Chandler (2010) for the specialized references.

British subfamilies and genera:
Asteiinae: *Asteia*
Sigaloessinae: *Astiosoma, Leiomyza*

Fig. 19.29 *Clusia flava* (Clusiidae) (Photo: Roger Key)

Fig. 19.30 *Opomyza florum* (Opomyzidae) (Photo: Roger Key)

Family Aulacigastridae (1 genus, 1 species)

The only British species, *Aulacigaster leucopeza*, is a distinctive small brownish fly, 2–3 mm long, which has reddish eyes with iridescent banding. It is found only found in sap runs on a variety of trees, where the larvae presumably feed on micro-organisms in the decaying sap. The adult flies may hibernate in woodland.

Identification: see Chandler (2010) for notes and specialized references.

British genus: *Aulacigaster*

Family Clusiidae (4 genera, 10 species)

This family was previously known as the Heteroneuridae. All are small flies of various colours, some being mainly black, others pale yellow (Fig. 19.29), and the adults are usually found in damp deciduous woods though they are never conspicuous. The larvae are basically saproxylic, found in rotting wood or occasionally in sap runs.

Clusiodes geomyzinus (Strathspey clusiid fly) is on the UKBAP list.

Identification: Czerny's (1928) review is very out of date and Chandler (2010) should be consulted for recent references.

British genera: *Clusia, Clusiodes, Heteromeringia, Paraclusia*

Family Odiniidae (1 genus, 9 species)

The members of this family are all small greyish flies, which can be found in woodland near to tree-wounds or fungi, although none seems to be particularly common. Their larvae are saproxylic, and some are found with other insect larvae such as wood-boring beetles either in wood or in fungi, but the details of these associations are not clear; later instars may actually prey on these other insects. *Odinia pomona* has only been found under the bark of apple trees, and is apparently one of the few species of Diptera to be dependent on fruit trees.

Identification: see Chandler (2010) for the necessary specialized references.

British genus: *Odinia*

Family Opomyzidae (2 genera, 17 species)

These small flies are slender-bodied and of varying colours of yellow, brown or black (Fig. 19.30) but always with marked or spotted wings. Their phytophagous larvae all live in the stems of grasses of various kinds, including some cereal crops; several species are common and widespread, wherever the food plants are to be found.

Identification: Drake (1993) gave a key to the British species; see also Chandler (2010) for additional references.

British genera: *Geomyza, Opomyza*

Family Periscelididae (1 genus, 3 species)

This family has also been known as the Periscelidae; there are only four species in Europe, which were previously all on the British list, but it is now clear that there are no genuine records of *Periscelis annulipes*. The larvae are all saproxylic, and adults have been seen feeding on sap-runs, but none of the British species seems to be common or widespread.

Identification: Duda's (1934) key covered the British species.

British genus: *Periscelis*

Family Stenomicridae (1 genus, 2 species)
The members of this family have been variously included in the Aulacigastridae, the Periscelididae or the Anthomyzidae. There are only two species recorded from Britain, with just one more known in Europe. They are very small flies, less than 2 mm long, greyish-yellow in colour and usually found in wetlands or damp grassland. Although their biology has been little studied, the larvae apparently live in the water at the bases of *Carex* tussocks.

Identification: see Chandler (2010) for the necessary specialized references.

British genus: *Stenomicra*

SUPERFAMILY SCIOMYZOIDEA

Family Coelopidae (2 genera, 3 species)
These are known as seaweed-flies, kelp-flies or wrack-flies because the larvae feed on rotting seaweed on the shore. The adults are small and dark coloured, often seen in large numbers around the larval habitat and can breed all through the year. *Coelopa frigida* is one of the most widespread and common species. *Malacomyia sciomyzina* was previously in the Dryomyzidae, and authors who recognize the Helcomyzinae as a separate family would place it there.

Identification: see Chandler (2010) for the necessary specialized references.

British genera: *Coelopa, Malacomyia*

Family Dryomyzidae (4 genera, 5 species)
The Helcomyzinae are sometimes regarded as a separate family, the Helcomyzidae, and some authors suggest that *Heterocheila* should be in its own family, the Heterocheilidae. The Dryomyzinae are yellowish or brownish flies that breed in dung, carrion or rotting fungi (Fig. 19.31). The Helcomyzinae (including *Heterocheila buccata*, the only European species in the genus) have similar habits to the Coelopidae, breeding in seaweed although some species prefer slightly drier habitats further up the shore. At 6–10 mm they are also larger than the Coelopidae and are greyish in colour.

Identification: see Chandler (2010) for the necessary specialized references.

Fig. 19.31 *Neuroctena anilis* (Dryomyzidae) (Photo: Robin Williams)

British subfamilies and genera:
Dryomyzinae: *Dryomyza, Neuroctena*
Helcomyzinae: *Helcomyza, Heterocheila*

Family Phaeomyiidae (1 genus, 2 species)
This group was previously placed within the Sciomyzidae, but the morphological separation from that family has been confirmed by the discovery that the phaeomyiids are parasitoids on millipedes. The larva develops internally in the host, which remains alive until the fully grown larva emerges to pupate in the soil. The adults are occasionally seen visiting woodland flowers; they are yellowish-brown species with brownish wings.

Identification: see Chandler (2010) for the necessary specialized references.

British genus: *Pelidnoptera*

Family Sciomyzidae (23 genera, 68 species)
The dramatic common name of snail-killing flies indicates the habits of this moderately large family; all are parasitoids or predators on snails and slugs, sometimes on their eggs. They are always found in damp or marshy places, and many larvae are truly aquatic; these species tend to be active predators on freshwater snails. The terrestrial species are often parasitoids, developing inside the slug or snail hosts from which they emerge to pupate. There is still much work to do on the biology of this group, particularly in studying how host-specific the sciomyzids are. Some adults are quite striking species with mottled wings (Fig. 19.32) and these resemble Platystomatidae.

Salticella fasciata (Dune snail-killing fly) is on the UKBAP list.

Fig. 19.32 *Coremacera marginata* (Sciomyzidae) (Photo: Roger Key)

Fig. 19.33 *Sepsis punctum* (Sepsidae) (Photo: Roger Key)

Identification: Rozkosny (1984) covers most of the British species, but see Chandler (2010) for the necessary additional references.

British subfamilies and genera:

Salticellinae: *Salticella*

Sciomyzinae: *Anticheta, Colobaea, Coremacera, Dichetophora, Dictya, Ditaeniella, Ectinocera, Elgiva, Euthycera, Hydromya, Ilione, Limnia, Pherbellia, Pherbina, Psacadina, Pteromicra, Renocera, Sciomyza, Sepedon, Tetanocera, Tetanura, Trypetoptera*

Family Sepsidae (6 genera, 29 species)
This family has no common name; the flies are small and rather ant-like, metallic and mostly black in colour and *Sepsis* species have a dark spot at the wing-tip (Fig. 19.33). They are most often seen running around on vegetation, sometimes in large

numbers, constantly 'waving' their wings as part of their courtship display. They are generally found in damp places where the larvae feed on various organic materials such as rotting vegetation, fungi and dung. Some can be found in wet mud at the side of freshwater bodies and they may be tolerant of some organic pollution. *Orygma luctuosum* is unusual in being found on the seashore where its larvae live in seaweed; it was formerly included in the Coelopidae.

Identification: Pont's (1979) handbook covers most of the British species, but see also Chandler (2010) for additional references.

British genera: *Meroplius, Nemopoda, Orygma, Saltella, Sepsis, Themira*

SUPERFAMILY SPHAEROCEROIDEA

Family Chyromyidae (3 genera, 11 species)
Next to nothing is known about the lifecycles of these flies. The adults are small, often yellowish, and some species such as *Chryromya flava* can be very common. They have been found in numerous habitats, both dry and damp, including saltmarshes and sand dunes, and sometimes on flowers. The larvae are associated with rotting organic matter, including bird guano in communal nesting sites, but there is clearly much scope for further work on the larval stages of this family.

Identification: Czerny's (1927) review is very out of date; see Chandler (2010) for the necessary specialized references.

British genera: *Aphaniosoma, Chyromya, Gymnochiromyia*

Family Heleomyzidae (17 genera, 63 species)
This family was previously known as the Helomyzidae or Heteromyzidae; some authors propose that it should be split into several smaller families, including the Borboropsidae, Chiropteromyzidae, Cnemospathidae and Trixoscelididae. It is certainly a diverse family in terms of both morphology and life histories; like many other families in this group the larvae are associated with decaying vegetation, fungi, carrion and dung. Some species have been found in caves, though not exclusively, and some adults hibernate in such habitats. *Neossos nidicola* is associated with bird and bat guano, and *Trixoscelis* species have been reared from birds' nests in dry coastal areas. Most members of the genus *Suillia* are found on fungi, but one or two species seem to be phytophagous. A few members of this

Fig. 19.34 *Tephrochlamys* sp. (Heleomyzidae) (Photo: Roger Key)

Fig. 19.35 A flightless sphaerocerid (Photo: Roger Key)

family regularly enter houses, such as species of *Tephrochlamys* (Fig. 19.34).

A recently introduced species, *Prosopantrum flavifrons*, of South American origin, is sometimes placed in the separate family Cnemospathidae, but this group is also treated as a tribe of the Heleomyzinae.

Identification: see the detailed references in Chandler (2010) (also under Borboropsidae, Chiropteromyzidae, Cnemospathidae and Trixoscelididae).

British subfamilies and genera:

Chiropteromyzinae: *Neossos*

Heleomyzinae: *Borboropsis, Eccoptomera, Gymnomus, Heleomyza, Morpholeria, Neoleria, Oecothea, Oldenbergiella, Prosopantrum, Schroederella, Scoliocentra*

Heteromyzinae: *Heteromyza, Tephrochlaena, Tephrochlamys*

Suilliinae: *Suillia*

Trixoscelidinae: *Trixoscelis*

Family Sphaeroceridae (36 genera, 137 species)
This family, whose members are known as lesser dung-flies, was previously known as the Cypselidae or Borboridae; some authors have suggested that they should be combined with the Heleomyzidae (*sensu stricto*). At 1–5 mm in length, dark brown or black in colour, they are often not noticed around cow dung because the observer's eye is caught by the larger Scatophagidae, but they can be very common and many species can be seen together on a single cow-pat. Like many other dung-feeding insects they are often seen to be carrying mites. Other species are found on fungi and various kinds of rotting organic material, including seaweed; a few are very common on riverine sediments left after receding floodwater. All species are reluctant to fly long distances, and they often merely 'jump' a few centimetres or run rapidly like Phoridae. Wing reduction is common in this family, with some species being completely flightless (Fig. 19.35).

Identification: Pitkin's (1988) handbook covered most of the British species; see Chandler (2010) for many additional references.

British subfamilies and genera:

Copromyzinae: *Alloborborus, Borborillus, Copromyza, Crumomyia, Lotophila, Norrbomia*

Limosininae: *Apteromyia, Archicollinella, Chaetopodella, Coproica, Elachisoma, Gigalimosina, Gonioneura, Herniosina, Leptocera, Limosina, Minilimosina, Opacifrons, Opalimosina, Paralimosina, Philocoprella, Phthitia, Pseudocollinella, Pteremis, Pullimosina, Puncticorpus, Spelobia, Spinilimosina, Telomerina, Terrilimosina, Thoracochaeta, Trachyopella, Xenolimosina*

Sphaerocerinae: *Ischiolepta, Lotobia, Sphaerocera*

SUPERFAMILY TEPHRITOIDEA

Family Lonchaeidae (5 genera, 46 species)
Until recently this family had no common name, but authors have now adopted the American name of lance-flies, based on the long, pointed ovipositor of the female (MacGowan & Rotheray, 2008). All the British species are black or bluish-black in colour, and often shiny, around 3–6 mm long. Although several species are common and widespread, it is

usually only the females that are seen because males form swarms high in the tree canopy. Lonchaeids are unusual amongst acalyptrate families in mating in the air like many Muscoidea, rather than sedentarily on a surface. Larvae of the largest genus, *Lonchaea*, are saproxylic and are most frequently found under the bark of recently fallen trees. Of the other genera whose larvae are known, some are found in decaying vegetation and others are phytophagous, but there is still much work to be done on the biology of this family. A few species seem to be necrophagous on dead or dying insect larvae, as well as being saproxylic. *Lonchaea chorea* and *L. mallochi* are among the most common species in Britain.

Lonchaea ragnari (Viking sword fly) is on the UKBAP list.

Identification: MacGowan and Rotheray's (2008) handbook also has keys to the known early stages.

British subfamilies and genera:
Dasiopinae: *Dasiops*
Lonchaeinae: *Earomyia, Lonchaea, Protearomyia, Setisquamalonchaea*

Family Pallopteridae (2 genera, 13 species)
Although they have no common name, this small family consists of rather distinctive flies; many are pale coloured and all have very long, strongly patterned wings (Fig. 19.36), which they frequently 'wave'. The Pallopteridae, Platystomatidae and Ulidiidae are sometimes loosely known as picture-winged flies. The adults are usually found in shaded damp places, sometimes on flowers, but the larval habitats are not clear for many species; some have been found under tree bark, but it is not known

Fig. 19.36 *Palloptera saltuum* (Pallopteridae) (Photo: Roger Key)

whether they are feeding on fungi or scavenging on other dead insects, and others are phytophagous on umbellifers and other plants, often in the flower-heads.

Identification: see Chandler (2010) for the necessary specialized references.

British genera: *Eurygnathomyia, Palloptera*

Family Piophilidae (12 genera, 14 species)
These were previously known as Neottiophilidae or Thyreophoridae, both now regarded as subfamilies, although some authors would still place the Neottiophilinae as a separate family. All are small, shiny black flies, usually with no wing markings. Most piophilid larvae are found on dung, carrion or fungi, but *Neottiophilum praeustum* lives in birds' nests where it sucks the blood of nestlings. The larva of *Piophila casei* is the Cheese-skipper, sometimes found on cheese and dried meats, but perhaps not as commonly as before with modern food storage. Like the larvae of a few other families it can leap by holding its hind end with its mouth-hooks and suddenly letting go.

Identification: see Chandler (2010) for the necessary specialized references.

British subfamilies and genera:
Neottiophilinae: *Actenoptera, Neottiophilum*
Piophilinae: *Allopiophila, Liopiophila, Mycetaulus, Parapiophila, Piophila, Prochyliza, Protopiophila, Pseudoseps, Stearibia*
Thyreophorinae: *Centrophlebomyia*

Family Platystomatidae (2 genera, 2 species)
The two species in this family are usually found in damp habitats, and are known as signal-flies; they have strongly patterned wings which they 'wave' like several related families. Along with the Pallopteridae and Ulidiidae they are sometimes loosely known as picture-winged flies. The two British species are easily separated: *Platystoma seminationis* has dark wings with small clear spots (Fig. 19.37), whereas *Rivellia syngenesiae* has clear wings with dark cross-bands. The larvae are associated with fungi and rotting vegetation.

Identification: see Chandler (2010) for the necessary specialized references.

British genera: *Platystoma, Rivellia*

Family Tephritidae (34 genera, 76 species)
This family has also been known as Trypetidae, Euribiidae or Trupaneidae; they are sometimes known as fruit-flies, though this can cause confusion with

Fig. 19.37 *Platystoma seminationis* (Platystomatidae) (Photo: Roger Key)

Fig. 19.39 *Oxyna parietina* (Tephritidae) (Photo: Roger Key)

Fig. 19.38 *Platyparaea discoidea* (Tephritidae) (Photo: Roger Key)

Fig. 19.40 *Urophora cardui* gall on thistle (Tephritidae) (Photo: Roger Key)

the Drosophilidae, and the neutral name tephritids is perhaps safer. Like several other families in this group they have strongly patterned wings (Figs. 19.38 & 19.39) but the most significant feature of this family is that the larvae are all phytophagous, some causing gall formation, and with a small number of species being serious agricultural and horticultural pests. Different species develop in many parts of the plant, with some in seed-heads or fruits, others mining in stems and leaves, and some in roots; the commonest plant hosts are composites (Asteraceae). Because the feeding of tephritid larvae can seriously reduce seed production of their hosts, some species have been tested as potential biocontrol agents of composite weeds, such as *Urophora cardui* on thistles (Fig. 19.40), though not in Britain. The pests of spe-

cific plants often have individual common names, including *Euleia heraclei*, the Celery fly and *Ceratitis capitata*, the Mediterranean fruit-fly.

Identification: White's (1988) handbook covers nearly all the British species, but see Chandler (2010) for additional references. The economically important species were described by White and Elson-Harris (1992).

British subfamilies and genera:

Tephritinae: *Acanthiophilus, Acinia, Campiglossa, Chaetorellia, Chaetostomella, Dioxyna, Dithryca, Ensina,*

Merzomyia, Myopites, Noeeta, Orellia, Oxyna, Sphenella, Tephritis, Terellia, Trupanea, Urophora, Xyphosia

Trypetinae: *Acidia, Anomoia, Ceratitis, Chetostoma, Cornutrypeta, Cryptaciura, Euleia, Euphranta, Goniglossum, Philophylla, Platyparea, Plioreocepta, Rhagoletis, Stemonocera, Trypeta*

Family Ulidiidae (11 genera, 20 species)

This family has formerly been known as the Otitidae or Ortalidae. Its members have no common name, though they are sometimes loosely linked with the Pallopteridae and Platystomatidae as picture-winged flies. Not all have the strong banding of typical picture-winged flies, though they exhibit the typical wing-waving of this group of families. They are often to be found in damp areas and the few known larvae are found under tree bark or associated with dung; a few are phytophagous, but the larvae of several species are completely unknown.

Dorycera graminum (Phoenix fly) is on the UKBAP list.

Identification: see Chandler (2010) for the necessary specialized references.

British subfamilies and genera:

Otitinae: *Ceroxys, Dorycera, Herina, Melieria, Myennis, Otites, Seioptera, Tetanops*

Ulidiinae: *Homalocephala, Physiphora, Ulidia*

Infraorder Muscomorpha (Schizophora Calyptratae)

This group is distinguished from the acalyptrates by the presence of a complete mesonotal suture on the thorax, and a dorsal cleft on the antennal pedicel; they usually have well-developed calypters or squamae behind the bases of the wings. Many are hairy or bristly, and this group includes the familiar house-flies (Muscidae), blowflies, bluebottles and greenbottles (Calliphoridae), and dung-flies (Scatophagidae). There are many unusual life-styles, however, with several parasitic groups, both internal and external. Only the Anthomyiidae and Scatophagidae have developed phytophagous larvae, although this habit is much more common in the acalyptrates.

In contrast to the acalyptrates, most of the British families in this group are reasonably well covered by recent literature, although the Anthomyiidae are the main exception.

SUPERFAMILY HIPPOBOSCOIDEA

All members of this group are external parasites on birds or mammals. They are unusual in not laying

Fig. 19.41 Slide preparation of a hippoboscid (Photo: Peter Barnard)

eggs; the single larva develops inside the female's abdomen and pupates immediately after being laid (hence the older name of Pupipara for this superfamily).

Family Hippoboscidae (10 genera, 14 species)

These are generally known as flat-flies or louse-flies (when parasitic on birds) or as keds (when on mammals). The most extremely adapted examples are scarcely recognizable as Diptera, as they are wingless, greatly flattened dorso-ventrally, and with large claws for remaining attached to their hosts (Fig. 19.41). All feed on their host's blood and although they are generally restricted to a small group of host species, few are entirely monoxenous. Some adults are fully winged, at least when newly emerged, and these species fly to locate their hosts; these include the common bird parasites in *Ornithomyia* such as *O. avicularia* (Fig. 19.42) as well as *Hippobosca equina*, the Forest-fly, which used to be common on horses. Other species may have reduced wings, such as *Crataerina* and *Stenepteryx*, again parasites on birds. Some may shed their wings once they have found the right host, and a common example is *Lipoptena cervi*, the Deer-fly (Fig. 19.43), while others such as *Melophagus ovinus*, the Sheep ked are permanently apterous. Several species on the British list are occasional vagrants, and are not established.

Fig. 19.42 *Ornithomya avicularia* (Hippoboscidae)
(Photo: Roger Key)

Fig. 19.44 *Anthomyia* sp. (Anthomyiidae) (Photo: Roger
Key)

Fig. 19.43 *Lipoptena cervi* (Hippoboscidae) (Photo:
Robin Williams)

Identification: Hutson's (1984) handbook covers
all except one species; see Chandler (2010) for additional references.

British subfamilies and genera:
  Hippoboscinae: *Hippobosca*
  Lipopteninae: *Lipoptena, Melophagus*
  Ornithomyinae: *Crataerina, Icosta, Olfersia,
Ornithomya, Ornithophila, Pseudolynchia, Stenepteryx*

Family Nycteribiidae (3 genera, 3 species)
These are known as bat-flies and have been found
on most species of British bats. The adults are even
more bizarre than the wingless hippoboscids, being
spider-like in appearance. All are small and apterous, and the dorsal surface of the thorax is reduced
and membranous, such that the long legs seem to
be attached to the dorsal side of the body; the small

head is bent backwards on the thorax, except when
feeding, which adds to the spidery appearance.

Identification: Hutson's (1984) handbook covers
the British species.

British genera: *Basilia, Nycteribia, Phthiridium*

SUPERFAMILY MUSCOIDEA

Family Anthomyiidae (29 genera, 242 species)
This large family was formerly included within the
Muscidae; most species are grey, black or brownish, sometimes with distinctive markings but apart
from a few common species this is not an easy
group to identify, mainly because there are no
recent British works covering this family (Fig.
19.44). Adult anthomyiids are often seen on
flowers; the larvae of some species are saprophagous in habitats such as dung and fungi, but many
are phytophagous with a few becoming leaf-
miners. Some of these phytophagous species are
pests of economic significance, including *Delia
radicum*, the Cabbage root fly; *D. coarctata*, the
Wheat bulb fly; and *D. antiqua*, the Onion fly. The
larvae of *Eustalomyia* and some other genera are
cleptoparasites of wasps and bees.
  Botanophila fonsecai (Fonseca's seed fly) is on the
UKBAP list.

Identification: Hennig (1966–76) covered the
Palaearctic species, but see Chandler (2010) for
numerous additional references.

British subfamilies and genera:
  Anthomyiinae: *Adia, Anthomyia, Botanophila,
Chiastocheta, Chirosia, Delia, Egle, Eustalomyia,
Fucellia, Heterostylodes, Hydrophoria, Hylemya,*

Hylemyza, Lasiomma, Leucophora, Paregle, Phorbia, Strobilomyia, Subhylemyia, Zaphne

Pegomyinae: *Alliopsis, Calythea, Emmesomyia, Eutrichota, Mycophaga, Myopina, Paradelia, Pegomya, Pegoplata*

Family Fanniidae (2 genera, 60 species)

This family was formerly treated as a subfamily of the Muscidae. Most species are greyish or yellowish in colour and some are frequently seen indoors; these include *Fannia canicularis*, the Lesser house-fly, which has probably overtaken *Musca domestica* (Muscidae) as the commonest synanthropic species, and *F. scalaris*, the Latrine fly. Indoors, *F. canicularis* flies in overlapping three- or four-sided shapes below the ceiling, usually around a central light-fitting; the same behaviour in seen in the swarms of other species of *Fannia* outside, but in all cases the flight is centred on some kind of visual marker. The larvae in this family are saprophagous and are found in a wide variety of habitats wherever suitable micro-organisms are developing, including rotting vegetation, fungi, dung and carrion.

Identification: Fonseca's (1968) handbook is still useful, but see Chandler (2010) for additional references.

British genera: *Fannia, Piezura*

Family Muscidae (40 genera, 282 species)

Generally known as house-flies, although only a few species are synanthropic, most muscids are grey or black, often with a tessellated or chequered pattern. The best known is *Musca domestica*, the Common house-fly, which in Britain is now less common than before, at least in urban areas. It is a cosmopolitan species that is implicated in carrying pathogens from dung and rotting refuse, where its larvae develop, onto unprotected human food. It can still be common in rural parts of Britain where horses or cattle are numerous, but in urban areas it is less frequently seen in houses than *Fannia canicularis* (Fanniidae). Another member of this genus, *M. autumnalis*, is known as the Face-fly from its attraction to sweat, both human and animal; similar behaviour is shown by *Hydrotaea irritans*, the Sweat-fly, which lives up to its specific name when present in large numbers in wooded areas. A more disturbing muscid is *Stomoxys calcitrans*, the Stable-fly, which can bite people persistently when near cattle or horses. Other species are more benign and more attractive, such as *Mesembrina meridiana* (Fig. 19.45). Muscid larvae can be found in wide variety of habitats; many are saprophagous in rotting vegetation or dung, but in some of these the later instar larvae

Fig. 19.45 *Mesembrina meridiana* (Muscidae) (Photo: Robin Williams)

become carnivorous. Other species are entirely carnivorous, and some, such as *Limnophora*, are truly aquatic.

Phaonia jaroschewskii (Hairy canary) is on the UKBAP list.

Identification: Fonseca's (1968) handbook is still useful but out of date; see Chandler (2010) for many additional references. Skidmore (1985) covers many larvae, as well as being a good general account of the biology.

British subfamilies and genera:

Achanthipterinae: *Achanthiptera*

Coenosiinae: *Coenosia, Dexiopsis, Limnophora, Limnospila, Lispe, Lispocephala, Macrorchis, Neolimnophora, Orchisia, Pseudocoenosia, Schoenomyza, Spanochaeta, Spilogona, Villeneuvia*

Muscinae: *Azelia, Drymeia, Eudasyphora, Haematobia, Haematobosca, Hydrotaea, Mesembrina, Morellia, Musca, Muscina, Neomyia, Polietes, Potamia, Pyrellia, Stomoxys, Thricops*

Mydaeinae: *Brontaea, Graphomya, Hebecnema, Mydaea, Myospila*

Phaoniinae: *Atherigona, Helina, Lophosceles, Phaonia*

Family Scathophagidae (23 genera, 54 species)

This family has variously been known as Scatophagidae, Cordyluridae or Scopeumatidae; they are commonly called the dung-flies. The adults vary in coloration, but all are noticeably bristly or hairy and the long wings often have some faint marks or tinges of colour. The species of *Cordilura* are narrow-bodied black flies (Fig. 19.46); their larvae live in damp places such as marshes and bogs, especially on sedges. The best-known species is probably *Scathophaga stercoraria*, the Yellow dung-fly, which is

Fig. 19.46 *Cordilura ciliata* (Scathophagidae) (Photo: Roger Key)

Fig. 19.48 *Lucilia* sp. (Calliphoridae) (Photo: Roger Key)

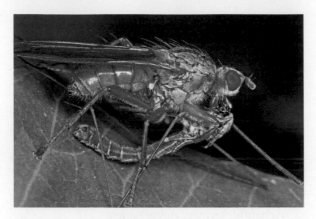

Fig. 19.47 *Scathophaga* eating crane-fly (Scathophagidae) (Photo: Colin Rew)

frequently seen in large numbers on cow-pats. The adults are predators with a preference for Dipteran prey (Fig. 19.47).

Identification: see Chandler (2010) for the necessary specialized references.

British subfamilies and genera:

Delininae: *Delina, Leptopa, Parallelomma*

Scathophaginae: *Acanthocnema, Ceratinostoma, Chaetosa, Cleigastra, Conisternum, Cordilura, Cosmetopus, Ernoneura, Gimnomera, Gonatherus, Hydromyza, Megaphthalma, Microprosopa, Nanna, Norellia, Norellisoma, Pogonota, Scathophaga, Spaziphora, Trichopalpus*

SUPERFAMILY OESTROIDEA

Family Calliphoridae (14 genera, 38 species)
Even with the removal of the Rhinophoridae and Sarcophagidae as separate families, there is still

doubt about the monophyly of the Calliphoridae, commonly known as the blow-flies. Many species are metallic green or blue with a dense covering of bristles. Most larvae in this group are carrion-feeders and many are attracted to vertebrate bodies; hence the larvae of some species, especially *Calliphora vicina,* are of great importance in forensic entomology, where the known rates of development of blowfly species on corpses can be an indicator of time of death. Members of the genus *Calliphora* are known generally as bluebottles, while those of *Lucilia* are the greenbottles (Fig. 19.48); it should be noted that there are similar metallic green flies in other groups, and Colyer and Hammond (1968) provided a small key to distinguish the main genera. The propensity of blowfly larvae to feed on decaying flesh has led to their use in 'maggot therapy' where they are used to clean up wounds; *Lucilia sericata* is the commonest species used in this way. Anglers are also familiar with blowfly larvae, using these 'gentles' as bait.

A rather different calliphorid is *Pollenia rudis,* the Cluster-fly, which has a mat of yellow hairs over the thorax; it frequently enters buildings in large numbers during late autumn in order to hibernate, and its larvae are parasitoids of earthworms.

Identification: Emden's (1954) handbook is still useful, but a more recent treatment by Rognes (1991) covers most of the British species. A useful general guide to the group is by Erzinclioglu (1996).

British subfamilies and genera:

Calliphorinae: *Bellardia, Calliphora, Cynomya*

Chrysomyiinae: *Phormia, Protocalliphora, Protophormia*

Helicoboscinae: *Eurychaeta*

Luciliinae: *Lucilia*
Melanomyinae: *Angioneura, Eggisops, Melanomya, Melinda*
Polleniinae: *Pollenia*
Rhiniinae: *Stomorhina*

Family Oestridae (5 genera, 11 species)

There have always been disputes about the validity of this family, with some authors treating each of the three subfamilies as separate families, but recent evidence seems to support the monophyly of the Oestridae (see Chandler, 2010). They constitute the warble-flies and bot-flies, and all are parasitic on mammals. Females of the Cephenemyiinae such as *Oestrus ovis*, the Sheep nostril-fly, lay larvae rather than eggs and these are ejected into the nostrils of the host, usually sheep or goats; the larvae develop in the nasal cavities or pharynx, leaving via the nostrils to pupate in the soil. The best-known species of *Gasterophilus* is *G. intestinalis*, the Bot-fly, which lays its eggs on the skin of its host, usually a horse. The larvae burrow through the skin and make their way to the intestinal tract usually via the mouth; after development they leave via the rectum to pupate in the soil.

The Hypodermatinae were previously treated as a subgroup of the Calliphoridae; a typical species is *Hypoderma bovis*, the Warble-fly. Adults of this groups are quite large, 10 mm or more in length, rather hairy and almost bee-like; their eggs are laid on the skin of cattle and the larvae burrow through the host body, eventually congregating along the back of the cow and forming swellings called 'warbles' with their posterior spiracles poking through a perforation in the host skin. The associated infection causes distress to the animal and the holes in the cow-hide damage its value for leather production; strict control has now made this group scarce in commercial farming.

Identification: Emden's (1954) handbook covered all except the Gasterophilinae (under Tachinidae); see Chandler (2010) for additional references. Also useful are works by Grunin (1964–9) and Colwell et al. (2004)

British subfamilies and genera:
  Cephenemyiinae: *Cephenemyia, Oestrus, Pharyngomyia*
  Gasterophilinae: *Gasterophilus*
  Hypodermatinae: *Hypoderma*

Family Rhinophoridae (6 genera, 8 species)

Some earlier authors treated this group as a subfamily of the Calliphoridae; they are undistinguished bristly black or grey flies. The larvae are remarkable, however, in that the known species are all internal parasitoids of woodlice. The eggs are not laid on the host, but in suitable sites for the woodlice to live, and the newly hatched larvae actively seek out their hosts. The rhinophorid larva consumes the host from within and then pupates in the empty skin of the woodlouse.

Identification: Emden's (1954) handbook covered all except one British species (under the Calliphoridae); see Chandler (2010) for additional references.

British genera: *Melanophora, Paykullia, Phyto, Rhinophora, Stevenia, Tricogena*

Family Sarcophagidae (15 genera, 60 species)

This group was previously treated by some authors as a subfamily of the Calliphoridae; they are generally known as flesh-flies. Members of the largest genus, *Sarcophaga*, are dark grey with tessellated patterns on the abdomen, noticeably large 'feet' and bright red eyes (Fig. 19.49). Their larvae develop in carrion, dung and similar habitats and the adult flies are commonly seen on flowers. Apart from the saprophagous species there are many other lifestyles in this family, with several species being cleptoparasites of wasps and bees, and others being parasitoids of various insects, spiders and snails. Many sarcophagids are larviparous, rather than oviparous, and the parasitoid species have a modified 'larvipositor' to place their larva in or on the host.

Identification: Emden's (1954) handbook covered this group (under the Calliphoridae) but is now out of date; for a more recent treatment see Pape (1987), which includes most of the British species, and

Fig. 19.49 *Sarcophaga* sp. (Sarcophagidae) (Photo: Roger Key)

Fig. 19.50 Tachinid (Photo: Roger Key)

Fig. 19.51 Tachinid puparium in cocoon (Photo: Roger Key)

Verves (1982–93). Chandler (2010) gives additional references. A useful website for information abou this family is http://www.zmuc.dk/entoweb/sarcoweb/sarcweb/sarc_web.htm.

British subfamilies and genera:

Miltogramminae: *Amobia, Macronychia, Metopia, Miltogramma, Oebalia, Pterella, Senotainia*

Paramacronychiinae: *Agria, Angiometopa, Brachi-coma, Nyctia, Sarcophila*

Sarcophaginae: *Blaesoxipha, Ravinia, Sarcophaga*

Family Tachinidae (141 genera, 261 species)

Earlier authors used the name Larvaevoridae for this group. The classification and identification of species within this large and complex family is not easy and, despite much recent work on the British fauna, there are likely to be many changes in the accepted British list as studies continue. Most adult tachinids are black or grey and very bristly (Fig. 19.50); all their larvae are endoparasitoids on a wide range of other insects. There are four main modes of getting the larvae into the hosts: some female flies lay their eggs directly on the host's integument and the newly hatched larvae burrow into the insect's body; others puncture the host skin and lay their eggs internally; some species lay eggs very near the host and the larvae actively seek them out; the most curious group lays huge numbers of very tiny eggs on leaves, where they will be eaten by suitable Lepidopteran caterpillars. Obviously the latter group cannot be sure exactly which host will pick up the eggs, so host specificity is quite varied throughout the family. In all cases the larvae gradually eat the host from within and eventually pupate, either in the dead host's skin or in the silk cocoon of a lepidopteran larva, which is not killed until its pupal cocoon is finished (Fig. 19.51).

Fig. 19.52 *Tachina grossa* (Tachinidae) (Photo: Roger Key)

Tachina grossa is a particularly large and striking species (Fig. 19.52); others such as species of *Siphona* are smaller and brown, often seen on flower-heads (Fig. 19.53); some other groups are quite brightly coloured (Fig. 19.54). *Gymnocheta viridis* is a bright green species that could be mistaken for the greenbottles of the genus *Lucilia* (Calliphoridae), though it is more hairy.

Identification: Emden's (1954) handbook was superseded by that of Belshaw (1993), but several species have been added since; see Chandler (2010) for additional references.

British subfamilies and genera:

Dexiinae: *Athrycia, Billaea, Blepharomyia, Campylocheta, Cyrtophleba, Dexia, Dinera, Dufouria, Eriothrix, Estheria, Freraea, Microsoma, Periscepsia, Phyllomya,*

Fig. 19.53 *Siphona* sp. (Tachinidae) (Photo: Roger Key)

Fig. 19.55 *Oxycera trilineata* (Stratiomyidae) (Photo: Roger Key)

Fig. 19.54 *Eriothrix rufomaculata* (Tachinidae) (Photo: Roger Key)

Prosena, Ramonda, Rondania, Thelaira, Trixa, Voria, Wagneria

Exoristinae: *Admontia, Aplomya, Bactromyia, Belida, Bessa, Blondelia, Brachicheta, Cadurciella, Carcelia, Chetogena, Clemelis, Compsilura, Cyzenis, Diplostichus, Drino, Elodia, Epicampocera, Erycia, Erycilla, Erynnia, Eumea, Eurysthaea, Exorista, Frontina, Gastrolepta, Gonia, Gymnosoma, Hebia, Hemimacquartia, Huebneria, Leiophora, Ligeria, Lydella, Medina, Meigenia, Myxexoristops, Nemorilla, Nilea, Ocytata, Oswaldia, Pales, Parasetigena, Phebellia, Phorocera, Phryno, Phryxe, Platymya, Policheta, Pseudoperichaeta, Rhaphiochaeta, Senometopia, Smidtia, Thecocarcelia, Thelymorpha, Timavia, Tlephusa, Townsendiellomyia, Vibrissina, Winthemia, Xylotachina, Zaira, Zenillia*

Phasiinae: *Catharosia, Cinochira, Cistogaster, Clytiomya, Cylindromyia, Dionaea, Hemyda, Labigastera,*

Leucostoma, Litophasia, Lophosia, Opesia, Phania, Phasia, Redtenbacheria, Subclytia

Tachininae: *Actia, Anthomyiopsis, Aphantorhaphopsis, Aphria, Appendicia, Bithia, Ceranthia, Ceromya, Chrysosomopsis, Cleonice, Demoticus, Dexiosoma, Eloceria, Entomophaga, Ernestia, Eurithia, Fausta, Germaria, Goniocera, Graphogaster, Gymnocheta, Hyalurgus, Leskia, Linnaemya, Loewia, Lydina, Lypha, Macquartia, Mintho, Neaera, Nemoraea, Nowickia, Pelatachina, Peleteria, Peribaea, Phytomyptera, Siphona, Solieria, Tachina, Triarthria, Zophomyia*

Infraorder Tabanomorpha

Within the suborder Brachycera, this group of families is considered to be among the most primitive. Many larvae live in damp or aquatic habitats, and many are predators though some are found in decaying plant material and rotting wood. Many adults are large and brightly coloured so that some families such as Stratiomyidae (soldier-flies), Tabanidae (horse-flies) and Rhagionidae (snipe-flies) are quite well known even to the non-entomologist. The book by Stubbs and Drake (2001) covers all the British families and supersedes Oldroyd's (1969) handbook.

SUPERFAMILY STRATIOMYOIDEA

Family Stratiomyidae (16 genera, 48 species)
The members of this family are commonly known as soldier-flies, presumably because their bright patterns are reminiscent of ceremonial military uniforms (Fig. 19.55). All the British species have been given common names (Stubbs & Drake, 2001). Their

Fig. 19.56 *Eupachygaster tarsalis* (Stratiomyidae) (Photo: Roger Key)

Fig. 19.57 *Xylomya maculata* larva (Xylomyidae) (Photo: Roger Key)

larvae feed on algae or rotting vegetable matter in damp conditions or in running water, and the adults are found in waterside habitats or damp woodland, often visiting flowers for nectar. In some species even the larvae have distinct patterns and colours (Fig. 19.56) and because of the relative popularity of this attractive group of flies most of the larvae and life histories are known.

Although still included on the British list, *Clitellaria ephippium* is almost certainly extinct in this country. *Odontomyia hydroleon* (Barred green colonel) is on the UKBAP list.

Identification: the book by Stubbs and Drake (2001) is the key reference for all the British species. Rozkosny (1982, 1983) covered the European species.

British subfamilies and genera:
Beridinae: *Beris, Chorisops*
Clitellariinae: *Clitellaria, Nemotelus, Oxycera, Vanoyia*
Pachygasterinae: *Eupachygaster, Neopachygaster, Pachygaster, Zabrachia*
Sarginae: *Chloromyia, Microchrysa, Sargus*
Stratiomyinae: *Odontomyia, Oplodontha, Stratiomys*

Family Xylomyidae (2 genera, 3 species)
This family was previously called the Xylomyiidae or Solvidae, and they have also been included within the Stratiomyidae. The British species have all been given common names (Stubbs & Drake, 2001). They are known as wood soldier-flies from their close similarity to the Stratiomyidae, coupled with the fact that their larvae are found in dead wood. Most species are commonest in southern

Britain, where *Xylomya maculata* larvae live in rot-holes in ancient woodland (Fig. 19.57).

Identification: Stubbs and Drake (2001) cover the British species.

British genera: *Solva, Xylomya*

SUPERFAMILY TABANOIDEA

Family Athericidae (3 genera, 3 species)
This group was formerly included within the Rhagionidae; they are known as water snipe-flies from their similarity to the rhagionids coupled with the aquatic larvae. All three British species have been given common names (Stubbs & Drake, 2001). One species, *Atherix ibis* presents an unusual phenomenon in that hundreds of females can gather in a cluster on a branch overhanging a river; they lay their eggs in a gelatinous matrix and then die, so the end result is a 'ball' of dead flies and egg-masses (Fig. 19.58). The newly hatched larvae drop into the water, where they are active predators, feeding on other aquatic insects such as mayfly nymphs and chironomid larvae, although they leave the water in order to pupate.

Identification: Stubbs and Drake (2001) cover the British species.

British genera: *Atherix, Atrichops, Ibisia*

Family Rhagionidae (3 genera, 12 species)
The members of this family have long been known as snipe-flies, although the origin of the name is not clear. The most distinctive species are in the genus *Rhagio*; they have yellow patterned bodies and legs, and long wings with a dark pterostigma (Fig. 19.59).

Fig. 19.58 *Atherix ibis* 'ball' (Athericidae) (Photo: Roger Key)

Fig. 19.60 *Chrysops relictus* (Tabanidae) (Photo: Robin Williams)

Fig. 19.59 Rhagionid (Photo: Colin Rew)

Characteristically they rest with the head facing downwards. The larvae are predators on other insects and invertebrates, living in damp soil, mosses or rotting wood.

Identification: Stubbs and Drake (2001) cover the British species.

British genera: *Chrysopilus, Rhagio, Symphoromyia*

Family Spaniidae (2 genera, 3 species)
The separation of this family from the Rhagionidae is not accepted by all authors, but the phytophagous larvae are leaf-miners on bryophytes and mosses, quite different to the predatory larvae of the true snipe-flies. All three British species have been given common names (Stubbs & Drake, 2001).

Identification: Stubbs and Drake (2001) cover the British species (under Rhagionidae).

British genera: *Ptiolina, Spania*

Family Tabanidae (5 genera, 30 species)
For many people the horse-flies or clegs have negative associations because of their reputation for biting both humans and animals. The family includes some very large flies, up to 25 mm in length, and their larvae can be aquatic, semi-aquatic or terrestrial, living in damp soil or leaf-litter; most are predatory on other insects, worms and molluscs, but larvae of *Chrysops* species are saprophagous.

All the British species have been given common names (Stubbs & Drake, 2001) but the layman would at most distinguish between the smaller clegs (*Haematopota*) and the larger horse-flies. There are three main groups of species, which can be based on the wing patterns: *Chrysops* species have banded wings, sometimes mainly dark, and with fairly bright body patterns (Fig. 19.60); *Haematopota* species have essentially dark or 'clouded' wings with patterns of pale spots; and the other genera such as *Tabanus* have clear wings (Fig. 19.61). All groups have strikingly coloured eyes with iridescent or metallic bands across them, apparently indicating different kinds of lenses to aid visual acuity; Stubbs and Drake (2001) illustrate several examples.

Fig. 19.61 *Tabanus sudeticus* (Tabanidae) (Photo: Roger Key)

Fig. 19.62 *Xylophagus ater* (Xylophagidae) (Photo: Roger Key)

It is only the female horse-flies that bite mammals, and only after mating do they need a blood meal for the eggs to develop, similar to mosquitoes. The larger *Tabanus* species usually target horses and cattle; *Chrysops* species are sometimes known as deer-flies although they are not confined to that group; and the common cleg, *Haematopota pluvialis*, is the species that causes most distress to humans when its numbers are very large in damp woodland.

Identification: Stubbs and Drake's (2001) book is the key reference for the British species. Chvála et al. (1972) covered the European species, and Chvála and Jezek (1997) described the aquatic groups.

British subfamilies and genera:
Chrysopsinae: *Chrysops*
Tabaninae: *Atylotus*, *Haematopota*, *Hybomitra*, *Tabanus*

Infraorder Xylophagomorpha

This group contains the single family Xylophagidae.

SUPERFAMILY XYLOPHAGOIDEA

Family Xylophagidae (1 genus, 3 species)
This group was previously included in the Rhagionidae. They have been given the name awl-flies, and each of the three British species has been given a common name (Stubbs & Drake, 2001); there are only two more species in Europe. The adults are narrow, dark flies which resemble ichneumonid wasps both in appearance and in their behaviour. The larvae live in rotting wood though

it is not clear whether they are saprophagous, nec-rophagous or predatory; they are easily recognized by the prominent awl-shaped head that gives the family their common name (Fig. 19.62).

Identification: Stubbs and Drake (2001) cover the British species.

British genus: *Xylophagus*

SUBORDER NEMATOCERA

These are generally considered to be the most primitive suborder of the Diptera but their common characteristics are plesiomorphic, which strongly suggests that the group is paraphyletic. Nonetheless they form a convenient grouping, recognized by their elongate, multi-segmented antennae, a rather complete wing venation and often a slender, long-legged appearance. Unlike the Brachycera their larvae have a more-or-less complete head capsule, though again this is clearly a plesiomophy.

Infraorder Bibionomorpha

SUPERFAMILY BIBIONOIDEA

Family Bibionidae (2 genera, 18 species)
These are sometimes known as March-flies, though this is an American name, and most British species are on the wing in April or later, so the alternative name of St Mark's flies is preferred because St Marks' Day (25th April) is approximately when common species like *Bibio marci* start to fly. These are black, very hairy flies (Fig. 19.63), though the females of some species have red markings; they are bulky insects that fly quite slowly, though some can

Fig. 19.63 *Bibio marci* (Bibionidae) (Photo: Roger Key)

Fig. 19.65 Cecidomyiidae larvae on peas (Photo: Roger Key)

Fig. 19.64 *Bibio marci* pupa (Bibionidae) (Photo: Roger Key)

hover. Bibionid larvae are usually found in the soil where they are mainly saprophagous, though some attack the roots of plants and can even reach pest proportions; pupation occurs in the soil (Fig. 19.64). *Bibio* and *Dilophus* are the only two genera in Europe.

Identification: the handbook by Freeman and Lane (1985) covers all the British species, but see the references in Chandler (2010) for important updates.

British genera: *Bibio, Dilophus*

SUPERFAMILY SCIAROIDEA

Within this group, the larvae of those families that are fungus-feeders secrete mucus, which may help to trap fungal spores on which they feed.

Family Bolitophilidae (1 genus, 17 species)
This group was previously treated as a subfamily of the Mycetophilidae, and they share the common name of fungus gnats. They are small, delicate grey or brownish flies that are found in damp habitats, and their larvae feed gregariously in soft fungi.

Bolitophila is the only European genus.

Identification: the handbook by Hutson et al. (1980) covers most of the British species, but see the references in Chandler (2010) for important updates and additions.

British genus: *Bolitophila*

Family Cecidomyiidae (ca 140 genera, 652 species)
These were previously known as the Itonididae or Itonidae. The gall midges form a very distinct and very large family, though the adult flies are usually less than 5 mm long and can be difficult to identify to species; critical examination often depends on making microscope slide preparations. They are relatively easy to recognize as a family, however; the wings are broad, rounded and often fringed, and the antennae are relatively long with whorls of hairs or loop-shaped sensilla on each segment. Most cecidomyiid larvae are gall-makers, with a few leaf-miners, and this is the largest group of phytophagous Diptera. Unlike most nematoceran families, the larvae have a very small unsclerotized head capsule and are often yellow, orange or red in colour (Fig. 19.65). A few groups are endoparasitoids of other insects; at least two genera are known to prey on aphids, and others prey on mites, with some species feeding on fungi, occasionally on rotting wood. The great majority are phytophagous and the

Fig. 19.66 *Contarinia nasturtii* gall and larvae (Cecidomyiidae) (Photo: Roger Key)

larvae, or their effects, are usually more noticeable than the adults. Many records of British species are therefore based on larvae and the plant associations are very important; Chandler's checklist (1998) listed the known host associations. Many larvae can be identified using the works on galls by Redfern and Askew (1998) and Redfern and Shirley (2002). Like so many other groups of gall-makers, they are very prone to parasitism by Hymenoptera, especially the Platygastridae. A few specialized larvae create 'ambrosia' galls, in which the larvae feed on specialized fungi that grow inside the gall, rather than on the gall tissue itself.

There are numerous pest species with common names, including *Sitodiplosis mosellana* and *Contarinia tritici* (Wheat blossom midges), *C. pisi* (Pea midge, see Fig. 19.65), *C. nasturtii* (Swede midge, Fig. 19.66) *Haplodiplosis marginata* (Saddle gall midge), *Dasineura tetensi* (Blackcurrant leaf midge), and many more.

Identification: There are no works covering the entire British fauna, but Barnes (1946–56) and Nijveldt (1969) can be consulted for the economically important species, with Möhn (1966–71) for fairly recent reviews of the European species. Many additional references are listed in Chandler (2010). The European Porricondylinae were reviewed by Panelius (1965).

British subfamilies and genera:

Cecidomyiinae: *Acodiplosis, Amerhapha, Ametrodiplosis, Anabremia, Anisostephus, Antichiridium, Aphidoletes, Arceuthomyia, Arnoldiola, Arthrocnodax, Asphondylia, Atrichosema, Baldratia, Bayeriola, Blastodiplosis, Blastomyia, Brachineura, Brachyneurina, Bremiola, Camptodiplosis, Cecidomyia, Clinodiplosis,* *Contarinia, Craneiobia, Cystiphora, Dasineura, Dichodiplosis, Didymomyia, Diodaulus, Drisina, Endaphis, Endopsylla, Fabomyia, Feltiella, Geocrypta, Geodiplosis, Gephyraulus, Giardomyia, Giraudiella, Hadrobremia, Haplodiplosis, Harmandiola, Hartigiola, Hybolasioptera, Hygrodiplosis, Iteomyia, Jaapiella, Janetiella, Kaltenbachiola, Kiefferia, Lasioptera, Lathyromyza, Ledomyia, Lestodiplosis, Loewiola, Macrodiplosis, Macrolabis, Massalongia, Mayetiola, Mikiola, Mikomya, Misospatha, Monarthropalpus, Monobremia, Monodiplosis, Mycocecis, Mycodiplosis, Myricomyia, Neomikiella, Octodiplosis, Oligotrophus, Ozirhincus, Parallelodiplosis, Pemphigocecis, Phegomyia, Physemocecis, Placochela, Planetella, Polystepha, Plemeliella, Prolauthia, Psectrosema, Putoniella, Rabdophaga, Resseliella, Rhopalomyia, Rondaniola, Sackenomyia, Schizomyia, Schmidtiella, Semudobia, Silvestriola, Sitodiplosis, Spurgia, Stefaniella, Stenodiplosis, Taxomyia, Thecodiplosis, Tricholaba, Trotteria, Wachtliella, Xenodiplosis, Xylodiplosis, Zeuxidiplosis, Zygiobia*

Lestremiinae: *Anarete, Anaretella, Aprionus, Bryomyia, Campylomyza, Catocha, Corinthomyia, Heterogenella, Lestremia, Micromya, Monardia, Mycophila, Neurolyga, Peromyia, Polyardis, Stenospatha, Strobliella, Trichopteromyia, Xylopriona*

Porricondylinae: *Asynapta, Brittenia, Camptomyia, Cassidoides, Claspettomyia, Dicerura, Dirhiza, Henria, Heteropeza, Heteropezula, Holoneurus, Leptosyna, Miastor, Parepidosis, Porricondyla, Stackelbergiella, Tetraneuromyia, Winnertzia*

Family Diadocidiidae (1 genus, 3 species)

These were previously treated as a subfamily of the Mycetophilidae, and are also known as fungus gnats. They are small yellowish brown flies, around 3–4 mm long, and their larvae feed on fungal mycelia in rotting wood. *Diadocidia* is the only genus in Europe, with just two more included species.

Identification: this group is covered in the handbook by Hutson et al. (1980).

British genus: *Diadocidia*

Family Ditomyiidae (2 genera, 3 species)

These were previously regarded as a subfamily of the Mycetophilidae; at 6–8 mm long they are among the larger fungus gnats, and some have patterned wings. The larvae are found in fungi in rotten wood, and also bracket fungi. There is only one further species (of *Ditomyia*) in Europe.

Identification: this group is covered in the handbook by Hutson et al. (1980) with one additional species; see Chandler (2010) for the reference.

British genera: *Ditomyia, Symmerus*

Family Keroplatidae (15 genera, 52 species)

The keroplatids were previously treated as a subfamily of the Mycetophilidae and are hence known as fungus gnats; they have also been known as Macroceridae, which is now regarded as a subfamily. The larvae are associated with fungi on dead wood, some feeding on fungal spores, while others are predatory.

Asindulum nigrum (Black fungus gnat) is on the UKBAP list.

Identification: most species can be identified using the handbook by Hutson et al. (1980), but see also the additions listed in Chandler (2010).

British subfamilies and genera:

Keroplatinae: *Antlemon, Asindulum, Cerotelion, Isoneuromyia, Keroplatus, Macrorrhyncha, Monocentrota, Neoplatyura, Orfelia, Platyura, Pyratula, Rocetelion, Rutylapa, Urytalpa*
Macrocerinae: *Macrocera*

Family Mycetophilidae (61 genera, 471 species)

The main family of fungus gnats was previously known as the Fungivoridae; although some groups have been split off into new families the mycetophilids still remain a large family, not always easy to identify especially because there is currently no publication covering the largest subfamily, the Mycetophilinae. Adults are usually found in damp, shady places; they are small, slender flies with long legs and a domed thorax that give a humpbacked appearance. Most mycetophilid larvae live up their name and feed on fungi, sometimes on rotting wood; a few are confined to myxomycete slime-moulds, but there is still much work to do on the larval stages and life histories of this family.

Neoempheria lineola (Giant wood-gnat) is on the UKBAP list.

Identification: Hutson et al. (1980) covered all the known British species except for the large subfamily Mycetophilinae; with the division of the family into several smaller families, the tribes of the Sciophilinae in this handbook are now placed at subfamily level. See the many additional references in Chandler (2010).

British subfamilies and genera:

Eudicraninae: *Eudicrana*
Gnoristinae: *Apolephthisa, Boletina, Coelosia, Creagdhubhia, Dziedzickia, Ectrepesthoneura, Gnoriste, Grzegorzekia, Palaeodocosia, Saigusaia, Synapha, Syntemna, Tetragoneura*

Fig. 19.67 *Sciara* sp. (Sciaridae) (Photo: Roger Key)

Leiinae: *Clastobasis, Docosia, Leia, Megophthalmidia, Rondaniella*
Manotinae: *Manota*
Mycetophilinae: *Allodia, Allodiopsis, Anatella, Brachypeza, Brevicornu, Cordyla, Dynatosoma, Epicypta, Exechia, Exechiopsis, Mycetophila, Myrosia, Phronia, Platurocypta, Pseudexechia, Pseudobrachypeza, Pseudorymosia, Rymosia, Sceptonia, Stigmatomeria, Synplasta, Tarnania, Trichonta, Zygomyia*
Mycomyinae: *Mycomya, Neoempheria*
Sciophilinae: *Acnemia, Allocotocera, Anaclileia, Azana, Coelophthinia, Leptomorphus, Megalopelma, Monoclona, Neuratelia, Notolopha, Paratinia, Phthinia, Polylepta, Sciophila, Speolepta*

Family Sciaridae (19 genera, 266 species)

The members of this family are generally known as black fungus gnats; most species are entirely black although a few have striking yellow markings (Fig. 19.67). In older works they were sometimes treated as a subfamily of the Mycetophilidae. They are generally very small flies, and their larvae are extremely common in damp soil and leaf-litter where they feed on fungal mycelia; they can also be found associated with the fungi in rotting wood, carrion and dung. Even non-entomologists may be familiar with the one or two species that occur frequently in houses when they breed in pot-plants. The adult flies are sometimes seen on flowers (Fig. 19.68).

Identification: Freeman's (1983) handbook has been updated by various specialist references and a revision of the British species is needed; see Chandler (2010) for details.

British genera: *Bradysia, Bradysiopsis, Camptochaeta, Corynoptera, Cratyna, Ctenosciara, Epidapus,*

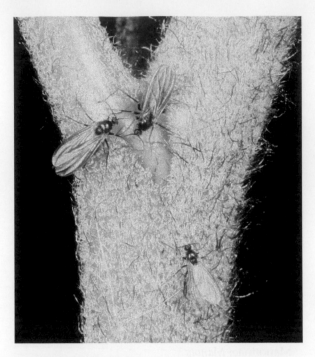

Fig. 19.68 Sciarid (Photo: Roger Key)

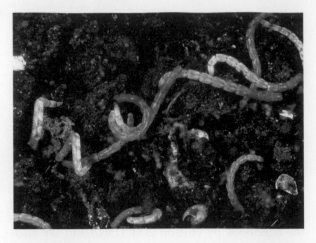

Fig. 19.69 *Dasyhelea saxicola* (Ceratopogonidae) (Photo: Roger Key)

Leptosciarella, Lycoriella, Phytosciara, Pnyxia, Pseudolycoriella, Scatopsciara, Schwenckfeldina, Sciara, Scythropochroa, Trichosia, Xylosciara, Zygoneura

Infraorder Culicomorpha

The families of this group are generally united by having aquatic larvae, although in some families the young stages live in damp terrestrial habitats. Many female flies are well-known blood-feeders, such as the Ceratopogonidae (biting midges), Culicidae (mosquitoes) and Simuliidae (blackflies). The males feed on nectar but in some families, especially the Chironomidae, the adults do not feed at all.

SUPERFAMILY CHIRONOMOIDEA

Family Ceratopogonidae (20 genera, 170 species)
The biting midges are well-known for just a few species that bite humans, but overall it is a large family of which the members of only one genus, *Culicoides*, feed on the blood of mammals. Adult ceratopogonids are very small, inconspicuous flies (hence the name 'no-see-ums' in the USA) and their larvae develop in a range of aquatic or damp habitats. Some are found deep in lakes, some live in rivers, while others are in rot-holes in trees. The semi-aquatic species live in bogs, rotting vegetation, dung, fungi and similar damp habitats (Fig. 19.69).

The impact of biting midges on human activity is best known in the north and west of Britain, and the west Highlands of Scotland are notorious for this problem on summer evenings. *Culicoides impunctatus* is the main culprit, and its habits can make any outdoor activity almost impossible at certain times. There can be few other tiny insects that have merited semi-popular books such as Hendry (2003). It is only the females that bite, and they need a blood meal for their eggs to develop. They secrete a 'recruitment' pheromone, which attracts others to the same host; they can afford to be altruistic as each is only taking a minute amount of blood. Some biting midges are known to feed on other insects, and species of *Forcipomyia* can be found on the wing-veins of insects such as dragonflies and lacewings. Apart from their nuisance value, species of *Culicoides* also transmit bluetongue virus in cattle and sheep.

Identification: there are no recent works covering the British species; see Chandler (2010) for the many references which need to be consulted. Boorman (1993) gives an overview of the medically important groups, and Szadziewski et al. (1997) helps to identify the aquatic stages of the north European species.

British subfamilies and genera:

Ceratopogoninae: *Allohelea, Alluaudomyia, Bezzia, Brachypogon, Ceratopogon, Clinohelea, Culicoides, Kolenohelea, Mallochohelea, Neurohelea, Palpomyia, Phaenobezzia, Probezzia, Schizohelea, Serromyia, Sphaeromias, Stilobezzia*

Dasyheleinae: *Dasyhelea*

Forcipomyiinae: *Atrichopogon, Forcipomyia*

Fig. 19.70 Male chironomid (Photo: Roger Key)

Family Chironomidae (139 genera, 608 species)
The non-biting midges form another very large family whose larvae are mainly aquatic. Adult chironomids are small and delicate flies with a strongly humped thorax that almost hides the head in dorsal view; the male antennae are noticeably plumose (Fig. 19.70). Their mouthparts are not developed and they do not feed as adults. A few species have terrestrial larvae, living in damp soil, rotting vegetation and so on, but most are genuinely aquatic and may build small tubes in which they live, rather like caddisfly larvae. Many species have such precise environmental requirements that the species profile of this one family can be a good indicator of freshwater quality. The few species that can tolerate much organic pollution include the well-known bloodworms whose red colour is due to haemoglobin that enables them to survive in oxygen-poor sediments. Adult chironomids are most noticeable when they perform mating swarms over the water bodies in which they breed.

Identification: Langton and Pinder (2007) give keys to the adults of the British species; Cranston (1982) is a guide to the larvae of the Orthocladiinae. A generic guide to pupal exuviae is by Wilson and Ruse (2005), and Wiederholm (1983, 1986, 1989) is an important review of the larvae, pupae and adult males of the Holarctic species. There is a useful CD-ROM covering larval identification by Klink and Moller Pillott (2003). The following website contains much information about the family worldwide: http://insects.ummz.lsa.umich.edu/~ethanbr/chiro/index.html.

British subfamilies and genera:
Buchonomyiinae: *Buchonomyia*

Chironominae: *Chironomus, Cladopelma, Cladotanytarsus, Corynocera, Cryptochironomus, Cryptotendipes, Demeijerea, Demicryptochironomus, Dicrotendipes, Einfeldia, Endochironomus, Glyptotendipes, Graceus, Harnischia, Kiefferulus, Lauterborniella, Microchironomus, Micropsectra, Microtendipes, Neozavrelia, Nilothauma, Omisus, Pagastiella, Parachironomus, Paracladopelma, Paralauterborniella, Parapsectra, Paratanytarsus, Paratendipes, Phaenopsectra, Polypedilum, Pseudochironomus, Rheotanytarsus, Saetheria, Sergentia, Stempellina, Stempellinella, Stenochironomus, Stictochironomus, Synendotendipes, Tanytarsus, Tribelos, Virgatanytarsus, Xenochironomus, Zavrelia, Zavreliella*

Diamesinae: *Diamesa, Potthastia, Protanypus, Pseudodiamesa, Pseudokiefferiella, Sympotthastia, Syndiamesa*

Orthocladiinae: *Acamptocladius, Acricotopus, Brillia, Bryophaenocladius, Camptocladius, Cardiocladius, Chaetocladius, Clunio, Corynoneura, Corynoneurella, Cricotopus, Diplocladius, Epoicocladius, Eukiefferiella, Eurycnemus, Euryhapsis, Georthocladius, Gymnometriocnemus, Halocladius, Heleniella, Heterotanytarsus, Heterotrissocladius, Krenosmittia, Limnophyes, Mesosmittia, Metriocnemus, Nanocladius, Orthocladius, Paracladius, Paracricotopus, Parakiefferiella, Paralimnophyes, Parametriocnemus, Paraphaenocladius, Paratrichocladius, Paratrissocladius, Parorthocladius, Psectrocladius, Pseudorthocladius, Pseudosmittia, Rheocricotopus, Rheosmittia, Smittia, Synorthocladius, Thalassosmittia, Thienemannia, Thienemanniella, Tokunagaia, Trissocladius, Tvetenia, Zalutschia*

Podonominae: *Lasiodiamesa, Paraboreochlus, Parochlus*

Prodiamesinae: *Monodiamesa, Odontomesa, Prodiamesa*

Tanypodinae: *Ablabesmyia, Anatopynia, Apsectrotanypus, Arctopelopia, Clinotanypus, Conchapelopia, Guttipelopia, Hayesomyia, Krenopelopia, Labrundinia, Larsia, Macropelopia, Monopelopia, Natarsia, Nilotanypus, Paramerina, Procladius, Psectrotanypus, Rheopelopia, Schineriella, Tanypus, Telmatopelopia, Thienemannimyia, Trissopelopia, Xenopelopia, Zavrelimyia*

Telmatogetoninae: *Telmatogeton, Thalassomya*

Family Simuliidae (3 genera, 35 species)
These are commonly known as blackflies (but this name is not to be confused with the black aphids, sometimes called blackfly). As their name suggests they are small and black, and the wings are unusually broad with an anal lobe. Like the

Ceratopogonidae, some species are vicious biters of both humans and other animals; in Britain they do not normally transmit any specific diseases, though they are vectors of serious diseases such as river-blindness in the tropics. As with any insect bite or sting the host can suffer allergic reactions to the fly's saliva or the wound can become secondarily infected. *Simulium posticatum* has even been given the common name of Blandford fly because of its severe nuisance value around the River Stour in Dorset.

Blackfly larvae are aquatic, and are adapted to living in fast-flowing water; they attach themselves to the substrate with a silk thread, and filter-feed on particles in the current.

Identification: the work by Davies (1968) keys the adults, with updates by Bass (1998); the latter covers the larvae and pupae. Additions and updates are provided by Chandler (2010). There is a useful overview of the whole family by Crosskey (1990) and their medical significance is covered by Crosskey (1993). For an overview of the northern European species see Jenssen (1997); the larvae and pupae are also covered by Lechthaler and Car (2005).

British genera: *Metacnephia, Prosimulium, Simulium*

Family Thaumaleidae (1 genus, 3 species)
As their common name of trickle midges suggests, the larvae of this small family live in shallow trickles and splash zones of fast-running water, usually in upland areas in the north and west of Britain. The adult flies are small and yellowish or brownish, and they do not fly far from the larval habitats. Although there are only three species in Britain, there are around 75 in Europe.

Identification: keys to the adults and larvae are provided by Disney (1999); see also Wagner (1997c) for the north European species.

British genus: *Thaumalea*

SUPERFAMILY CULICOIDEA

The three families in this group were all treated as Culicidae by early authors.

Family Chaoboridae (2 genera, 6 species)
The phantom midges get their name from the remarkably transparent larvae that 'hangs' motionless in the water waiting for its prey of small invertebrates. The adults can look rather similar to some Chironomidae, and the males have plumed antennae like Culicidae; earlier authors treated them as a subgroup of the latter family.

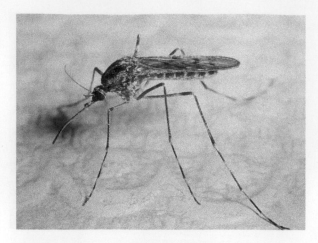

Fig. 19.71 *Culiseta annulata* (Culicidae) (Photo: Roger Key)

Identification: the handbook by Coe et al. (1950) covered the British species (within the Culicidae); see also Saether (1997) for the northern European species and Chandler (2010) for additional references.

British genera: *Chaoborus, Mochlonyx*

Family Culicidae (6 genera, 34 species)
The mosquitoes are well-known to everyone, and they are easily recognized from related families by the elongate mouthparts (Fig. 19.71). These are non-piercing in the males, which drink nectar and other liquids, whereas the females pierce vertebrate skin to drink blood, which is necessary for egg development. Mosquitoes are of course serious disease vectors in the tropics and much study has gone into the recognition of cryptic species groups, which are common in this family; in Britain the species previously called *Culex pipiens* actually consists of two morphologically almost identical species, which have completely different life-styles and which prefer different hosts. Mosquito larvae are all aquatic detritivores or predators, and are found in a wide variety of water types, from rivers to tree-holes.

Some adult mosquitoes have bright patterns formed from scales; a large and common example is *Culiseta annulata*, which commonly bites humans (Fig. 19.72). Male Culicidae have plumose antennae (Fig. 19.73) whereas the antennae of the blood-feeding females are relatively unadorned.

A world overview of the mosquitoes is give by Eldridge (2009).

Identification: the works by Cranston et al. (1987) and Snow (1990) provide keys to the adults and

Fig. 19.72 *Culiseta annulata* feeding (Culicidae) (Photo: Robin Williams)

Fig. 19.74 *Sylvicola* sp. (Anisopodidae) (Photo: Roger Key)

larval habitat of slow or still freshwater. The larvae stay at the surface of the water in a characteristic inverted U-shape, which gives the family its common name of meniscus midges. They were previously regarded as a subfamily of the Culicidae, but the males do not have plumed antennae and both sexes have short mouthparts.

Identification: Disney (1999) covers the adults and larvae; see also Wagner (1997b) for the north European species.

British genera: *Dixa*, *Dixella*

Infraorder Psychodomorpha

Within this group most larvae live in damp habitats, rather than being aquatic.

SUPERFAMILY ANISOPODOIDEA

Family Anisopodidae (1 genus, 4 species)
The members of this family are delicate yellowish or brownish flies, usually found in damp habitats where their larvae live. They were formerly known as Phryneidae, Rhyphidae or Sylvicolidae, and they have the common name of window gnats, or window midges. The larvae feed in various kinds of organic matter, including rotting wood, dung and fungi; the adults can be seen on flowers and sap-runs or resting on tree-trunks (Fig. 19.74); some species are frequently seen on house windows or walls, hence their common name.

Indentification: this group was covered in the handbook by Coe et al. (1950); there are some useful additional references in Chandler (2010).

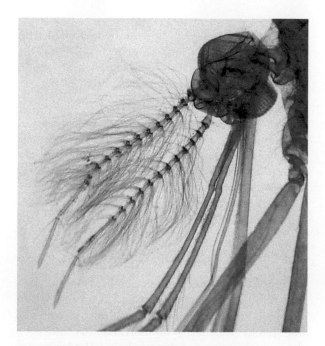

Fig. 19.73 Male mosquito head, slide preparation (Culicidae) (Photo: Peter Barnard)

juvenile stages; see Chandler (2010) for additional references. The medically important groups are covered by Service (1993), and there is a European guide on CD-ROM by Schaffner et al. (2001).

British subfamilies and genera:
Anophelinae: *Anopheles*
Culicinae: *Aedes*, *Coquillettidia*, *Culex*, *Culiseta*, *Orthopodomyia*

Family Dixidae (2 genera, 15 species)
The members of this family are small brownish flies with very long antennae, always found close to the

Fig. 19.75 Mycetobiid larva in sap-run (Photo: Roger Key)

Fig. 19.76 Psychodid (Photo: Colin Rew)

British genus: *Sylvicola*

Family Mycetobiidae (1 genus, 3 species)
This small group was originally in the Mycetophilidae, and was then moved to the Anisopodidae. The small dark flies are found near their larval habitats in woodland, and are commonly known as wood gnats. The larvae are usually found in rotting wood, rot-holes or sap-runs (Fig. 19.75).

Identification: see the specialist references listed in Chandler (2010).

British genus: *Mycetobia*

SUPERFAMILY PSYCHODOIDEA

Family Psychodidae (20 genera, 99 species)
Although many people are familiar with the tiny moth-flies running around near sinks, public toilets and on windows in houses, it is often a suprise to learn how large a family this is. The members of the family are easily recognized by the wing-shape, and the general hairiness of the wings and body (Fig. 19.76) but species recognition is not easy and usually needs microscopic examination of genitalia. The larvae live in a variety of semi-aquatic habitats, including damp soil, mosses, splash zones of streams, tree rot-holes, waterside mud, fungi and sewage. Most adults do not feed, but those of *Sycorax* feed on the blood of amphibians.

The subfamily Phlebotominae does not occur in Britain, but some of its members are vectors of disease in warmer countries, where they are known as sand-flies. *Phlebotomus* occurs in France, and there is an early but isolated record from Jersey in the Channel Islands.

Identification: most species can be identified with Withers (1989); see also Wagner (1997a) for the north European aquatic Psychodinae. Many additional references are listed in Chandler (2010).

British subfamilies and genera:
  Psychodinae: *Bazarella, Boreoclytocerus, Brunettia, Feuerborniella, Mormia, Panimerus, Paramormia, Pericoma, Peripsychoda, Philosepedon, Psychoda, Szaboiella, Telmatoscopus, Threticus, Tinearia, Tonnoiriella, Trichopsychoda, Vaillantodes*
    Sycoracinae: *Sycorax*
    Trichomyiinae: *Trichomyia*

SUPERFAMILY SCATOPSOIDEA

Family Scatopsidae (16 genera, 46 species)
This family was originally treated as a subgroup of the Bibionidae, which they superficially resemble in being stout black flies, with short antennae; however, the scatopsids are not densely hairy. The larvae are found in all kinds of rotting organic matter including dung, and the adults of some species are often seen in large numbers on adjacent vegetation (Fig. 19.77).

Identification: the handbook by Freeman and Lane (1985) covers most of the British species, but there are important additions and updates in Chandler (2010).

British subfamilies and genera:
    Aspistinae: *Aspistes*
    Ectaetiinae: *Ectaetia*
    Plectrosciarinae: *Anapausis*
    Scatopsinae: *Apiloscatopse, Coboldia, Colobostema, Efcookella, Ferneiella, Holoplagia, Neorhegmoclemina,*

Fig. 19.77 *Anapausis soluta* (Scatopsidae) (Photo: Roger Key)

Fig. 19.79 *Ptychoptera contaminata* (Ptychopteridae) (Photo: Roger Key)

Fig. 19.78 *Trichocera regelationis* (Trichoceridae) (Photo: Roger Key)

Parascatopse, Reichertella, Rhexoza, Scatopse, Swammerdamella, Thripomorpha

SUPERFAMILY TRICHOCEROIDEA

Family Trichoceridae (2 genera, 10 species)
This family was formerly known as the Petauristidae. They are one of the few groups of insects to be seen flying in cold weather in the winter, even when snow is on the ground, and this has earned them the common name of winter gnats. They are slender, long-legged flies that superficially resemble small crane-flies (Fig. 19.78), and they often form large conspicuous swarms in the sunshine. Trichocerid larvae are saprophagous, found in damp soil, under bark, in fungi, dung and carrion.

Identification: this group was covered in the handbook by Coe et al. (1950).

British genera: *Diazosma, Trichocera*

Infraorder Ptychopteromorpha

This group contains the single family Ptychopteridae.

SUPERFAMILY PTYCHOPTEROIDEA

Family Ptychopteridae (1 genus, 7 species)
These were formerly known as the Liriopeidae. They resemble tipuloid crane-flies but have a different wing-venation, and the thoracic transverse suture is looped, not V-shaped; they are sometimes known as fold-winged crane-flies. Many species have dark wing-markings, and the adults are usually found resting on vegetation in damp habitats (Fig. 19.79); the larvae live in wet mud or marshy pools and they breathe by means of a posterior respiratory siphon, an interesting parallel with the similar structure developed in aquatic hoverfly larvae.

Ptychoptera is the only European genus, and some species show sexual dimorphism (Fig. 19.80).

Identification: males can be identified using the handbook by Coe et al. (1950). Peus (1958) monographed the European species; see also Andersson (1997) and Stubbs (1993). Additional references are given by Chandler (2010), and there is draft key on the following website: http://www.dipteristsforum.org.uk.

British genus: *Ptychoptera*

Fig. 19.80 *Ptychoptera* sp. (Ptychopteridae) (Photo: Roger Key)

Fig. 19.82 *Neolimonia dumetorum* (Limoniidae) (Photo: Roger Key)

Fig. 19.81 Tipulid mouthparts (Photo: Colin Rew)

Infraorder Tipulomorpha

All the members of this group are known as crane-flies, and all were previously included in the single family Tipulidae. However, the relationships of the four families that are now recognized are still open to further revision, and there is some indication that at least one family is not monophyletic. Functionally the adult mouthparts are reduced, though they are sometimes elongated with a distinct rostrum (Fig. 19.81). All members of this group have a V-shaped thoracic suture.

The handbook by Coe et al. (1950) is still useful for some of these families (all included under Tipulidae), but there are also several draft keys on the following website: http://www.dipteristsforum. org.uk.

SUPERFAMILY TIPULOIDEA

Family Cylindrotomidae (4 genera, 4 species)
The members of this small family of crane-flies are all yellow or pale brown. The larvae are unusual in being phytophagous; they are bright green in colour, and three species feed on mosses while the fourth is a generalized phytophage on low vegetation in woodland.

Identification: Coe et al. (1950) handbook covered this group (within the Tipulidae); see also Brinkmann (1997) and specialized references listed by Chandler (2010).

British genera: *Cylindrotoma, Diogma, Phalacrocera, Triogma*

Family Limoniidae (49 genera, 214 species)
These were previously known as the Limnobiidae, and they form the largest family of crane-flies. They are particularly delicate, long-legged flies that fold their wings over each other at rest (Fig. 19.82) and the males are noted for flying in large swarms. Most larvae in this group feed in rotting vegetable material, fungi and so on, in terrestrial or semi-aquatic habitats; some are aquatic predators that live in silken tubes, while others feed on algae.

Gnophomyia elsneri (Royal splinter cranefly), *Idiocera sexguttata* (Six-spotted cranefly), *Lipsothrix ecucullata* (Scottish yellow splinter), *Lipsothrix errans* (Northern yellow splinter), *Lipsothrix nervosa* (Southern yellow splinter), *Lipsothrix nigristigma* (Scarce yellow splinter, Fig. 19.83) and *Rhabdomastix japonica* (River-shore cranefly) are on the UKBAP list.

Identification: the handbook by Coe et al. (1950) is still useful for most species (within the Tipulidae);

Fig. 19.83 *Lipsothrix nigristigma* (Limoniidae) (Photo: Roger Key)

Fig. 19.84 *Pedicia rivosa* (Pediciidae) (Photo: Roger Key)

Reusch and Oosterbroek (1997) cover the early stages of the aquatic species; for other specialized references see Chandler (2010).

British subfamilies and genera:

Chioneinae: *Arctoconopa, Cheilotrichia, Crypteria, Ellipteroides, Erioconopa, Erioptera, Gnophomyia, Gonempeda, Gonomyia, Hoplolabis, Idiocera, Ilisia, Molophilus, Neolimnophila, Ormosia, Rhabdomastix, Rhypholophus, Scleroprocta, Symplecta, Tasiocera, Trimicra*

Dactylolabinae: *Dactylolabis*

Limnophilinae: *Austrolimnophila, Eloeophila, Epiphragma, Euphylidorea, Hexatoma, Idioptera, Limnophila, Neolimnomyia, Paradelphomyia, Phylidorea, Pilaria, Pseudolimnophila*

Limoniinae: *Achyrolimonia, Antocha, Atypophthalmus, Dicranomyia, Dicranoptycha, Discobola, Geranomyia, Helius, Limonia, Lipsothrix, Metalimnobia, Neolimonia, Orimarga, Rhipidia, Thaumastoptera*

Family Pediciidae (4 genera, 19 species)

These were previously treated as a subfamily within the Limoniidae, but they differ in the wing venation, and many species have distinctive wing-markings; like Tipulidae they hold their wings open at rest (Fig. 19.84). Their larvae live in semi-aquatic habitats or small streams; a few species feed on fungi, but most are predatory on small invertebrates.

Identification: the handbook by Coe et al. (1950) covers most of the British species (within the Tipulidae); Reusch and Oosterbroek (1997) describe the early stages of the aquatic species. For other specialized references see Chandler (2010).

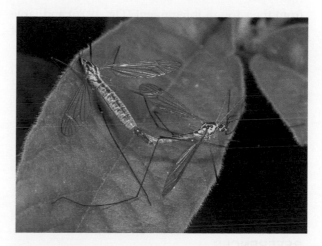

Fig. 19.85 *Nephrotoma* mating pair (Tipulidae) (Photo: Roger Key)

British subfamilies and genera:

Pediciinae: *Dicranota, Pedicia, Tricyphona*

Ulinae: *Ula*

Family Tipulidae (8 genera, 87 species)

This family includes the largest of the crane-flies, especially in the genus *Tipula* where *T. maxima* has the largest wing-span of any British dipteran, at over 60mm. Most crane-flies are grey or brown, sometimes with conspicuous wing-markings, but *Nephrotoma* species are yellow and black (Fig. 19.85). The larvae in this family are not usually aquatic, but are often found in damp soil bordering streams or ponds; some species live in mosses and liverworts, while others are in decayed wood. Some are significant pests on grassland, where they eat the plant roots, and these are principally *Tipula paludosa* and *T. oleracea*. These larvae are known as leatherjackets,

Fig. 19.86 *Tipula paludosa* larva (Tipulidae) (Photo: Roger Key)

from their tough integument (Fig. 19.86) and large populations can represent a significant source of food for many birds.

Identification: Coe et al.'s (1950) handbook is still important though outdated; see also Theowald (1973–80) and Stubbs (1992); more specialized references are listed in Chandler (2010).

British subfamilies and genera:

Ctenophorinae: *Ctenophora, Dictenidia, Tanyptera*
Dolichopezinae: *Dolichopeza*
Tipulinae: *Nephrotoma, Nigrotipula, Prionocera, Tipula*

REFERENCES

ANDERSSON, H. 1977. Taxonomic and phylogenetic studies of Chloropidae (Diptera) with special reference to old world genera. *Entomologica Scandinavica* (Suppl.) 8: 200 pp.

ANDERSSON, H. 1997. Diptera Ptychopteridae, phantom crane flies. In: NILSSON, A. (ed.) *Aquatic insects of north Europe. Vol. 2. Odonata–Diptera.* Apollo Books, Stenstrup, pp. 193–207.

BÄCHLI, G., VILELA, C.R., ANDERSSON ESCHER, S. & SAURA, A. 2004. The Drosophilidae (Diptera) of Fennoscandia and Denmark. *Fauna Entomologica Scandinavica* 39: 362 pp.

BARNES, H.F. 1946–56. *Gall midges of economic importance.* 7 vols. Crosby Lockwood & Son, London.

BASS, J. 1998. Last-instar larvae and pupae of the Simuliidae of Britain and Ireland. *Scientific Publications of the Freshwater Biological Association* 55: 102 pp.

BECKER, T. 1926. 56a. Ephydridae. In: LINDNER, E. (ed.) *Die Fliegen der paläarktischen Region* 6(1): 115 pp.

BELSHAW, R. 1993. Tachinid flies. Diptera: Tachinidae. *Handbooks for the identification of British insects* 10(4ai): 169 pp.

BOORMAN, J. 1993. Biting midges (Ceratopogonidae). In: LANE, R.P. & CROSSKEY, R.W. (eds.) *Medical insects and arachnids.* Chapman & Hall, London, pp. 288–309.

BRINKMANN, R. 1997. Diptera Cylindrotomidae. In: NILSSON, A. (ed.) *Aquatic insects of north Europe. Vol. 2. Odonata–Diptera.* Apollo Books, Stenstrup, pp. 99–104.

CHANDLER, P. (ed.) 1998. Checklists of insects of the British Isles (new series). Part 1: Diptera. *Handbooks for the identification of British insects* 12(1): 234 pp.

CHANDLER, P.J. 2001. The flat-footed flies (Diptera: Opetiidae and Platypezidae) of Europe. *Fauna Entomologica Scandinavica* 36: 276 pp.

CHANDLER, P. (ed.) 2010. *A dipterist's handbook.* Amateur Entomologists' Society, London.

CHVÁLA, M. 1975. The Tachydromiinae (Dipt. Empididae) of Fennoscandia and Denmark. *Fauna Entomologica Scandinavica* 3: 336 pp.

CHVÁLA, M. 1983. The Empidoidea (Diptera) of Fennoscandia and Denmark. II General Part. The families Hybotidae, Alelestidae and Microphoridae. *Fauna Entomologica Scandinavica* 12: 279 pp.

CHVÁLA, M. 1994. The Empidoidea (Diptera) of Fennoscandia and Denmark. III. Genus *Empis. Fauna Entomologica Scandinavica* 29: 192 pp.

CHVÁLA, M. 2005. The Empidoidea (Diptera) of Fennoscandia and Denmark. IV. Genus *Hilara. Fauna Entomologica Scandinavica* 40: 233 pp.

CHVÁLA, M. & JEZEK, J. 1997. Diptera Tabanidae, horse flies. In: NILSSON, A. (ed.) *Aquatic insects of north Europe. Vol. 2. Odonata–Diptera.* Apollo Books, Stenstrup, pp. 295–309.

CHVÁLA, M., LYNEBORG, L. & MOUCHA, J. 1972. *The horse flies of Europe (Diptera, Tabanidae).* Entomological Society of Copenhagen, 499 pp.

COE, R.L. 1966. Diptera Pipunculidae. *Handbooks for the identification of British insects* 10(2c): 83 pp.

COE, R.L., FREEMAN, P. & MATTINGLY, P.F. 1950. Diptera 2. Nematocera: families Tipulidae to Chironomidae. *Handbooks for the identification of British insects* 9(2): 216 pp.

COLLIN, J.E. 1961. *British flies.* Vol. VI. Empididae. Cambridge University Press, Cambridge, 782 pp.

COLWELL, D.D., HALL, M.J.R. & SCHOLL, P.J. 2004. *The oestrid flies.* CAB International Publishing, Wallingford, 359 pp.

COLYER, C.N. & HAMMOND, C.O. 1968. *Flies of the British Isles* (2nd edn.). Warne & Co., London.

CRANSTON, P.S. 1982. A key to the larvae of the British Orthocladiinae (Chironomidae). *Scientific Publications of the Freshwater Biological Association* 45: 152 pp.

CRANSTON, P.S., RAMSDALE, C.D., SNOW, K.R. & WHITE, G.B. 1987. Adults, larvae and pupae of British mosquitoes (Culicidae). *Scientific Publications of the Freshwater Biological Association* 48: 152 pp.

CROSSKEY, R.W. 1990. *The natural history of blackflies.* Wiley, Chichester.

CROSSKEY, R.W. 1993. Blackflies (Simuliidae). In: LANE, R.P. & CROSSKEY, R.W. (eds.) *Medical insects and arachnids.* Chapman & Hall, London, pp. 241–87.

CZERNY, L. 1927. Chiromyidae. In: LINDNER, E. (ed.) *Die Fliegen der paläarktischen Region* 5(1): 51–54.

CZERNY, L. 1928. Clusiidae. In: LINDNER E. (ed.) *Die Fliegen der paläarktischen Region* 6(1): 112 pp.

DAHL, R.G. 1959. Studies on Scandinavian Ephydridae (Diptera Brachycera). *Opuscula Entomologica* (Suppl.) 15: 224 pp.

DAVIES. L. 1968. A key to the species of Simuliidae (Diptera) in the larval, pupal and adult stages. *Scientific Publications of the Freshwater Biological Association* 24: 126 pp.

DISNEY, R.H.L. 1983. Scuttle flies. Diptera, Phoridae (except *Megaselia*). *Handbooks for the identification of British insects* 10(6): 81 pp.

DISNEY, R.H.L. 1989. Scuttle flies. Diptera, Phoridae Genus *Megaselia*. *Handbooks for the identification of British insects* 10(8): 155 pp.

DISNEY, R.H.L. 1994. *Scuttle flies: the Phoridae*. Chapman & Hall, London.

DISNEY, R.H.L. 1999. British Dixidae (meniscus midges) and Thaumaleidae (trickle midges). *Scientific Publications of the Freshwater Biological Association* 56: 128 pp.

DRAKE, C.M. 1993. A review of the British Opomyzidae (Diptera). *British Journal of Entomology and Natural History* 6(Suppl.): 18 pp.

DUDA, O. 1934. Periscelididae. In: LINDNER, E. (ed.) *Die Fliegen der paläarktischen Region* 6(1): 13 pp.

ELDRIDGE, B.F. 2009. Mosquitoes. In: RESH, V.H. & CARDÉ, R.T. (eds.) *Encyclopedia of insects* (2nd edn.). Academic Press/Elsevier, San Diego & London, pp. 658–63.

EMDEN, F.I. van 1954. Diptera Cyclorrhapha. Calyptrata (I) section (a). Tachinidae and Calliphoridae. *Handbooks for the identification of British insects* 10(4a): 133 pp.

ERZINCLIOGLU, Z. 1996. *Blowflies*. Naturalists' Handbooks no. 23, Richmond Publishing, Slough.

FALK, S. 1991. *A review of the scarce and threatened flies of Great Britain. Part 1*. Research and Survey in Nature Conservation 39. Joint Nature Conservation Committee, Peterborough, 194 pp.

FALK, S. & CHANDLER, P.J. 2005. *A review of the scarce and threatened flies of Great Britain. Part 2: Nematocera and Aschiza not dealt with by Falk (1991)*. Joint Nature Conservation Committee, Peterborough, 189 pp.

FALK, S. & CROSSLEY, R. 2005. *A review of the scarce and threatened flies of Great Britain. Part 3: Empidoidea*. Joint Nature Conservation Committee, Peterborough, 134 pp.

FONSECA, E.C.M. d'Assis 1968. Diptera Cyclorrhapha Calyptrata Section (b) Muscidae. *Handbooks for the identification of British insects* 10(4b): 119 pp.

FONSECA, E.C.M. d'Assis 1978. Diptera: Orthorrhapha, Brachycera, Dolichopodidae. *Handbooks for the identification of British insects* 9(5): 90 pp.

FREEMAN, P. 1983. Sciarid flies. Diptera, Sciaridae. *Handbooks for the identification of British insects* 9(6): 68 pp.

FREEMAN, P. & LANE, R.P. 1985. Bibionid and scatopsid flies. Diptera: Bibionidae and Scatopsidae. *Handbooks for the identification of British insects* 9(7): 74 pp.

GILBERT, F. 1986. *Hoverflies*. Naturalists' Handbooks no. 5. Cambridge University Press, Cambridge.

GOOT, V.S. van der & VEEN, M.P. van 1996. De spillebeenvliegen *[Calobatidae, Micropezidae & Tanypezidae]*, wortelvliegen *[Psilidae]* en wolzwevers *[Bombyliidae]* van Noordwest-Europa (2nd edn.). Jeugdbondsuitgeverij, Utrecht, 57 pp.

GRIMALDI, D. & ENGEL, M.S. 2005. *Evolution of the insects*. Cambridge University Press, Cambridge.

GRUNIN, K.Y. 1966–9. Oestridae, Gasterophilidae, Hypodermatidae. In: LINDNER, E. (ed.) *Die Fliegen der paläarktischen Region* 8(1): 160 pp.

HENDRY, G. 2003. *Midges in Scotland*. Mercat Press, Edinburgh.

HENNIG, W. 1937. Milichiidae, Carnidae. In: LINDNER, E. (ed.) *Die Fliegen der paläarktischen Region* 6(1): 91 pp.

HENNIG, W. 1966–76. Anthomyiidae. In: LINDNER, E. (ed.) *Die Fliegen der paläarktischen Region* 7(1): 974 pp.

HUTSON, A.M. 1984. Keds, flat-flies and bat-flies. Diptera, Hippoboscidae and Nyctcribiidae. *Handbooks for the identification of British insects* 10(7): 40 pp.

HUTSON, A.M., ACKLAND, D.M. & KIDD, L.N. 1980. Mycetophilidae (Bolitophilinae, Ditomyiinae, Diadocidiinae, Keroplatinae, Sciophilinae and Manotinae). Diptera, Nematocera. *Handbooks for the identification of British insects* 9(3): 111 pp.

JENSEN, F. 1997. Diptera Simuliidae, Blackflies. In: NILSSON, A. (ed.) *Aquatic insects of north Europe. Vol. 2. Odonata–Diptera*. Apollo Books, Stenstrup, pp. 209–41.

KLINK, A.G. & Moller PILLOTT, H.K.M. 2003. *Chironomidae larvae: key to the higher taxa and species of the lowlands of northwestern Europe*. CD-ROM. World Biodiversity Database, ETI, Amsterdam.

KLOET, G.S. & HINCKS, W.D. 1976. A checklist of British insects (2nd edn., completely revised). Part 5: Diptera and Siphonaptera. *Handbooks for the identification of British insects* 11(5): 139 pp.

LANE, R.P. & CROSSKEY, R.W. (eds.) 1993. *Medical insects and arachnids*. Chapman & Hall, London, 723 pp.

LANGTON, P.H. & PINDER, L.C.V. 2007. Keys to the adult male Chironomidae of Britain and Ireland. *Scientific Publications of the Freshwater Biological Association* 64: (in 2 vols) 239 pp. & 168 pp.

LECHTHALER, W. & CAR, M. 2005. *Simuliidae. Key to larvae and pupae from central and western Europe*. CD-ROM. Eutaxa, Wien.

MacGOWAN, I. & ROTHERAY, G. 2008. British Lonchaeidae. Diptera, Cyclorrhapha, Acalyptratae. *Handbooks for the identification of British insects* 10(15): 142 pp.

MERRITT, R.W., COURTNEY, G.W. & KEIPER, J.B. 2009. Diptera (flies, mosquitoes, midges, gnats). In: RESH, V.H. & CARDÉ, R.T. (eds.) *Encyclopedia of insects* (2nd edn.). Academic Press/Elsevier, San Diego & London, pp. 284–97.

MÖHN, E. 1966–71. Cecidomyiidae (= Itonididae). Cecidomyiinae (part). In: LINDNER, E. (ed.) *Die Fliegen der paläarktischen Region* 2(2): 248 pp.

NIJVELDT, W. 1969. *Gall midges of economic importance*. Crosby Lockwood & Son, London.

O'GRADY, P.M. 2009. Drosophila melanogaster. In: RESH, V.H. & CARDÉ, R.T. (eds.) *Encyclopedia of insects* (2nd edn.). Academic Press/Elsevier, San Diego & London, pp. 301–3.

OLAFSSON, E. 1991. Taxonomic revision of western palae-arctic species of the genera *Scatella* R.-D. and *Lampro-scatella* Hendel, and studies on their phylogenetic positions within the subfamily Ephydrinae (Diptera, Ephydridae). *Entomologica Scandinavica* (Suppl.) 37: 100 pp.

OLDROYD, H. 1964. *The natural history of flies*. Weidenfeld & Nicholson, London.

OLDROYD, H. 1969. Diptera Brachycera Section (a) Taba-noidea and Asiloidea. *Handbooks for the identification of British insects* 9(7): 74 pp.

OLDROYD, H. 1970. Diptera 1. Introduction and key to families (3rd edn., rewritten and enlarged). *Handbooks for the identification of British insects* 9(1): 104 pp.

OOSTERBROEK, P. 2006. *The European families of the Diptera: identification, diagnosis, biology*. KNNV Publishing, Utrecht.

PANELIUS, S. 1965. A revision of the European gall midges of the subfamily Porricondylinae (Diptera: Itonidae). *Acta Zoologica Fennica* 113: 157 pp.

PAPE, T. 1987. The Sarcophagidae (Diptera) of Fennoscandia and Denmark. *Fauna Entomologica Scandinavica* 19: 203 pp.

PEUS, F. 1958. Liriopeidae. In: LINDNER, E. (ed.) *Die Fliegen der paläarktischen Region* 3(1): 44 pp.

PITKIN, B.R. 1988. Lesser dung flies. Diptera: Sphaero-ceridae. *Handbooks for the identification of British insects* 10(5e): 175 pp.

PONT, A.C. 1979. Sepsidae Diptera Cyclorrhapha, Acaly-ptrata. *Handbooks for the identification of British insects* 10(5c): 35 pp.

REDFERN, M. & ASKEW, R.R. 1998. *Plant galls*. Naturalists' Handbooks no. 17 (2nd edn.). Richmond Publishing, Slough.

REDFERN, M. & SHIRLEY, P. 2002. *British plant galls: identification of galls on plants and fungi*. Field Studies Council, OP270.

REUSCH, H. & OOSTERBROEK, P. 1997. Diptera Limoniidae and Pediciidae, Short-palped crane flies. In: NILSSON, A. (ed.) *Aquatic insects of north Europe. Vol. 2. Odonata–Diptera*. Apollo Books, Stenstrup, pp. 105–32.

ROGNES, K. 1991. The Calliphoridae (Diptera) of Fenno-scandia and Denmark. *Fauna Entomologica Scandinavica* 24: 272 pp.

ROTHERAY, G.E. 1993. Colour guide to hoverfly larvae (Diptera, Syrphidae). *Dipterist's Digest* 9: 155 pp.

ROZKOSNY, R. 1982. *A biosystematic study of the European Stratiomyidae (Diptera). Vol. 1. Introduction, Beridinae, Sarginae and Stratiomyinae*. Junk, Den Haag.

ROZKOSNY, R. 1983. *A biosystematic study of the European Stratiomyidae (Diptera). Vol. 2. Clitellariinae, Hermetiinae, Pachygasterinae and bibliography*. Junk, Den Haag.

ROZKOSNY, R. 1984. The Sciomyzidae (Diptera) of Fenno-scandia and Denmark. *Fauna Entomologica Scandinavica* 14: 224 pp.

SAETHER, O.A. 1997. Diptera Chaoboridae, Phantom midges. In: NILSSON, A. (ed.) *Aquatic insects of north Europe. Vol. 2. Odonat–Diptera*. Apollo Books, Stenstrup, pp. 149–61.

SCHAFFNER, F., ANGEL, G., GEOFFROY, B., HERVY, J.P., RHAIEM, A. & BRUNHES, J. 2001. *Les moustiques d'Europe: the mosquitoes of Europe*. CD-ROM. Institut de Recherche pour le Developpement, Montpellier.

SERVICE, M.W. 1993. Mosquitoes (Culicidae). In: LANE, R.P. & CROSSKEY, R.W. (eds.) *Medical insects and arach-nids*. Chapman & Hall, London, pp. 120–240.

SHORROCKS, B. 1972. *Invertebrate types: Drosophila*. Ginn, London.

SKIDMORE, P. 1985. The biology of the Muscidae of the world. *Series Entomologica* 29: 550 pp.

SMITH, K.G.V. 1969a. Diptera Conopidae. *Handbooks for the identification of British insects* 10(3a): 19 pp.

SMITH, K.G.V. 1969b. Diptera Lonchopteridae. *Handbooks for the identification of British insects* 10(2ai): 9 pp.

SMITH, K.G.V. 1989. An introduction to the immature stages of British flies. Diptera larvae, with notes on eggs, puparia and pupae. *Handbooks for the identification of British insects* 10(14): 280 pp.

SNOW, K.R. 1990. *Mosquitoes*. Naturalists' Handbooks no. 14. Richmond Publishing, Slough.

SOÓS, Á. & PAPP, L. (eds.) 1984–93. *Catalogue of palaearctic Diptera*. 13 vols. Elsevier, Budapest.

SPENCER, K.A. 1972. Diptera Agromyzidae. *Handbooks for the identification of British insects* 10(5g): 136 pp.

SPENCER, K.A. 1976. The Agromyzidae (Diptera) of Fenno-scandia and Denmark. *Fauna Entomologica Scandinavica* 5(1): 1–304; 5(2): 305–606.

STACKELBERG, A.A. & NEGROBOV, O.P. 1930–79. Dolichopo-didae. In: LINDNER, E. (ed.) *Die Fliegen der paläarktischen Region* 4(5): 530 pp.

STUBBS, A.E. 1992. *Provisional atlas of the long-palped crane-flies (Diptera: Tipulinae) of Britain and Ireland*. Biological Records Centre, Huntingdon.

STUBBS, A.E. 1993. *Provisional atlas of the ptychopterid crane-flies (Diptera: Ptychopteridae) of Britain and Ireland*. Biological Records Centre, Huntingdon.

STUBBS, A.E. & DRAKE, M. 2001. *British soldierflies and their allies*. British Entomological and Natural History Society, Reading.

STUBBS, A.E. & FALK, S. 2002. *British hoverflies, an illustrated identification guide* (2nd edn.). British Entomological and Natural History Society, Reading.

SZADZIEWSKI, R., KRZYWINSKI, J. & GILKA, W. 1997. Diptera Ceratopogonidae, biting midges. In: NILSSON, A. (ed.) *Aquatic insects of north Europe. Vol. 2. Odonata–Diptera*. Apollo Books, Stenstrup, pp. 243–63.

THEOWALD, B. 1973–80. Tipulidae. In: LINDNER, E. (ed.) *Die Fliegen der paläarktischen Region* 3(5): 538 pp.

UNWIN, D.M. 1981. *A key to the families of British Diptera*. Field Studies Council (AIDGAP) no. 143: 40 pp. (reprinted with minor revisions, 1984).

VEEN, M.P. van 2004. *Hoverflies of northwest Europe*. KNNV Publishing, Utrecht.

VERVES, Y.G. 1982–93. Sarcophaginae. In: LINDNER, E. (ed.) *Die Fliegen der paläarktischen Region* 11(64h): 235–504.

WAGNER, R. 1997a. Diptera Psychodidae, moth flies. In: NILSSON, A. (ed.) *Aquatic insects of north Europe. Vol. 2. Odonata–Diptera.* Apollo Books, Stenstrup, pp. 133–44.

WAGNER, R. 1997b. Diptera Dixidae, meniscus midges. In: NILSSON, A. (ed.) *Aquatic insects of north Europe. Vol. 2. Odonata–Diptera.* Apollo Books, Stenstrup, pp. 145–8.

WAGNER, R. 1997c. Diptera Thaumaleidae. In: NILSSON, A. (ed.) *Aquatic insects of north Europe. Vol. 2. Odonata–Diptera.* Apollo Books, Stenstrup, pp. 187–90.

WHITE, I.M. 1988. Tephritid flies. Diptera: Tephritidae. *Handbooks for the identification of British insects* 10(5a): 134 pp.

WHITE, I.M. & ELSON-HARRIS, M.M. 1992. *Fruitflies of economic significance: their identification and bionomics.* CAB International Publishing, Wallingford, 601 pp.

WIEDERHOLM, T. (ed.) 1983. Chironomidae of the holarctic region. Keys and diagnoses. Part 1 Larvae. *Entomologica Scandinavica* (Suppl.) 19: 457 pp.

WIEDERHOLM, T. (ed.) 1986. Chironomidae of the holarctic region. Keys and diagnoses. Part 2 Pupae. *Entomologica Scandinavica* (Suppl.) 28: 482 pp.

WIEDERHOLM, T. (ed.) 1989. Chironomidae of the holarctic region. Keys and diagnoses. Part 3 Adult males. *Entomologica Scandinavica* (Suppl.) 34: 532 pp.

WILSON, R.S. & RUSE, L.P. 2005. *A guide to the identification of genera of chironomid pupal exuviae occurring in Britain and Ireland.* Freshwater Biological Association, Ambleside, Special Publication no.13.

WITHERS, P. 1989. Moth flies. Diptera: Psychodidae. *Dipterist's Digest* 4: 83 pp.

WYATT, N.P. & CHAINEY, J.E. 1999. Diptera: the flies. In: BARNARD, P.C. (ed.) *Identifying British insects and arachnids: an annotated bibliography of key works.* Cambridge University Press, Cambridge, pp. 171–93.

ZATWARNICKI, T. 1997. Diptera Ephydridae, shore flies. In: NILSSON, A. (ed.) *Aquatic insects of north Europe. Vol. 2. Odonata–Diptera.* Apollo Books, Stenstrup, pp. 383–99.

WEBSITES

http://www.dipteristsforum.org.uk/
The essential site for anyone interested in British Diptera, which also has some draft keys for testing.

http://www.ukflymines.co.uk/
An important site for information on all the leaf-mining Diptera.

http://www.leafmines.co.uk/index.htm
A site containing photos and information on all groups of leaf-mining insects.

http://www.brc.ac.uk/DBIF/homepage.aspx
A general site listing records of the foodplants of British insects.

http://www.sel.barc.usda.gov/Diptera/biosys.htm
The world checklist of Diptera.

http://www.diptera.info/news.php
A worldwide site about Diptera in general.

20 Order Hymenoptera: the ants, bees and wasps

c. 7000 species in 57 families

This group is arguably the largest and most complex of all the orders of insects and they seem to be only distantly related to the other endopterygotes. There is evidence that they are closer to the panorpid groups rather than the Coleoptera but the picture is far from clear; see Grimaldi and Engel (2005) for a detailed analysis of the current knowledge. Although their monophyly is not in doubt, it is hard to give defining characters for the whole group that are not plesiomorphic. In the winged species the fore wings are larger than the hind ones, and the wings on each side are linked during flight by rows of hooks called hamuli; some authors consider this the origin of the word Hymenoptera, from the Greek god of marriage 'Hymen', referring to the 'married' wings, instead of 'hymen' meaning a membrane. Most people are familiar with the common groups of ants, bees and wasps, and perhaps with the sawflies and parasitic wasps. The larger and more conspicuous groups are often important pollinators of plants, and a few are pests in domestic settings, but the great majority of Hymenoptera belong in families little known except to the specialists.

Understanding the biology of the Hymenoptera depends on two major points, the method of sex determination, and the different kinds of parasitism; both of these have shaped the evolution of the order as a whole. All are haplodiploid, in that females develop from diploid (fertilized) eggs, while males develop from haploid (unfertilized) ones, though this method of controlling sex ratios is not confined to the Hymenoptera. Although the group as a whole exhibits a wide range of life styles, from phytophagy to predation, the development of highly sophisticated types of parasitism has contributed enormously to the success of the group. Many families are loosely termed parasitic wasps, but true parasites do not kill their hosts; most Hymenoptera are actually parasitoids, either internal or external, and they eventually kill the host, though not before they have finished their own development. Some are even hyperparasitoids, which attack a different parasitoid that is already feeding inside or on its host. Within the aculeate Hymenoptera, the evolution of complex social behaviour has been another major influence on their success.

Traditionally the Hymenoptera are divided into two main groups, the Symphyta (sawflies) and the Apocrita, which are in turn divided into the Aculeata (including the stinging and social groups) and the Parasitica (the largest section of the order). Only the Aculeata can be considered as monophyletic, and they are probably derived from some of the 'Parasitica', which means that the latter group is paraphyletic; the Symphyta are considered the most basal group of all and, again, cannot be monophyletic. Nonetheless these convenient groupings are retained for the time being, until a more robust phylogeny is uncovered.

In general, the entomologist wanting to identify many groups of British Hymenoptera will need to consult specialist papers. Although there are a few recent guides to popular groups such bumblebees, even the aculeates are not well covered by comprehensive works. There are several useful guides listed on the Bees, Wasps and Ants Recording Society (BWARS) website (http://www.bwars.com) and the single most valuable book is the comprehensive guide to all families by Gauld and Bolton

The Royal Entomological Society Book of British Insects, First Edition. Peter C. Barnard.
© 2011 Royal Entomological Society. Published 2011 by Blackwell Publishing Ltd.

(1988), which contains many important references. The newcomer to this group will also find Betts (1986) useful, though it is becoming out of date. Other broadly-based works that contain valuable information are Askew (1971) on parasitic insects in general, Godfray (1994) on parasitoid biology and Quicke (1997) on the parasitic wasps. A useful overview of the parasites of Lepidoptera is given by Shaw and Askew (1976). A key to families was provided by Richards (1977) but it is now superseded by that of Gauld and Bolton (1988). A simple guide to the whole group is Zahradník (1998); although basically a translation of a central European book it has British species indicated and includes illustrations of a few groups not usually covered in general insect guides. Step's (1932) book was a classic of its time, but is not recommended for accurate identifications, even in the aculeates; other works are listed under the appropriate sections below. Identification of many groups will need access to specialist literature, and frequent reference to Noyes et al. (1999) will be essential.

There are about 16,000 species in 66 families in Europe, with around 150,000 species in 90 families worldwide; a brief world overview of the Hymenoptera is given by Quicke (2009).

HIGHER CLASSIFICATION OF BRITISH HYMENOPTERA

Suborder Apocrita–Aculeata
  Superfamily Apoidea
    Family Apidae (31 genera, 256 species)
    Family Crabronidae (31 genera, 118 species)
    Family Sphecidae (2 genera, 4 species)
  Superfamily Chrysidoidea
    Family Bethylidae (10 genera, 22 species)
    Family Chrysididae (11 genera, 33 species)
    Family Dryinidae (7 genera, 33 species)
    Family Embolemidae (1 genus, 1 species)
  Superfamily Vespoidea
    Family Formicidae (20 genera, c. 60 species)
    Family Mutillidae (3 genera, 3 species)
    Family Pompilidae (14 genera, 42 species)
    Family Sapygidae (2 genera, 2 species)
    Family Tiphiidae (2 genera, 3 species)
    Family Vespidae (12 genera, 33 species)
Suborder Apocrita–Parasitica
  Superfamily Ceraphronoidea
    Family Ceraphronidae (3 genera, 26 species)
    Family Megaspilidae (5 genera, 61 species)
  Superfamily Chalcidoidea
    Family Aphelinidae (12 genera, 38 species)
    Family Chalcididae (6 genera, 8 species)
    Family Elasmidae (1 genus, 4 species)
    Family Encyrtidae (77 genera, c. 200 species)
    Family Eucharitidae (1 genus, 1 species)
    Family Eulophidae (c. 50 genera, c. 500 species)
    Family Eupelmidae (5 genera, 14 species)
    Family Eurytomidae (5 genera, 91 species)
    Family Mymaridae (19 genera, 87 species)
    Family Ormyridae (1 genus, 3 species)
    Family Perilampidae (2 genera, 9 species)
    Family Pteromalidae (c. 150 genera, c. 600 species)
    Family Signiphoridae (2 genera, 2 species)
    Family Tetracampidae (5 genera, 7 species)
    Family Torymidae (7 genera, 75 species)
    Family Trichogrammatidae (17 genera, 29 species)
  Superfamily Cynipoidea
    Family Cynipidae (18 genera, 91 species)
    Family Figitidae (30 genera, 93 species)
    Family Ibaliidae (1 genus, 2 species)
  Superfamily Evanioidea
    Family Aulacidae (1 genus, 1 species)
    Family Evaniidae (2 genera, 2 species)
    Family Gasteruptiidae (1 genus, 5 species)
  Superfamily Ichneumonoidea
    Family Braconidae (c. 140 genera, c. 1045 species)
    Family Ichneumonidae (c. 400 genera, c. 2100 species)
  Superfamily Mymarommatoidea
    Family Mymarommatidae (1 genus, 1 species)
  Superfamily Platygastroidea
    Family Platygastridae (15 genera, c. 157 species)
    Family Scelionidae (14 genera, c. 102 species)
  Superfamily Proctotrupoidea
    Family Diapriidae (38 genera, c. 300 species)
    Family Heloridae (1 genus, 3 species)
    Family Proctotrupidae (11 genera, 40 species)
  Superfamily Trigonaloidea
    Family Trigonalidae (1 genus, 1 species)
Suborder Symphyta
  Superfamily Cephoidea
    Family Cephidae (5 genera, 12 species)
  Superfamily Orussoidea
    Family Orussidae (1 genus, 1 species)
  Superfamily Pamphilioidea
    Family Pamphiliidae (4 genera, 19 species)
  Superfamily Siricoidea
    Family Siricidae (4 genera, 11 species)
  Superfamily Tenthredinoidea
    Family Argidae (3 genera, 15 species)
    Family Blasticotomidae (1 genus, 1 species)
    Family Cimbicidae (4 genera, 14 species)
    Family Diprionidae (5 genera, 9 species)
    Family Tenthredinidae (c. 75 genera, c. 400 species)
  Superfamily Xiphrydioidea
    Family Xiphydriidae (1 genus, 3 species)
  Superfamily Xyeloidea
    Family Xyelidae (1 genus, 2 species)

SPECIES OF CONSERVATION CONCERN

No species of Hymenoptera have any legal protection but the following 35 species are on the UKBAP list: *Andrena ferox, Andrena tarsata, Anthophora retusa, Bombus distinguendus, Bombus humilis, Bombus muscorum, Bombus ruderarius, Bombus ruderatus, Bombus subterraneus, Bombus sylvarum, Colletes floralis, Colletes halophilus, Eucera longicornis, Lasioglossum angusticeps, Nomada armata, Nomada errans, Osmia inermis, Osmia parietina, Osmia uncinata* and *Osmia xanthomelana* (Apidae); *Chrysis fulgida* and *Chrysura hirsuta* (Chrysididae); *Cerceris quadricincta* and *Cerceris quinquefasciata* (Crabronidae); *Anergates atratulus, Formica exsecta, Formica pratensis, Formica rufibarbis, Formicoxenus nitidulus, Tapinoma erraticum* and *Temnothorax interruptus* (Formicidae); *Homonotus sanguinolentus* (Pompilidae); *Odynerus melanocephalus, Odynerus simillimus* and *Pseudepipona herrichii* (Vespidae).

The following species were removed from the UKBAP list in 2007 for various reasons: *Andrena gravida, Andrena lathyri* and *Nomada ferruginata* (Apidae); *Formica aquilonia, Formica candida, Formica lugubris* and *Formica rufa* (Formicidae); *Evagetes pectinipes* (Pompilidae)

Falk (1991) gave a useful overview of threatened species of ants, bees and wasps.

Fig. 20.1 Apidae with pollen baskets (Photo: Peter Barnard)

The Families of British Hymenoptera

SUBORDER APOCRITA–ACULEATA

In the aculeate Hymenoptera the female ovipositor is modified to form a sting. Some families have parasitoid larvae, others are free-living and solitary, but the true social families are also in this group. Willmer (1985) provided a simplified key to the genera of the aculeates.

SUPERFAMILY APOIDEA

Family Apidae (31 genera, 256 species)
These are known generally as the bees, although this is a large and complex family; the current subfamilies have all been treated as separate families in the past. Although *Apis mellifera*, the Honey bee, is the iconic member of this family, most species are important pollinators of angiosperms and they have special collecting areas for storing pollen in flight, either on the abdomen or the hind legs (Fig. 20.1). If these structures are not present the bee will eat the pollen and nectar, then later regurgitate

Fig. 20.2 *Hylaeus communis* (Apidae) (Photo: Roger Key)

them into the cells of the nest to feed the young; an example of this group is *Hylaeus* (Fig. 20.2).

The social structure of honeybees and some bumblebees is well known, with most nests being populated by worker bees, which are sterile females. However, most bee species are solitary but they often build nests close to each other, which can give a false impression of sociality. Nests may consist of varying numbers of cells, and these are usually provisioned before the eggs hatch, so there is no direct feeding of the larvae by their mother. The social nests of honeybees and bumblebees are built from wax, but other groups use a wide variety of materials, from mud to petals and leaves (Fig. 20.3); some use hollow stems (Figs. 20.4 & 20.5).

Fig. 20.3 *Megachile versicolor* (Apidae) (Photo: Robin Williams)

Fig. 20.4 *Osmia rufa* and nest (Apidae) (Photo: Robin Williams)

Fig. 20.5 *Osmia rufa* cell (Apidae) (Photo: Robin Williams)

Fig. 20.6 *Bombus (Psithyrus) vestalis* (Apidae) (Photo: Robin Williams)

Fig. 20.7 *Bombus terrestris* nest (Apidae) (Photo: Robin Williams)

Some bees are cleptoparasites on other Apidae, laying their eggs in the host's nest where the young larvae will destroy the host egg or larva and then live on the food provided by the host bee. One group of bumblebees in the subgenus *Psithyrus* are social parasites on other species of *Bombus*; they take over the nest and after killing the queen they use the host workers to raise their own brood. They often resemble their host species quite closely but lack any pollen storage structures. *Bombus (Psithyrus) vestalis* (Fig. 20.6) is a parasite on *B. terrestris* (Fig. 20.7).

Genera that are commonly seen include *Andrena* (Fig. 20.8), *Anthophora* (Fig. 20.9), *Megachile* (a

Fig. 20.8 *Andrena clarkella* (Apidae) (Photo: Roger Key)

Fig. 20.9 *Anthophora plumipes* (Apidae) (Photo: Robin Williams)

Fig. 20.10 *Osmia aurulenta* (Apidae) (Photo: Roger Key)

Fig. 20.11 *Osmia rufa* with phoretic mites (Apidae) (Photo: Robin Williams)

Fig. 20.12 *Eucera longicornis* (Apidae) (Photo: Roger Key)

leaf-cutter bee) and *Osmia* (Figs. 20.10 & 20.11). A striking species is *Eucera longicornis*, with its very long antennae in the male (Fig. 20.12). One species that may be becoming established in Britain is the southern European carpenter bee *Xylocopa violacea* (Fig. 20.13); there have been several recent sightings.

Bumblebees (Fig. 20.14) are a familiar sight, but there is concern about the status of several species. However, it should be noted that some guides to bumblebees will omit *Bombus hypnorum* (the Tree

Fig. 20.13 *Xylocopa* (Apidae) (Photo: Peter Barnard)

Fig. 20.14 *Bombus hortorum* (Apidae) (Photo: Robin Williams)

bumblebee) as it was not found in the UK until 2004, though it has rapidly spread northwards through England and is quite a common garden insect in some areas

The following species are on the UKBAP list: *Andrena ferox* (Oak mining bee), *Andrena tarsata* (Tormentil mining bee), *Anthophora retusa* (Potter flower bee), *Bombus distinguendus* (Great yellow bumblebee), *Bombus humilis* (Brown-banded carder-bee), *Bombus muscorum* (Moss carder-bee), *Bombus ruderarius* (Red-shanked carder-bee), *Bombus ruderatus* (Large garden bumblebee), *Bombus subterraneus* (Short-haired bumble-bee), *Bombus sylvarum* (Shrill carder-bee), *Colletes floralis* (The northern Colletes),

Colletes halophilus (Sea-aster Colletes bee), *Eucera longicornis* (Long-horned bee), *Lasioglossum angusticeps*, *Nomada armata*, *Nomada errans*, *Osmia inermis*, *Osmia parietina* (Wall mason bee), *Osmia uncinata* and *Osmia xanthomelana* (Large mason bee).

Identification: despite the popularity of this group there are currently no comprehensive identification guides to all the British Apidae; Edwards and Jenner (2009) reliably cover the bumblebees, and Koster (1986) includes a key (in English) to the British species of *Hylaeus*. A work in German by Westrich (1989) covers many British species with good photos. Noyes et al. (1999) should be consulted for specialist references.

The last work to cover the entire group was Saunders (1896) and while it is a classic work it is obviously out of date. On a world scale, the well-illustrated guide to this family by O'Toole and Raw (1991) is useful; and for an overview of honey bees see Crane (2009). There is an online guide to the bumblebees (http://www.nhm.ac.uk/research-curation/research/projects/bombus/key_british_colour.html) and a useful general website to the same group (http://www.bumblebee.org). The following site has information on solitary bees: http://www.insectpix.net/index.htm.

British subfamilies and genera:
Andreninae: *Andrena, Panurgus*
Anthophorinae: *Anthophora, Epeolus, Eucera, Melecta, Nomada*
Apinae: *Apis, Bombus*
Colletinae: *Colletes, Hylaeus*
Halictinae: *Dufourea, Halictus, Lasioglossum, Rophites, Sphecodes*
Megachilinae: *Anthidium, Chalicodoma, Chelostoma, Coelioxys, Heriades, Hoplitis, Hoplosmia, Megachile, Osmia, Stelis*
Melittinae: *Dasypoda, Macropis, Melitta*
Xylocopinae: *Ceratina, Xylocopa*

Family Crabronidae (31 genera, 118 species)
These were formerly included as a subfamily in the Sphecidae. Adult crabronids are often seen feeding on flowers, but the females use a wide range of prey species to provision their nests, taking spiders, Coleoptera, Hemiptera and many other insect groups. *Nysson* species are cleptoparasites in the nests of other crabronids. Most species are restricted to a very narrow range of prey species, but *Crossocerus* and *Lindenius* are known to prey on many insect groups; nearly always the nest is provisioned with many small insects, rather than a small number of large specimens.

Fig. 20.15 *Crossocerus megacephalus* (Crabronidae) (Photo: Robin Williams)

Fig. 20.17 *Pemphredon lugubris* (Crabronidae) (Photo: Roger Key)

Fig. 20.16 *Ectemnius* sp. (Crabronidae) (Photo: Roger Key)

Fig. 20.18 *Philanthus triangulum* (Crabronidae) (Photo: Robin Williams)

Common genera include *Crabro*, *Crossocerus* (Fig. 20.15), *Ectemnius* (Fig. 20.16), *Pemphredon* (Fig. 20.17), and *Trypoxylon*. One species, *Philanthus triangulum*, has the dramatic name of Bee killer (Fig. 20.18); it seems to be declining in Britain, though this may be due to reduction in numbers of honeybees.

Cerceris quadricincta (Four-banded weevil-wasp) and *Cerceris quinquefasciata* (Five-banded weevil-wasp) are on the UKBAP list.

Identification: the British species were keyed by Richards (1980) under the Sphecidae; a useful simplified guide is by Yeo and Corbet (1995).

British subfamilies and genera:

Astatinae: *Astata*, *Dinetus*

Crabroninae: *Crabro*, *Crossocerus*, *Ectemnius*, *Entomognathus*, *Lestica*, *Lindenius*, *Nitela*, *Oxybelus*, *Rhopalum*

Larrinae: *Miscophus*, *Tachysphex*

Mellininae: *Mellinus*

Nyssoninae: *Alysson*, *Argogorytes*, *Gorytes*, *Lestiphorus*, *Nysson*

Pemphredoninae: *Diodontus*, *Mimesa*, *Mimumesa*, *Passaloecus*, *Pemphredon*, *Psen*, *Psenulus*, *Spilomena*, *Stigmus*

Philanthinae: *Cerceris*, *Philanthus*

Trypoxylinae: *Trypoxylon*

Family Sphecidae (2 genera, 4 species)

Now that most members of this family have been removed to the Crabronidae, there are just four species remaining in the Sphecidae. All take Lepidopteran caterpillars as prey to provision their

Fig. 20.19 *Ammophila sabulosa* with caterpillar prey (Sphecidae) (Photo: Roger Key)

Fig. 20.20 *Chrysis viridula* (Chrysididae) (Photo: Roger Key)

nests and they usually catch a small number of very large specimens, which are often larger than the wasp itself (Fig. 20.19). After closing the nest entrance some species are reported as tamping down the soil with their head; some (non-British) species even use a small stone held in the mandibles, an interesting example of tool use in the insects.

Identification: the British species were keyed by Richards (1980) under the Sphecidae; a useful simplified guide is by Yeo and Corbet (1995).

British genera: *Ammophila, Podalonia*

SUPERFAMILY CHRYSIDOIDEA

Family Bethylidae (10 genera, 22 species)
Bethylids are all predators on the larvae of Coleoptera or Lepidoptera, which are often taken under bark or in similar hidden situations. The large prey is subdued by the female's powerful sting and is then dragged away; female bethylids are often brachypterous or wingless, though the males are usually fully winged. Eggs are laid on the prey, and the larvae then feed externally on it. Bethylid stings can penetrate human skin, with painful results.

Identification: the handbook by Perkins (1976) is still useful though slightly outdated.

British subfamilies and genera:
Bethylinae: *Bethylus, Goniozus*
Epyrinae: *Allepyris, Cephalonomia, Epyris, Holepyris, Plastanoxus*
Pristocerinae: *Pseudisobrachium*

Fig. 20.21 *Trichrysis cyanea* (Chrysididae) (Photo: Robin Williams)

Family Chrysididae (11 genera, 33 species)
Because they are cleptoparasites on other Hymenoptera, chrysidids are known as cuckoo wasps, but their metallic colours also lead to names such as ruby-tailed or jewel wasps. As well as being brightly coloured, the cuticle is also very thick and heavily sculptured. The adults are usually seen flying in sunshine investigating likely holes, which might indicate the nests of their hosts; they also feed at flowers.

Common genera include *Chrysis* (Fig. 20.20), *Cleptes* and *Trichrysis* (Fig. 20.21).

Chrysis fulgida (Ruby-tailed wasp) and *Chrysura hirsuta* (Northern osmia ruby-tailed wasp) are on the UKBAP list.

Identification: Morgan's (1984) handbook is still the key work, though some nomenclature has changed slightly; Kimsey and Bohart (1990) give a world overview of the family.

British subfamilies and genera:
   Chrysidinae: *Chrysis, Chrysogona, Chrysura, Pseudospinolia, Trichrysis*
   Cleptinae: *Cleptes*
   Elampinae: *Elampus, Hedychridium, Hedychrum, Omalus*

Family Dryinidae (7 genera, 33 species)

Dryinid wasps usually capture nymphs of Hemipteran Cicadelloidea or Fulgoroidea; the prey is paralysed by a sting from the female who then lays her eggs on its abdomen and the larvae develop in a kind of cyst projecting from the host, whilst feeding on its internal tissues. By the time the dryinid larva emerges to pupate, the host is usually dead or moribund. Pupation takes place in a silk cocoon, and at this stage they are frequently parasitized by other Hymenoptera such as Diapriidae or Encyrtidae.

Identification: Perkins (1976) is still useful but see Noyes et al. (1999) for specialist literature. Olmi (1984) gives a valuable world overview of the family, which includes keys to the Palaearctic species.

British genera: *Anteon, Aphelopus, Dryinus, Gonatopus, Haplogonatopus, Lonchodryinus, Mystrophorus*

Family Embolemidae (1 genus, 1 species)

This small family includes only one British species, *Embolemus ruddii*, and little is known of its biology although it seems to be fairly widespread in England at least. Females have been found throughout the months of winter and spring, and are therefore assumed to overwinter as adults. The host of *E. ruddii* is unknown, but it is likely to be an auchenorrynchan Hemipteran.

Identification: Perkins (1976)

British genus: *Embolemus*

SUPERFAMILY VESPOIDEA

One family, the Scoliidae, is represented by a single species from the Channel Islands, and is not included here.

Family Formicidae (20 genera, c. 60 species)

The number of ant species in Britain is small, with around 10 of the 60 species being introductions that are usually confined to heated buildings. Ants usually attract attention when they invade houses in search of food supplies (Fig. 20.22) or when they

Fig. 20.22 *Lasius niger* foraging in house (Formicidae) (Photo: Roger Key)

Fig. 20.23 *Lasius niger* swarming (Formicidae) (Photo: Roger Key)

form mating swarms of 'flying ants' in the summer (Fig. 20.23). Most species nest in the soil, often beneath stones, although some such as *Lasius flavus* build mounds, which can be prominent features, around 30 cm high, in heath or meadowland. The large nests of the Wood ant, *Formica rufa* (or a related species) are impressive structures, which can contain a colony of up to half a million individuals. Ant colonies are usually permanent, and most species have a wingless worker caste; the much larger queen sheds her wings after mating (Fig. 20.24). A few species are social parasites on other species and one of these has no worker caste, instead using host workers to create its nest and maintain its brood.

   The so-called 'ants eggs' used as bait by anglers are actually the ant pupae (Fig. 20.25).

Fig. 20.24 *Lasius niger* queen, workers and pupae (Formicidae) (Photo: Peter Barnard)

Fig. 20.26 *Mutilla europaea* (Mutillidae) (Photo: Roger Key)

Fig. 20.25 *Myrmica* sp. with pupae (Formicidae) (Photo: Roger Key)

The following species are on the UKBAP list: *Anergates atratulus* (Dark guest ant), *Formica exsecta* (Narrow-headed ant), *Formica pratensis* (Black-backed meadow ant), *Formica rufibarbis* (Red barbed ant), *Formicoxenus nitidulus* (Shining guest ant), *Tapinoma erraticum* (Erratic ant) and *Temnothorax interruptus* (Long-spined ant).

Identification: Bolton and Collingwood (1975) and Skinner and Allen (1996) give keys to the British species; there is a useful simplified guide by Skinner (1987). Bolton (1994, 1995) gives important reviews at a world level, and the biology of the family is covered by Hölldobler and Wilson (1990); for a world overview of ants see Franks (2009). The worldwide Antbase website has much useful information (http://antbase.org).

British subfamilies and genera:
Dolichoderinae: *Iridomyrmex, Tapinoma*
Formicinae: *Formica, Lasius, Paratrechina*
Myrmicinae: *Anergates, Crematogaster, Formicoxenus, Leptothorax, Monomorium, Myrmica, Myrmecina, Pheidole, Sifolinia, Solenopsis, Stenamma, Strongylognathus, Temnothorax, Tetramorium*
Ponerinae: *Hypoponera, Ponera*

Family Mutillidae (3 genera, 3 species)
The members of this small family are known as velvet ants from the bands of silvery hairs on their body; the wingless females are particularly ant-like in appearance, but the males are fully winged. The three British species are *Mutilla europaea* (Fig. 20.26), *Smicromyrme rufipes* and *Myrmosa atra*; all are uncommon and are restricted to sandy soils such as heathland and sand dunes. They are parasitoids on other aculeate Hymenoptera such as bumblebees or solitary wasps; they lay their eggs on the host larva or pupa and the mutillid larvae then attack and eat the host.

Identification: Richards (1980) with the Myrmosinae included in the Tiphiidae; Yeo and Corbet (1995).

British subfamilies and genera:
Mutillinae: *Mutilla, Smicromyrme*
Myrmosinae: *Myrmosa*

Family Pompilidae (14 genera, 42 species)
Like some other families, the spider wasps are basically a warm-climate group; many species are found

Fig. 20.27 *Cryptocheilus notatus* (Pompilidae) with spider prey (Photo: Robin Williams)

Fig. 20.29 *Anoplius infuscatus* (Pompilidae) (Photo: Roger Key)

Fig. 20.28 *Priocnemis exaltata* with prey (Pompilidae) (Photo: Robin Williams)

only in the southern half of Britain and there are several more in the Channel Islands. As their common name suggests, they provision their nests with spiders, placing just one large spider in each cell (Figs. 20.27 & 20.28). Some species are cleptoparasites on other pompilids.

Adult spider wasps are rather easy to recognize in the field, as they spend more time running erratically on the ground than in flight; their antennae and wings are constantly vibrating, which has led some authors to dub them the 'Hymenoptera Neurotica'. Most of the British species have red markings on the abdomen (Fig. 20.29), though a few are completely black.

Homonotus sanguinolentus (Bloody spider-hunting wasp) is on the UKBAP list.

Identification: Day's (1988) handbook covers the British species.

British subfamilies and genera:
Ceropalinae: *Ceropales*
Pepsinae: *Auplopus, Caliadurgus, Cryptocheilus, Dipogon, Priocnemis*
Pompilinae: *Agenioideus, Anoplius, Aporinellus, Aporus, Arachnospila, Episyron, Evagetes, Homonotus, Pompilus*

Family Sapygidae (2 genera, 2 species)
The two solitary wasps in this family, *Monosapyga clavicornis* and *Sapyga quinquepunctata*, are brightly coloured with black, yellow and sometimes red bands; neither species is common. Their larvae are cleptoparasites in the nests of bees in the Megachilinae (Apidae). Unusually in the aculeates, the sting keeps its original function as an ovipositor.

Identification: Richards (1980) or Yeo and Corbet (1995).

British genera: *Monosapyga, Sapyga*

Family Tiphiidae (2 genera, 3 species)
This small family is represented in Britain by two species of *Tiphia, T. femorata* and *T. minuta*, with one

Fig. 20.30 *Methoca ichneumonides* (Tiphiidae) (Photo: Roger Key)

Fig. 20.31 *Dolichovespula media* queen (Vespidae) (Photo: Roger Key)

species of Methocinae, *Methocha ichneumonides*. All three are uncommon species that are most likely to be seen in southern Britain, and all lay their eggs on beetle larvae, *Tiphia* on scarabaeids and *Methoca* on Cicindelinae (Carabidae). *M. ichneumonides* shows remarkable sexual dimorphism, as the male is fully winged, with a narrow black body, while the wingless female has a red thorax and legs, and is very ant-like in appearance (Fig. 20.30).

Identification: Richards (1980) or Yeo and Corbet (1995).

British subfamilies and genera:
Methocinae: *Methoca*
Tiphiinae: *Tiphia*

Family Vespidae (12 genera, 33 species)
Until recently the Eumenidae (the potter and mason wasps) were regarded as a distinct family, but they are now placed as a subfamily of the Vespidae (the social wasps), both groups can fold their wings longitudinally when at rest (Fig. 20.31).

The Eumeninae are solitary wasps which provision their mud nests with insect prey, often caterpillars of Lepidoptera (Fig. 20.32) or various beetle larvae. *Odynerus* species are black and yellow wasps (Fig. 20.33) that build subterranean nests, which have a characteristic chimney-shaped entrance (Fig. 20.34). *Eumenes coarctatus*, the aptly named potter wasp, builds a flask or pot-shaped nest that is usually attached to heather plants (Fig. 20.35).

The familiar social wasps build much more complex structured nests from 'carton' or wasp paper, made by chewing wood fibres; they contain hundreds of cells with an outer covering and the general shape of the nest varies between species

Fig. 20.32 *Ancistrocerus nigricornis* cell with prey (Vespidae) (Photo: Robin Williams)

Fig. 20.33 *Odynerus spinipes* (Vespidae) (Photo: Roger Key)

Fig. 20.34 *Odynerus spinipes* 'chimney' entrance to nest (Vespidae) (Photo: Roger Key)

Fig. 20.36 *Dolichovespula media* nest (Vespidae) (Photo: Roger Key)

Fig. 20.35 *Eumenes coarctatus* 'mud-pot' nest on heather (Vespidae) (Photo: Roger Key)

Fig. 20.37 *Vespula vulgaris* nest in roof-space (Vespidae) (Photo: Roger Key)

(Figs. 20.36 & 20.37). The internal structure is complex and organized (Fig. 20.38); eggs are laid individually in each cell (Fig. 20.39) so that each developing larva has a cell to itself (Fig. 20.40). Unlike the mass-provisioning seen in the Eumeninae, the social wasps continue to catch prey (Fig. 20.41), which is chewed up and fed to the developing larvae. Identification of British vespine wasps is not always as straightforward as some insect guides suggest, because the gross colour patterns can vary between castes as well as geographically; the patterns on the front of the head are often useful (Fig. 20.42).

One species, *Vespula austriaca*, is a social parasite on *V. rufa*, but this is the only British vespine wasp that does not have its own worker caste. *Polistes dominulus* has long been doubtfully included on the British list, but it may now be established in south-

ern England, as there have been several recent sightings. One alien species is a potential invader to this country: *Vespa velutina*, the Asian hornet, was introduced to France from China a few years ago; it is spreading through south-west France and may

Fig. 20.38 *Vespula germanica* nest showing interior structure (Vespidae) (Photo: Roger Key)

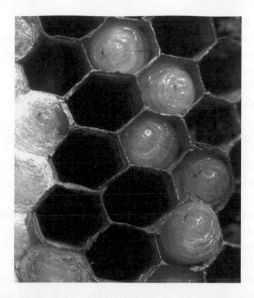

Fig. 20.40 *Vespula germanica* larvae in cells (Vespidae) (Photo: Roger Key)

Fig. 20.39 *Vespula germanica* eggs in cells (Vespidae) (Photo: Roger Key)

Fig. 20.41 Vespid wasp with fly prey (Photo: Colin Rew)

Fig. 20.42 Vespid wasp face (Photo: Colin Rew)

become an additional threat to honeybees, which are its main prey.

Odynerus melanocephalus (Black-headed mason-wasp), *Odynerus simillimus* (Fen mason-wasp) and *Pseudepipona herrichii* (Purbeck mason wasp) are on the UKBAP list.

Identification: Richards (1980) and Yeo and Corbet (1995) will work for the Eumeninae, but many books on the social wasps will omit two recent additions to the British fauna, *Dolichovespula media* and *D. saxonica*; see Noyes et al. (1999) for additional references. Two useful books on the biology of wasps are Spradbery (1973) and Edwards (1980).

British subfamilies and genera:

Eumeninae: *Ancistrocerus, Eumenes, Euodynerus, Gymnomerus, Microdynerus, Odynerus, Pseudepipona, Symmorphus*

Polistinae: *Polistes*

Vespinae: *Dolichovespula, Vespa, Vespula*

SUBORDER APOCRITA PARASITICA

Although the families of parasitic wasps are among the least well known, except to a relatively small number of specialist workers, they include over three-quarters of the British species of Hymenoptera. Although a few families have phytophagous larvae, principally in the Chalcidoidea and Cynipoidea, most are parasitoids on other invertebrates. Some species are widely used as control agents of insect pests on a commercial scale. Two useful general accounts are by Shaw (1997) and Shaw and Askew (1976). Two useful websites about the parasitic wasps are: http://chrisraper.org.uk/Html/parasitica.htm and http://hedgerowmobile.com/parasitica.html.

SUPERFAMILY CERAPHRONOIDEA

Some authors include this group within the Proctotrupoidea; see Muesebeck (1979) for a discussion. Identifications within this superfamily usually involve consulting specialist literature; see Noyes et al. (1999) for details. There is a checklist of the British species on the following website: http://www.nhm.ac.uk/resources-rx/files/superfamily-ceraphronoidea-27753.pdf.

Family Ceraphronidae (3 genera, 26 species)
The members of this family are assumed to be endoparasitoids on nematoceran flies, and possibly some other insect groups, but little is known of their biology.

Identification: there are no keys to the British ceraphronids, and it will be necessary to examine the specialized references in Noyes et al. (1999). The Russian work by Alekseev (1987a) is also useful.

British genera: *Aphanogmus, Ceraphron, Synarsis*

Family Megaspilidae (5 genera, 61 species)
Megaspilids are ectoparasitoids on a wide variety of insect hosts, which include Coccidae (Hemiptera), Neuroptera, Mecoptera and several families of Diptera. Several species are hyperparasitoids on Braconidae, Chalcidoidea and Cynipoidea, and some of these have an impact on agriculturally important pests such as aphids.

Identification: there are no guides to the British megaspilids, and it will be necessary to examine the specialized references in Noyes et al. (1999). The Russian work by Alekseev (1987b) is also useful.

British subfamilies and genera:

Lagynodinae: *Lagynodes*

Megaspilinae: *Conostigmus, Dendrocerus, Megaspilus, Trichosteresis*

SUPERFAMILY CHALCIDOIDEA

This large group of parasitic wasps attracts very little amateur attention because of a combination of several factors: there are obviously a great many species, most are very small and difficult to examine without a high-quality microscope, the higher classification is in some turmoil, and most of the relevant literature is in obscure journals in a variety of languages. It is therefore not surprising that the group is seen as firmly in the specialist's domain, even though they exhibit a wide variety of life styles. Most are parasitoids, but some are phytophagous on seeds, others cause stem-galls in grasses; the group as a whole probably includes more species used as biological control agents than any other.

Identification even to families can be very difficult, though the key in Gauld and Bolton (1988) is useful. Further identification will require specialist references; see Noyes et al. (1999) for details. It is also worth consulting Boucek (1988), as well as the Russian work by Nikol'skaya (1952) and the faunal work by Peck et al. (1964). There is a catalogue of world species with biological data on the following website: http://www.nhm.ac.uk/research-curation/research/projects/chalcidoids/index.html.

Family Aphelinidae (12 genera, 38 species)
Most aphelinids are parasitoids on scale insects, aphids and psyllids (Hemiptera), and they can be ectoparasitoids, endoparasitoids, egg parasitoids or egg predators. In some species the males and females develop on different hosts or as different forms of parasitoids. Most emerge from the dead host 'mummy' to pupate, though some form a pupation chamber inside the still-living host, where they tap into the host's air supply; they overwinter as mature larvae or pupae.

Several aphelinids are used in the biological control of insect pests. In Britain, *Encarsia formosa* is used in greenhouses to control *Trialeurodes vaporariorum*, the important whitefly pest, and *Aphelinus mali* has been used in the control of the woolly aphid *Eriosoma lanigerum* on apple trees.

Identification: there are no guides to the British species though many can be identified with Ferrière (1965) with care, but see also the specialized references detailed in Noyes et al. (1999).

British subfamilies and genera:

Aphelininae: *Aphelinus, Aphytis, Centrodora, Marietta, Mesidia*

Azotinae: *Ablerus*

Coccophaginae: *Coccobius, Coccophagoides, Coccophagus, Encarsia, Pteroptrix*

Eretmocerinae: *Eretmocerus*

Family Chalcididae (6 genera, 8 species)

This is only a small family in Britain, and they are endoparasitoids of Diptera, Coleoptera and sawflies (Hymenoptera), though many other host groups are known elsewhere in the world. The selected hosts are usually mature larvae or pupae, and the female chalcid oviposits directly into them. Although they are small species, some have bright coloration (Fig. 20.43).

Identification: the handbook by Ferrière and Kerrich (1958) covers the British species, but is not totally reliable; see Noyes et al. (1999) for further details. The work by Boucek (1952) is also worth consulting.

British subfamilies and genera:

Chalcidinae: *Brachymeria, Chalcis, Conura*

Haltichellinae: *Haltichella, Neochalcis, Psilochalcis*

Family Elasmidae (1 genus, 4 species)

The hosts of this small family are the caterpillars of Lepidoptera, on which the elasmids may be primary ectoparasitoids, or else hyperparasitoids on Braconidae or Ichneumonidae.

Identification: see Noyes et al. (1999) for the specialist literature.

British genus: *Elasmus*

Family Encyrtidae (77 genera, over 200 species)

Most members of this large family use scale insects (Hemiptera) as their hosts, either as endoparasitoids or occasionally as hyperparasitoids via other parasitic wasps. Some are polyembryonic parasitoids of Lepidopteran caterpillars and in some cases a single egg can give rise to several thousand larvae. Encyrtid eggs often retain a stalk protruding out of the host that supplies the young larva with air, though later larvae usually tap into the host's tracheal system; in these cases the host is kept alive until the encyrtid is ready to emerge. A few species are flightless (Fig. 20.44).

Some Encyrtidae are widely used in the biological control of insect pests, and in Britain several species are used to control mealybugs in greenhouses.

Identification: see Noyes et al. (1999) for the necessary specialized references, such as Sugonjaev (1964, 1965) and Trjapitzin (1978, 1989).

British subfamilies and genera:

Encyrtinae: *Adelencyrtus, Ageniaspis, Aphycoides, Aphycus, Arrhenophagus, Aschitus, Baeocharis, Blastothrix, Bothriothorax, Ceballosia, Cerapterocerus, Cerchysiella, Cerchysius, Cercobelus, Cheiloneurus, Choreia, Coelopencyrtus, Copidosoma, Discodes, Echthroplexiella, Echthroplexis, Ectroma, Encyrtus,*

Fig. 20.43 *Brachymeria minuta* (Chalcididae) (Photo: Roger Key)

Fig. 20.44 *Dinocarsis hemiptera* (Encyrtidae) (Photo: Roger Key)

Epitetracnemus, Eusemion, Ginsiana, Habrolepis, Helegonatopus, Heterococcidoxenus, Homalotyloidea, Homalotylus, Isodromus, Ixodiphagus, Lakshaphagus, Lamennaisia, Leiocyrtus, Mahencyrtus, Mayridia, Metaphycus, Microterys, Negeniaspidius, Ooencyrtus, Parablastothrix, Parablatticida, Parechthrodryinus, Prionomastix, Prionomitus, Prochiloneurus, Protynda-richoides, Pseudencyrtus, Pseudococcobius, Pseudor-hopus, Psyllaephagus, Sectiliclava, Stemmatosteres, Subprionomitus, Syrphophagus, Thomsonisca, Trechnites, Trichomasthus, Tyndarichus, Zaomma

Tetracneminae: *Aglyptus, Anagyrus, Anomalicornia, Anusia, Charitopus, Coccidoxenoides, Dinocarsis, Dusmetia, Ericydnus, Gyranusoidea, Leptomastidea, Leptomastix, Mira, Rhopus, Tetracnemus*

Family Eucharitidae (1 genus, 1 species)

The only British species in this family is *Eucharis adscendens*, known from a very few records. It is a parasitoid of the ant *Formica rufa*, but unusually the female wasp lays her eggs on flowers near the host's nest. The newly hatched planidial larvae attach themselves to worker ants, which carry them into the nest, where they feed on ant larvae.

Identification: the single British species was figured in the handbook by Ferrière and Kerrich (1958).

British genus: *Eucharis*

Family Eulophidae (c. 50 genera, c. 500 species)

This is one of the largest families of chalcidoids, with at least 500 species known in Britain, many of which are primary parasitoids of leaf-mining larvae of Lepidoptera, Diptera and Coleoptera. Some species attack gall-forming insects or mites, but there are many complications in the life histories of this large group, with numerous variations on host preferences and modes of parasitism.

Most eulophids are very small, with some less than 1 mm long, often yellowish or brownish, but some have distinct metallic colours (Fig. 20.45).

Identification: Askew's (1968) handbook has a key to the subfamilies and also covers most species in the Eulophinae and Euderinae, but it is also neces-sary to consult the specialist literature listed in Noyes et al. (1999). See also Boucek and Askew (1968), Hansson (1985) and Graham (1991).

British subfamilies and genera:

Entedontinae: *Achrysocharoides, Asecodes, Cera-nisus, Chrysocharis, Chrysonotomyia, Closterocerus, Derostenus, Desmatocharis, Entedon, Euderomphale, Eugerium, Grahamia, Holcopelte, Ionympha, Mesto-*

Fig. 20.45 *Aulogymnus skianeuros* **ex rose bedeguar (Eulophidae) (Photo: Robin Williams)**

charis, Neochrysocharis, Omphale, Pediobius, Pholema, Teleopterus

Euderinae: *Astichus, Euderus, Parasecodella*

Eulophinae: *Aulogymnus, Cirrospilus, Colpoclypeus, Dicladocerus, Diglyphus, Dimmockia, Elachertus, Eulophus, Euplectrus, Hemiptarsenus, Microlycus, Miotropis, Necremnus, Pnigalio, Ratzeburgiola, Stenomesius, Sympiesis, Trichoplectrus, Xanthellum*

Tetrastichinae: *Aceratoneuromyia, Aprostocetus, Baryscapus, Crataepus, Melittobia, Peckelachertus, Pronotalia, Tetrastichus*

Family Eupelmidae (5 genera, 14 species)

The members of this small family are often rather larger in size than other chalcidoids, with some species reaching about 7 mm in length. Even within this small group there is a great range of host preferences including the Coleoptera, Hemiptera, Hymenoptera, Lepidoptera, Neuroptera and Orth-optera; most species are either primary ectoparasi-toids or facultative hyperparasitoids; most are solitary though a few are gregarious. There are also a few egg predators or parasites on various insects or spiders. A peculiarity of the subfamily Eupe-lminae is that they are adapted for jumping; they have resilin in the thorax, which is the same elastic substance used by fleas; large muscles pull against this, distorting the thorax, and when released the energy is suddenly transferred to the mid legs.

Some eupelmids are yellowish in colour, but most are metallic (Fig. 20.46).

Identification: see the specialized references listed in Noyes et al. (1999). The works by Gibson (1989, 1995) are also useful.

Fig. 20.46 *Calosota aestivalis* (Eupelmidae) (Photo: Roger Key)

Fig. 20.48 *Sycophila biguttata* (Eurytomidae) ex *Andricus grossulariae* (Cynipidae) (Photo: Robin Williams)

Fig. 20.47 *Eurytoma brunniventris* ex knopper gall (Eurytomidae) (Photo: Robin Williams)

Fig. 20.49 *Sycophila biguttata* (Eurytomidae) on *Aulacidea hieracii* gall (Cynipidae) (Photo: Robin Williams)

British subfamilies and genera:
Calosotinae: *Calosota, Eusandalum*
Eupelminae: *Anastatus, Eupelmus, Merostenus*

Family Eurytomidae (5 genera, 91 species)
The members of this moderately large family show a wide range of life styles, particularly as there are two groups of phytophagous species. One of these groups feeds on plant seeds, and the others are stem-miners of grasses. However, most species are ectoparasitoids of insects that are feeding inside plant tissue, including leaf-miners and gall-makers, so that they may frequently be found emerging from a gall not of their own creation. For example, *Eurytoma brunniventris* (Fig. 20.47) can parasitize a cynipid, or its inquilines, other chalcidoid parasi-

toids, or can even feed on the gall itself. Similarly *Sycophila biguttata* (Figs. 20.48 & 20.49) is found in oak galls as an endoparasitoid of cynipids such as *Andricus* and *Aulacidea*. Most eurytomids are black, sometimes with noticeable sculpturing of the cuticle, but some are metallic.

Identification: see the specialized references listed in Noyes et al. (1999).

British genera: *Bruchophagus, Eurytoma, Sycophila, Systole, Tetramesa*

Family Mymaridae (19 genera, 87 species)
This is a large family of tiny wasps, often less than 1mm long, which are sometimes known as fairy flies. All are endoparasitoids of insect eggs, and in

general they do not seem confined to particular hosts, though preferred groups are various Hemiptera, Coleoptera and Psocoptera. Some species can even parasitize the eggs of aquatic beetles such as Dytiscidae; the adults can swim underwater using their wings like paddles, and mating has even been recorded under water.

Fairy flies have long been favourite insects for mounting as microscopic preparations because of their small size and delicate appearance; their fine details provide good tests of the resolution of quality lenses. However, they are a difficult group to identify to species because of their small size.

Identification: see the specialized references in Noyes et al. (1999), including Schauff (1984).

British genera: *Alaptus, Anagrus, Anaphes, Arescon, Camptoptera, Caraphractus, Cleruchus, Dicopus, Doryclytus, Erythmelus, Eustochus, Gonatocerus, Litus, Mymar, Ooctonus, Parallelaptera, Polynema, Stephanodes, Stethynium*

Family Ormyridae (1 genus, 3 species)
This small family contains just three species in the genus *Ormyrus*, which has previously been placed in the Pteromalidae or the Torymidae. They are small metallic wasps, but little is known of their biology; they are assumed to be parasitoids on gall-forming insects.

Identification: see the specialized references listed in Noyes et al. (1999).

British genus: *Ormyrus*

Family Perilampidae (2 genera, 9 species)
The members of this small family are medium-sized chalcidoids, up to 5mm long, and are often metallic with strong sculpturing. Most seem to be hyperparasitoids on various insect groups, but the details of most lifecycles have not been worked out and it is not always clear whether existing records show primary or hyperparasitism.

The exact relationships of the two subfamilies are not clear; some authors would place the Chrysolampinae in the family Pteromalidae, and the monophyly of the Perilampidae as currently constituted is in some doubt.

Identification: the British species can be identified using the handbook by Ferrière and Kerrich (1958).

British subfamilies and genera:
Chrysolampinae: *Chrysolampus*
Perilampinae: *Perilampus*

Family Pteromalidae (*c.* 150 genera, *c.* 600 species)
This is the largest chalcidoid family In Britain, with over 600 species known, though the family is probably paraphyletic and the relationships of the many subfamilies are not always clear. As one would expect in such a large group there is a wide variety of life histories, with ectoparasitoids and endoparasitoids, primary and secondary parasitoids and even predators known on a wide range of hosts including Diptera, Coleoptera, Hymenoptera, Lepidoptera and Siphonaptera. Many species seem to attack concealed hosts such as leaf-miners, stem-borers and gall-formers, and the species seen emerging from galls can be inquilines, parasitoids of the gall-causing insect, hyperparasitoids, or even feeding on the gall-tissue itself, a similar situation to that seen in the Eurytomidae.

One common endoparasitoid on butterfly pupae is *Pteromalus puparum*, a species often encountered by entomologists trying to rear butterflies in captivity. There are several commonly seen genera, including *Caenacis* (Fig. 20.50), *Cecidostiba* (Fig. 20.51), *Mesopolobus* (Fig. 20.52), *Pteromalus* (Fig. 20.53) and *Stenomalina* (Fig. 20.54).

Identification: the handbook of Ferrière and Kerrich (1958) includes keys to the genera and species of a few smaller families such as the Cleonyminae; otherwise it is necessary to consult the specialized references listed in Noyes et al. (1999). See also Boucek and Rasplus (1991).

British subfamilies and genera:
Asaphinae: *Asaphes, Hyperimerus*
Ceinae: *Cea, Spalangiopelta*

Fig. 20.50 *Caenacis inflexa* ex rose bedeguar (Pteromalidae) (Photo: Robin Williams)

Fig. 20.51 *Cecidostiba fungosa* (Pteromalidae) ex *Andricus grossulariae* (Cynipidae) (Photo: Robin Williams)

Fig. 20.54 *Stenomalina liparae* (Pteromalidae) (Photo: Roger Key)

Fig. 20.52 *Mesopolobus sericeus* ex oak marble gall (Pteromalidae) (Photo: Robin Williams)

Cerocephalinae: *Cerocephala, Theocolax*
Cleonyminae: *Cleonymus*
Colotrechninae: *Colotrechnus*
Cratominae: *Cratomus*
Diparinae: *Dipara*
Eunotinae: *Epicopterus, Eunotus*
Macromesinae: *Macromesus*
Miscogasterinae: *Ardilea, Callimerismus, Glyphognathus, Halticoptera, Halticopterina, Lamprotatus, Merismus, Miscogaster, Nodisoplata, Rhicnocoelia, Schimitschekia, Seladerma, Sphaeripalpus, Stictomischus, Telepsogina, Thinodytes, Tricyclomischus, Xestomnaster*
Neodiparinae: *Neodipara*
Ormocerinae: *Bugacia, Melancistrus, Ormocerus, Systasis*
Panstenoninae: *Panstenon*
Pireninae: *Ecrizotes, Gastrancistrus, Macroglenes, Spathopus, Stenophrus*
Pteromalinae: *Ablaxia, Acrocormus, Aggelma, Anisopteromalus, Anogmoides, Anogmus, Apelioma, Apsilocera, Arthrolytus, Atrichomalus, Bairamlia, Caenacis, Calliprymna, Callitula, Capellia, Catolaccus, Cecidostiba, Cheiropachus, Chlorocytus, Coelopisthia, Conomorium, Coruna, Cryptoprymna, Cyclogastrella, Cyrtogaster, Dibrachoides, Dibrachys, Diglochis, Dimachus, Dinarmus, Dinotiscus, Dinotoides, Dirhicnus, Endomychobius, Erdoesia, Erythromalus, Eulonchetron, Eumacepolus, Euneura, Gastracanthus, Gbelcia, Gyrinophagus, Habritys, Hemitrichus, Heteroprymna, Hobbya, Holcaeus, Homoporus, Isocyrtus, Janssoniella, Kaleva, Lampoterma, Lariophagus, Leptomeraporus, Meraporus, Merisus, Mesopolobus, Metacolus, Metastenus, Micradelus, Mokrzeckia, Muscidifurax, Nasonia, Nephelomalus, Pachycrepoideus, Pachyneuron, Pandelus, Pegopus,*

Fig. 20.53 *Pteromalus bedeguaris* (Pteromalidae) (Photo: Robin Williams)

245

Peridesmia, Perniphora, Phaenocytus, Platneptis, Platygerrhus, Plutothrix, Pseudocatolaccus, Psilocera, Psilonotus, Psychophagoides, Psychophagus, Pteromalus, Rakosina, Rhaphitelus, Rhopalicus, Rohatina, Roptrocerus, Sceptrothelys, Schizonotus, Semiotellus, Spaniopus, Sphegigaster, Spilomalus, Spintherus, Staurothyreus, Stenomalina, Stinoplus, Synedrus, Syntomopus, Termolampa, Tomicobia, Toxeuma, Trichomalopsis, Trichomalus, Trigonoderus, Tritneptis, Trychnosoma, Urolepis, Vrestovia, Xiphydriophagus

   Spalangiinae: *Spalangia*

Family Signiphoridae (2 genera, 2 species)

This family was previously known as the Thysanidae, and there are just two British species, *Chartocerus subaeneus* and *Thysanus ater*; both are uncommon species whose life histories have not been studied in detail. From what is known of exotic species in the family they are likely to be found on scale insects (Hemiptera) but whether as primary or hyperparasitoids is yet to be confirmed.

Identification: the two British species can be identified using Ferrière and Kerrich (1958), under the Thysanidae.

British genera: *Chartocerus, Thysanus*

Family Tetracampidae (5 genera, 7 species)

The members of this small family are less than 3 mm long and often metallic green in colour. They are generally found as endoparasitoids in the eggs of Coleoptera or Diprionidae (Hymenoptera), often of groups that mine plants, though the biologies of most species are not known.

Identification: see the specialist references listed in Noyes et al. (1999).

British subfamilies and genera:

   Platynocheilinae: *Platynocheilus*

   Tetracampinae: *Dipriocampe, Epiclerus, Foersterella, Tetracampe*

Family Torymidae (7 genera, 75 species)

Many members of this moderately large family are metallic green in colour, and the females have a characteristically long and upturned ovipositor (Fig. 20.55). Most torymids are entomophagous ectoparasitoids on the inhabitants of plant galls, although an appreciable number are phytophagous, either feeding on gall tissue of another species or feeding directly on seeds. Many of the latter group are in the genus *Megastigmus*, although *M. dorsalis* (Fig. 20.56) is an ectoparasitoid in cynipid galls. The largest genus is *Torymus* and some of its species, such as *T. auratus* (Fig. 20.57) are inquilines in

Fig. 20.55 Torymid (Photo: Roger Key)

Fig. 20.56 *Megastigmus dorsalis* **ex marble gall (Torymidae) (Photo: Robin Williams)**

Fig. 20.57 *Torymus auratus* (Torymidae) (Photo: Robin Williams)

plant galls while others are ectoparasitoids on gall-makers.

Some authors place *Monodontomerus* and some allied genera in a separate subfamily.

Identification: see the specialized references in Noyes et al. (1999).

British subfamilies and genera:
Megastigminae: *Megastigmus*
Toryminae: *Cryptopristus, Glyphomerus, Monodontomerus, Pseudotorymus, Torymoides, Torymus*

Family Trichogrammatidae (17 genera, 29 species)

The members of this fairly small family are all minute wasps, often less than 1 mm long, of various colours though not metallic. The known life histories show that they are generally primary endoparasitoids in the eggs of a wide range of insect groups. As in the Mymaridae, a few remarkable species such as *Prestwichia aquatica* attack the eggs of aquatic insects, swimming underwater using their wings. Several species of trichogrammatids are used as biological control agents, especially against pest Lepidoptera; inevitably with such tiny insects identification to species is not easy.

Identification: see the list of specialist references in Noyes et al. (1999).

British genera: *Aphelinoidea, Chaetostricha, Chaetostrichella, Epoligosita, Lathromeris, Mirufens, Monorthochaeta, Oligosita, Ophioneurus, Paracentrobia, Poropoea, Prestwichia, Trichogramma, Tumidiclava, Ufensia, Uscana, Xiphogramma*

SUPERFAMILY CYNIPOIDEA

The three families that constitute this group are the Cynipidae (gall-wasps), the Figitidae and Ibaliidae, the last two families being endoparasitoids of other insects. They are all small wasps, usually less than 5 mm long and generally black or brown in colour; they are therefore not easy to identify. However, the British fauna is well covered by a series of three handbooks by Eady and Quinlan (1963), Quinlan (1978) and Fergusson (1986); there is also a good general account of the group in Gauld and Bolton (1988). The work by Eady and Quinlan (1963) has a key to families and subfamilies of all groups. A useful website on this group is: http://hedgerowmobile.com/Cynipoidea.html.

Family Cynipidae (18 genera, 91 species)

The members of this moderately large family all have phytophagous larvae that form galls or are

Fig. 20.58 *Synergus gallaepomiformis* (Cynipidae) (Photo: Robin Williams)

Fig. 20.59 *Biorhiza pallida* sexual gall (Cynipidae) (Photo: Robin Williams)

inquilines in the galls produced by other insects. Although the female gall-wasp lays her eggs directly into the plant tissue it is the presence of the larva that causes gall formation. There are a great many different kinds of galls, some more conspicuous than others, and their form is directly related to the species of cynipid. The best-known group are probably the oak gall-wasps of the tribe Cynipini, which is the largest tribe in the family. Many species have alternation of generations, one being female only (agamic) and the other sexual; each generation can produce different types of galls. Members of the Synergini (Fig. 20.58) are inquilines in the galls of other cynipids, usurping their food supply; some of these kill the original gall-wasp larva, while others develop alongside without harming them.

One of the largest galls is the multi-chambered oak apple gall caused by the sexual generation of *Biorhiza pallida* (Fig. 20.59), which can be over 30 mm

Fig. 20.60 *Neuroterus quercusbaccarum* galls (Cynipidae) (Photo: Roger Key)

Fig. 20.62 *Neuroterus albipes* agg. (Cynipidae) (Photo: Robin Williams)

Fig. 20.61 *Neuroterus numismalis* and *N. quercusbaccarum* galls (Cynipidae) (Photo: Roger Key)

Fig. 20.63 *Cynips quercusfolii* agg. (Cynipidae) (Photo: Robin Williams)

in diameter. Some very common oak-galls are the spangle-galls found on the leaves, caused by species of *Neuroterus*. The common spangle-gall (Fig. 20.60) is caused *by N. quercusbaccarum*, while the silk button spangle is caused by *N. numismalis*; the two can frequently be seen on the same oak leaf (Fig. 20.61). *N. albipes* (Fig. 20.62) produces the smooth spangle gall, but it is a less gregarious species. Some species of *Cynips* produce large globular galls on oak leaves (Fig. 20.63), and another common group on oaks are species of *Andricus* (Fig. 20.64).

On roses, one of the most spectacular galls is the rose bedeguar, which is a large globular gall covered in long red hairs; it is caused by *Diplolepis rosae*, and there are other similar species.

Identification: Eady and Quinlan's (1963) handbook has keys to the adult gall wasps and to their

Fig. 20.64 *Andricus lucidus* agg. (Cynipidae) (Photo: Robin Williams)

galls. In many cases it is easier to identify the gall than its causer and this has led to several useful books on plant galls, such as Redfern and Askew (1998) and Redfern and Shirley (2002). Askew (1984) gave a more advanced overview of the biology of the group, and the book by Williams (2006) concentrates on oak-galls. There is much useful information on the website of the British Plant Gall Society (http://www.british-galls.org.uk).

British genera: *Andricus, Aphelonyx, Aulacidea, Aylax, Biorhiza, Callirhytis, Ceroptres, Cynips, Diastrophus, Diplolepis, Isocolus, Liposthenes, Neuroterus, Periclistus, Phanacis, Saphonecrus, Synergus, Trigonaspis, Xestophanes*

Family Figitidae (30 genera, 93 species)

Two subfamilies in this group, the Charipinae and Eucoilinae, were previously treated as distinct families. All species are small parasitic wasps, less than 4 mm long, and as endoparasitoids rather than gall-causers they have attracted less interest than the Cynipidae. Several groups of insects constitute the preferred hosts, but Diptera are the most commonly attacked. The five subfamilies of Figitidae have quite distinct life histories: the Charipinae includes very small species that are all hyperparasitoids of Aphididae or Psyllidae (Hemiptera), via the primary parasites Aphidiinae (Braconidae) or Aphelinidae (Chalcidoidea). The Aspiceratinae are parasitoids of Diptera, especially Syrphidae, while the Anacharitinae are parasitoids of the larvae of Neuroptera. The Figitinae is an ill-defined group whose members use Diptera hosts. Many Eucoilinae are endoparasitoids of dung or carrion-feeding flies, and the wasps are attracted to similar odours, which obviously aids in finding their hosts.

Identification: the handbook by Quinlan (1978) covers the Eucoilinae (as a separate family), and the remaining subfamilies (including the Charipinae as a distinct family) are included in Fergusson's (1986) handbook, which also has notes on the biology of the group.

British subfamilies and genera:
Anacharitinae: *Aegilips, Anacharis, Xyalaspis*
Aspicerinae: *Callaspidia*
Charipinae: *Alloxysta, Dilyta, Phaenoglyphis*
Eucoilinae: *Chrestosema, Cothonaspis, Diglyphosema, Disorygma, Episoda, Eucoila, Eutrias, Ganaspis, Glauraspidia, Hexacola, Kleidotoma, Leptopilina, Microstilba, Pseudopsichacra, Psichacra, Rhoptromeris, Trybliographa*
Figitinae: *Amphitectus, Figites, Lonchidia, Melanips, Sarothrus, Trischiza, Xyalophora, Zygosis*

Family Ibaliidae (1 genus, 2 species)

The two British species in this family, *Ibalia leucospoides* and *I. rufipes*, are easily recognized by their size; at 10 mm or more in length they are twice the size of other cynipoids. Both species are endoparasitoids of woodwasps of the family Siricidae. Woodwasp larvae live deep in the timber of coniferous trees, and *Ibalia* has a long coiled ovipositor that totals about 1.5 times the wasp's body length. This is uncoiled and passed down the woodwasp's oviposition shaft, so that a single egg can be laid inside the host egg or larva. The lifecycle of *Ibalia* can take up to three or four years; after pupation the adult wasp chews its way out of the wood using its mandibles.

Identification: the handbook by Fergusson (1986) covers the Ibaliidae and includes notes on their biology.

British genus: *Ibalia*

SUPERFAMILY EVANIOIDEA

The three families in this small grouping are rather different in their biology, and this superfamily may well be an artificial group.

Family Aulacidae (1 genus, 1 species)

There is just one species of this family in Britain, *Aulacus striatus*. This is a parasitoid on the woodwasp *Xiphydria camelus* (Xiphydriidae), whose eggs are laid under tree bark. The female *Aulacus* locates the host's oviposition shaft, inserts her ovipositor in it and lays individual eggs in as many of the host eggs as can be reached. The *Aulacus* larva can take a year to develop, and the host larva remains alive until the parasitoid is fully grown.

Identification: *Aulacus* can be identified using any of the standard keys to families.

British genus: *Aulacus*

Family Evaniidae (2 genera, 2 species)

The two British species in this family have a characteristic appearance, with a relatively large mesosoma to which the small metasoma is attached very high up by a long narrow petiole; the metasoma is often held up like a flag, giving rise to the common name of ensign flies. Both species are parasitoids in the egg-cases (oothecae) of cockroaches; *Brachygaster minutus* attacks the three native British species of *Ectobius* (Dictyoptera: Ectobiidae), while the cosmopolitan *Evania appendigaster* is associated with the domestic cockroaches *Blatta* and *Periplaneta*.

Fig. 20.65 *Gasteruption minutum* (Gasteruptiidae) (Photo: Roger Key)

Identification: see Noyes et al. (1999) for specialist references.

British genera: *Brachygaster, Evania*

Family Gasteruptiidae (1 genus, 5 species)
The members of this small family have a distinctive 'neck' formed from the propleura; they are elongate species, up to about 7 mm long (Fig. 20.65) and all five British species are in the genus *Gasteruption*. Earlier authors regarded them as ectoparasitoids on other Hymenoptera, but recent work suggests that they are all secondary cleptoparasites on food stored by solitary bees. Female *Gasteruption* species have very long ovipositors (Fig. 20.66).

Identification: see Noyes et al. (1999) for specialist references.

British genus: *Gasteruption*

SUPERFAMILY ICHNEUMONOIDEA

The two families in this group, the Ichneumonidae and Braconidae, include well over 3000 species, which is approaching half the total British Hymenoptera fauna. All are parasitoids of other insects, and in general the Braconidae are small and uniformly coloured, while many Ichneumonidae are much larger and often brightly coloured. Despite this, the ichneumonids have not attracted much attention from amateur workers, and they are still considered a 'difficult' group. Some members of this superfamily are nocturnal and are attracted to lights, where they may reach the attention of lepidopterists.

Keys to separate the two families, together with detailed descriptions of all the main subgroups, are provided by Gauld and Bolton (1988)

Fig. 20.66 *Gasteruption jaculator* (Gasteruptiidae) (Photo: Robin Williams)

Fig. 20.67 Aphidiinae (Braconidae) (Photo: Robin Williams)

Family Braconidae (*c.* 140 genera, *c.* 1045 species)
Despite the large size of this family, the British fauna is actually quite depauperate when compared with tropical regions. Most braconids are small and inconspicuous, usually black, brown or yellowish in colour (Fig. 20.67). There are a great many sub-

families although the monophyly of some is doubtful, but placing species into a subfamily is the only way to begin to identify the members of such a large group. For this reason, most accounts of this family such as that in Gauld and Bolton (1988) describe the characters of each subfamily separately, and brief outlines of these are given below.

The Adeliinae constitute a very small subfamily containing a single genus whose members were previously placed within the Microgastrinae; they are parasitoids of Nepticulidae (Lepidoptera). The Agathidinae include around 24 species, which are parasitoids on various families of small moths. The Alysiinae includes around 200 British species that are usually divided into two tribes, the Alysiini and Dacnusini, which differ in details of their wing venation; many species are parasitoids of Diptera, especially Agromyzidae, though identification to species is often difficult. The Aphidiinae was previously treated as a distinct family, and its members are all parasitoids of aphids (see Fig. 20.67); species identification can be difficult because there are many cryptic species, as well as variation induced by different hosts or environments. The Blacinae are small endoparasitoids of beetle larvae, though one is found in the larvae of *Boreus* (Mecoptera). The Braconinae contains the large genus *Bracon*, whose members are very difficult to identify at present; most are ectoparasitoids of Coleoptera or Lepidoptera larvae. The Cenocoeliinae are rarely recorded in Britain, and they are believed to be endoparasitoids of beetle larvae. The subfamily Charmontinae was recently erected for the single genus *Charmon*, which was previously included in the Helconinae or Orgilinae; it is an endoparasitoid of microlepidoptera. The Cheloninae are mainly endoparasitoids of Lepidopteran larvae, though identification to species level in the genus *Chelonus* is difficult. Members of the Doryctinae are ectoparasitoids on the larvae of wood-boring Coleoptera. The Euphorinae are endoparasitoids of a wide range of insect groups, including Coleoptera (Fig. 20.68) though the constitution of this subfamily varies with different authors. The Gnamptodontinae contain just one genus whose members are endoparasitoids of Nepticulidae (Lepidoptera). The Helconinae are mainly associated with wood-boring beetle larvae, as is the only British species in the Histeromerinae. The Homolobinae are all endoparasitoids on Lepidopteran larvae, especially Noctuidae and Geometridae. The three species in the Ichneutinae seem to be endoparasitoids on sawfly larvae (Hymenoptera). The Macrocentrinae are all endoparasitoids of Lepidopteran larvae,

Fig. 20.68 *Dinocampus coccinellae* (Braconidae) ex 7-spot ladybird (Photo: Roger Key)

while the two genera in the Meteorinae, which are sometimes placed in the Euphorinae, are found on Lepidoptera and Coleoptera larvae. The large subfamily Microgastrinae is extensively studied; all are endoparasitoids of Lepidopteran larvae, and they include the well-known *Apanteles glomeratus*, a common parasitoid of the large white butterfly, which was recently moved to the genus *Cotesia*. The single species of Microtypinae was only recently recorded from Britain; this group are endoparasitoids of microlepidoptera, as is the single British species in the Miracinae. The Neoneurinae are all apparently endoparasitoids of ants. The members of the large subfamily Opiinae are endoparasitoids of Diptera; most species are placed in the genus *Opius*. The Orgilinae contains a single genus whose members are endoparasitoids of Lepidopteran larvae, as are most members of the Rogadinae, but the exact constitution of this large subfamily varies with the opinions of different authors. The Sigalphinae are also endoparasitoids of Lepidopteran larvae.

Identification: an important introduction to the whole family is the handbook by Shaw and Huddleston (1991), which includes keys to the subfamilies. Noyes et al. (1999) gives a list of the numerous specialized works needed for the study of this family, such as Sharkey (1993), Starý (1966) and Taeger (1989). There is a checklist of the British species on the following website: http://www.nhm.ac.uk/resources-rx/files/braconidae-checklist-final-34139.pdf.

Although the current number of British genera is given as *c.* 140, there may well be more than 170 eventually recognized.

British subfamilies and genera:

Adeliinae: *Adelius*

Agathidinae: *Agathis, Bassus, Earinus*

Alysiinae: *Agonia, Alloea, Alysia, Amyras, Anisocyrta, Aphaereta, Aphanta, Aristelix, Asobara, Aspilota, Chaenusa, Chasmodon, Chorebus, Coelinidea, Coelinius, Coloneura, Cratospila, Dacnusa, Dapsilarthra, Dinotrema, Epimicta, Eudinostigma, Exotela, Grammospila, Heterolexis, Idiasta, Idiolexis, Laotris, Leptotrema, Mesocrina, Orthostigma, Panerema, Pentapleura, Phaenocarpa, Polemochartus, Protodacnusa, Pterusa, Sarops, Syncrasis, Synelix, Tanycarpa, Tates, Trachionus, Trachyusa*

Aphidiinae: *Adialytus, Aphidius, Areopraon, Binodoxys, Diaeretellus, Diaeretiella, Diaeretus, Dyscritulus, Ephedrus, Falciconus, Harkeria, Lysiphlebus, Monoctonus, Paralipsis, Pauesia, Praon, Toxares, Trioxys*

Blacinae: *Blacometeorus, Blacus, Dyscoletes, Taphaeus*

Brachistinae: *Eubazus, Foersteria, Schizoprymnus, Triaspis*

Braconinae: *Baryproctus, Bracon, Coeloides, Pigeria, Pseudovipio*

Cardiochilinae: *Cardiochiles*

Cenocoeliinae: *Cenocoelius, Lestricus*

Charmontinae: *Charmon*

Cheloninae: *Ascogaster, Chelonus, Phanerotoma*

Doryctinae: *Caenopachys, Dendrosoter, Doryctes, Ecphylus, Gildoria, Hecabolus, Heterospilus, Monolexis, Ontsira, Rhaconotus, Spathius, Wachsmannia*

Euphorinae: *Allurus, Centistes, Chrysopophthorus, Cosmophorus, Dinocampus, Euphorus, Leiophron, Meteorus, Myiocephalus, Neoneurus, Perilitus, Peristenus, Pygostolus, Rilipertus, Ropalophorus, Spathicopis, Syntretus, Townesilitus, Wesmaelia, Zele*

Exothecinae: *Colastes, Shawiana, Xenarcha*

Gnamptodoninae: *Gnamptodon*

Helconinae: *Aspigonus, Diospilus, Helcon, Helconidea, Wroughtonia*

Histeromerinae: *Histeromerus*

Homolobinae: *Homolobus*

Hormiinae: *Hormius*

Ichneutinae: *Ichneutes, Proterops*

Macrocentrinae: *Austrozele, Macrocentrus*

Microgastrinae: *Apanteles, Cotesia, Deuterixys, Diolcogaster, Hygroplitis, Microgaster, Microplitis, Paroplitis, Protapanteles*

Microtypinae: *Microtypus*

Miracinae: *Mirax*

Opiinae: *Ademon, Apodesmia, Atormus, Bathystomus, Biophthora, Biosteres, Bitomoides, Chilotrichia, Desmi-ostoma, Diachasma, Opius, Phaedrotoma, Utetes, Xynobius*

Orgilinae: *Orgilus*

Pambolinae: *Chremylus, Dimeris, Pambolus*

Rhysipolinae: *Rhysipolis*

Rhyssalinae: *Acrisis, Dolopsidea, Lysitermoides, Oncophanes, Pseudobathystomus, Rhyssalus*

Rogadinae: *Aleiodes, Chelonorhogas, Clinocentrus, Heterogamus, Rogas, Triraphis*

Sigalphinae: *Acampsis, Sigalphus*

Incertae sedis: *Streblocera*

Family Ichneumonidae (*c.* 400 genera, *c.* 2100 species)

Compared with the Braconidae, the ichneumons are often relatively large and brightly coloured species, and are noticed even by casual observers (Fig. 20.69). This is not only the largest family of British Hymenoptera, but also the largest family of any insect group and, like the Braconidae, it is divided into many subfamilies. Inevitably such a large group exhibits a wide range of life histories and behaviours, although one can generalize by saying that most ichneumonids are endoparasitoids on the larvae or pupae of insects such as Lepidoptera or sawflies (Hymenoptera). Some species are predators on spiders' eggs or on bee larvae but in general the range of host groups exploited by ichneumonids is quite narrow. Around two-thirds of the British species use Lepidoptera as hosts, with another quarter of the species using sawflies; it is noticeable that very few parasitize Coleoptera or Diptera. A few species of Mecoptera and Neuroptera are attacked by some Cryptinae and Campopleginae, while a small number of Trichoptera are parasitized

Fig. 20.69 Ophionine ichneumonid head (Photo: Roger Key)

by Cryptinae and *Agriotypus*. Some of the Hymenoptera Apocrita are also attacked, including ichneumonids themselves, as some species are hyperparasitoids. Spiders are often the hosts of ectoparasitic Pimplinae, while their egg sacs may be attacked by Cryptinae. Although most ichneumonids are very host-specific, the species that have more than one generation per year may use different hosts in each generation.

There are relatively few specialists working on the ichneumonids and most of the relevant literature is widely scattered in numerous journals and in many languages. Much of the European fauna is still little known and there are very few useful keys. However, a good start is the section in Gauld and Bolton (1988), which contains many useful references. The standard text for getting to subfamilies and many genera on a world basis is the multi-part work by Townes (1969, 1970a, 1970b, 1971); however, this is showing its age because the nomenclature has changed considerably in recent years. The keys are not easy to use because they are generally not illustrated although there are whole-insect figures for each genus. There are more recent works that will be useful even though they do not actually apply to the British fauna, including Gauld (1991) and the Russian work by Kasparyan (1981). Short (1978) is a valuable guide to the larvae in this family. Perkins' (1959) key to subfamilies can no longer be recommended, although his keys to the Ichneumoninae still work well. Very early works such as Morley (1903–15) are of historical value only. World catalogues are provided by Yu (1993) and Yu and Horstmann (1997).

As in the Braconidae, a brief account of each subfamily is given below.

The Acaenitinae is a small group, presumed to be endoparasitoids of Coleoptera larvae. The 15 species of Adelognathinae are all small wasps, less than 5 mm long and are easily overlooked; they are ectoparasitoids of sawfly larvae. The single species of Agriotypinae is unusual in attacking a few species of caddisfly larvae (Trichoptera); the female swims underwater to find the caddis case and the parasitoid produces a long ribbon-like structure protruding from the case that enables respiration. The two species of Alomyinae are endoparasitoids of Hepialidae (Lepidoptera); some authors place these in the Ichneumoninae. The Anomaloninae includes about 40 British species that are endoparasitoids of Lepidoptera, as are the Banchinae, which includes well over 100 species. The large subfamily Campopleginae includes more than 250 British species; they mostly attack Lepidoptera, though

Fig. 20.70 *Thaumatogelis audax* (Ichneumonidae) (Photo: Roger Key)

some are found on sawflies, Coleoptera and Raphidioptera. The two British species of Collyriinae are endoparasitoids of sawflies in the Cephidae. The Cremastinae are a small group that are endoparasitoids of Lepidoptera larvae. There are over 450 British species in the Cryptinae, which used to be known as the Phygadeuontinae. They use a wide range of hosts including spider egg-sacs, Dipteran pupae, bee larvae and even freshwater insects. Females of *Gelis* and related genera are wingless and very ant-like (Fig. 20.70).

Most of the 200 species of Ctenopelmatinae attack sawflies, though a few are known on Lepidoptera larvae. The small group of Cylloceriinae were previously placed in the Oxytorinae; their hosts are usually Diptera. The single species in the Diacritinae was formerly included in the Pimplinae. Diplazontinae are often seen around aphid colonies, where they attack aphidophagous hoverfly larvae (Syrphidae); there are about 50 British species. The three species of Eucerotinae are all uncommon, but have an unusual life history; their newly hatched larvae are carried on a host caterpillar but only develop if the host is parasitized, when the eucerotine becomes an endoparasitoid of the primary parasitoid. The subfamily Helictinae has variously been placed with the Oxytorinae and Orthocentrinae; their hosts are Diptera. The very large subfamily Ichneumoninae includes around 340 species; they are often brightly coloured and the large black and yellow species are commonly seen in summer. All are endoparasitoids of Lepidoptera. There is just a single rare species in the Lycorininae, which is probably a parasitoid of Lepidoptera. Most of the 50 species of Mesochorinae are small and

Fig. 20.71 Ophionine ichneumonid (Photo: Roger Key)

Fig. 20.72 *Orthopelma mediator* **(Ichneumonidae) ex rose bedeguar (Photo: Robin Williams)**

Fig. 20.73 Polysphinctine larva (Ichneumonidae) on spider (Photo: Roger Key)

inconspicuous, but species of *Cidaphus* look very like Ophioninae. They are obligatory hyperparasitoids, their larvae developing solitarily inside the bodies of primary parasitoid larvae, which include braconid, ichneumonid and tachinid (Diptera) endoparasitoids of Lepidopteran caterpillars and sawfly larvae.

Most of the 60 species of Metopiinae are black and inconspicuous, and buzz loudly if caught; they are endoparasitoids of Lepidoptera. The three species of Microleptinae are endoparasitoids of Stratiomyiidae (Diptera). The single species of Neorhacodinae is very small and was originally thought to be a braconid; it parasitizes species of Sphecidae. Most Ophioninae are nocturnal and brightly coloured (Fig. 20.71); they are frequently attracted to lights and are a familiar sight to lepidopterists using moth-traps. They are endoparasitoids of Lepidoptera caterpillars. There are over 40 species of Orthocentrinae; all are small and inconspicuous but can be abundant in damp shady places where they attack various Diptera larvae such as Mycetophilidae. Of the two British species of Orthopelmatinae, *Orthopelma mediator* (Fig. 20. 72) is an endoparasitoid of the cynipid *Diplolepis rosae* that causes bedeguar galls on roses. The biology of the single genus in the Oxytorinae is not known. The two species of Paxylommatinae seem to be associated with ants, but their biology is not known. Members of the small subfamily Phrudinae are presumed to be endoparasitoids of Coleoptera.

There are almost 100 species of British Pimplinae, and several are very common, on a wide range of hosts. Members of the tribe Polysphinctini are ectoparasitic on spiders (Fig. 20.73). The five species in the Poemeniinae were formerly included in the Pimplinae, as were the two species of Rhyssinae; *Rhyssa persuasoria* (Fig. 20.74) is one the largest ichneumonids that uses its extremely long ovipositor to reach the wood-boring larvae of the woodwasp *Urocerus*. Such long ovipositors are common in this subfamily (Fig. 20.75). One of the four species of Stilbopinae is commonly seen in woodland in the spring; they are all endoparasitoids of Incurvariidae (Lepidoptera). There are over 50 species of Tersilochinae; most are small species that usually attack phytophagous beetle larvae. The Tryphoninae include about 160 species, many of which are common; most attack sawflies though some specialize on Lepidopteran larvae. Xoridinae are not often seen and they seem to be ectoparasitoids on wood-boring beetles, and possibly also woodwasps.

Fig. 20.74 *Rhyssa persuasoria* (Ichneumonidae) (Photo: Robin Williams)

Fig. 20.75 *Ephialtes manifestator* ovipositing (Ichneumonidae) (Photo: Robin Williams)

Identification: there are a few key works that will help identification in certain subfamilies, as listed below, but in general it will be necessary to consult the extensive literature lists in Noyes et al. (1999). Agriotypinae, Perkins (1960); Alomyinae, Perkins (1960); Anomaloninae, Gauld and Mitchell (1977), Banchinae, Aubert (1978) and Townes (1970b); Campopleginae, Sanborne (1984) and Townes (1970b); Cryptinae, Townes (1970a, 1983); Ctenopelmatinae, Townes (1970b); Cylloceriinae, Townes (1971); Diacritinae, Fitton et al. (1988); Helictinae, Townes (1971); Ichneumoninae, Hilpert (1992) and Perkins (1959, 1960); Lycorininae, Perkins (1960); Metopiinae, Townes (1971); Orthocentrinae, Townes (1971); Orthopelmatinae, Gauld and Mitchell (1977); Pimplinae, Fitton et al. (1988);

Poemeniinae, Fitton et al. (1988), Rhyssinae, Fitton et al. (1988); Tryphoninae, Kasparyan (1973, 1990) and Townes (1969).

There is a checklist of the British species on the following website: http://www.brc.ac.uk/downloads/Ichneumonidae_checklist.pdf.

Tribes are also important in the division of some subfamilies, though they are not used here.

British subfamilies and genera:
Acaenitinae: *Acaenitus, Arotes, Coleocentrus, Leptacoenites, Phaenolobus*
Adelognathinae: *Adelognathus*
Agriotypinae: *Agriotypus*
Alomyinae: *Alomya*
Anomaloninae: *Agrypon, Anomalon, Aphanistes, Atrometus, Barylypa, Erigorgus, Gravenhorstia, Habrocampulum, Habronyx, Heteropelma, Parania, Perisphincter, Therion, Trichomma*
Banchinae: *Alloplasta, Apophua, Arenetra, Banchus, Cryptopimpla, Exetastes, Glypta, Lissonota, Syzeuctus, Rynchobanchus, Teleutaea*
Campopleginae: *Alcima, Bathyplectes, Campoletis, Campoplex, Casinaria, Charops, Clypeoplex, Cymodusa, Diadegma, Dolophron, Dusona, Echthronomas, Enytus, Eriborus, Gonotypus, Hyposoter, Lathroplex, Lathrostizus, Lemophagus, Macrus, Meloboris, Nemeritis, Nepiesta, Olesicampe, Phobocampe, Porizon, Pyracmon, Rhimphoctona, Scirtetes, Sinophorus, Synetaeris, Tranosema, Tranosemella, Venturia*
Collyriinae: *Collyria*
Cremastinae: *Cremastus, Dimophora, Pristomerus, Temelucha*
Cryptinae: *Aclastus, Aconias, Acrolyta, Acroricnus, Agasthenes, Agrothereutes, Amphibulus, Apsilops, Aptesis, Aritranis, Arotrephes, Ateleute, Atractodes, Bathythrix, Blapsidotes, Buathra, Caenocryptus, Ceratophygadeuon, Charitopes, Chirotica, Clypeoteles, Colocnema, Cratocryptus, Cremnodes, Cryptus, Cubocephalus, Demopheles, Diaglyptidea, Dichrogaster, Echthrus, Enclisis, Encrateola, Endasys, Ethelurgus, Eudelus, Fianoniella, Gambrus, Gelis, Giraudia, Glyphicnemis, Gnotus, Gnypetomorpha, Grasseiteles, Hedylus, Helcostizus, Hemiteles, Hidryta, Hoplocryptus, Idiolispa, Isadelphus, Ischnus, Javra, Leptocryptoides, Listrocryptus, Listrognathus, Lochetica, Lysibia, Mastrulus, Mastrus, Medophron, Megacara, Megaplectes, Meringopus, Mesoleptus, Mesostenus, Nematopodius, Neopimpla, Obisiphaga, Odontoneura, Oresbius, Orthizema, Parmortha, Phygadeuon, Platyrhabdus, Plectocryptus, Pleolophus, Pleurogyrus, Polyaulon, Polytribax, Pygocryptus, Rhembobius, Schenkia, Sphecophaga, Stibeutes, Stilpnus, Sulcarius, Thaumatogelis, Theroscopus, Thrybius, Tricholinum,*

Tropistes, Trychosis, Uchidella, Xenolytus, Xiphulcus, Xylophrurus, Zoophthorus

Ctenopelmatinae: *Absyrtus, Alcochera, Alexeter, Anisotacrus, Anoncus, Arbelus, Asthenara, Azelus, Barytarbes, Campodorus, Ctenopelma, Euryproctus, Glyptorhaestus, Gunomeria, Hadrodactylus, Himerta, Homaspis, Hypamblys, Hyperbatus, Hypsantyx, Ischyrocnemis, Labrossyta, Lagarotis, Lamachus, Lathiponus, Lathrolestes, Lethades, Lophyroplectus, Mesoleius, Mesoleptidea, Notopygus, Oetophorus, Olethrodotis, Opheltes, Otlophorus, Pantorhaestes, Perilissus, Perispuda, Phaestus, Phobetes, Pion, Priopoda, Protarchus, Rhaestus, Rhinotorus, Rhorus, Saotis, Scolobates, Scopesis, Semimesoleius, Smicrolius, Sympherta, Syndipnus, Synodites, Synoecetes, Synomelix, Syntactus, Trematopygodes, Trematopygus, Xenoschesis, Zaplethocornia, Zemiophora*

Cylloceriinae: *Allomacrus, Cylloceria*

Diacritinae: *Diacritus*

Diplazontinae: *Bioblapsis, Campocraspedon, Diplazon, Enizemum, Phthorima, Promethes, Sussaba, Syrphoctonus, Syrphophilus, Tymmophorus, Woldstedtius, Xestopelta*

Eucerotinae: *Euceros*

Hybrizontinae: *Ghilaromma, Hybrizon*

Ichneumoninae: *Achaius, Acolobus, Aethecerus, Amblyjoppa, Amblyteles, Anisobas, Aoplus, Apaeleticus, Asthenolabus, Baeosemus, Baranisobas, Barichneumon, Callajoppa, Centeterus, Chasmias, Coelichneumon, Coelichneumonops, Colpognathus, Cotiheresiarches, Cratichneumon, Crypteffigies, Crytea, Ctenichneumon, Ctenochares, Cyclolabus, Dentilabus, Deuterolabops, Diadromus, Dicaelotus, Dilleritomus, Diphyus, Dirophanes, Ectopius, Eparces, Epitomus, Eriplatys, Eristicus, Eupalamus, Eurylabus, Eutanyacra, Exephanes, Goedartia, Hemichneumon, Hepiopelmus, Heresiarches, Herpestomus, Heterischnus, Homotherus, Hoplismenus, Hybophorellus, Hypomecus, Ichneumon, Limerodes, Limerodops, Linycus, Listrodromus, Lymantrichneumon, Melanichneumon, Mevesia, Misetus, Nematomicrus, Neotypus, Notosemus, Obtusodonta, Oiorhinus, Oronotus, Orotylus, Paraethecerus, Phaeogenes, Platylabops, Platylabus, Platymischos, Poecilostictus, Pristicerops, Pristiceros, Probolus, Protichneumon, Psilomastax, Spilichneumon, Spilothyrateles, Stenaoplus, Stenichneumon, Stenobarichneumon, Stenodontus, Sycaonia, Syspasis, Thyrateles, Trachyarus, Tricholabus, Triptognathus, Trogus, Tycherus, Virgichneumon, Vulgichneumon*

Lycorininae: *Lycorina*

Mesochorinae: *Astiphromma, Cidaphus, Mesochorus*

Metopiinae: *Apolophus, Carria, Chorinaeus, Colpotrochia, Exochus, Hypsicera, Metopius, Periope, Stethoncus, Synosis, Triclistus, Trieces*

Microleptinae: *Microleptes*

Neorhacodinae: *Neorhacodes*

Ophioninae: *Enicospilus, Eremotylus, Ophion, Stauropoctonus*

Orthocentrinae: *Aniseres, Aperileptus, Apoclima, Batakomacrus, Catastenus, Dialipsis, Entypoma, Eusterinx, Gnathochorisis, Helictes, Hemiphanes, Megastylus, Neurateles, Orthocentrus, Pantisarthrus, Picrostigeus, Plectiscidea, Plectiscus, Proclitus, Stenomacrus, Symplecis*

Orthopelmatinae: *Orthopelma*

Oxytorinae: *Oxytorus*

Phrudinae: *Astrenis, Phrudus, Pygmaeolus*

Pimplinae: *Acrodactyla, Acropimpla, Afrephialtes, Apechthis, Clistopyga, Delomerista, Dolichomitus, Dreisbachia, Endromopoda, Ephialtes, Exeristes, Fredegunda, Gregopimpla, Iseropus, Itoplectis, Liotryphon, Oxyrrhexis, Paraperithous, Perithous, Pimpla, Piogaster, Polysphincta, Scambus, Schizopyga, Sinarachna, Theronia, Townesia, Tromatobia, Zaglyptus, Zatypota*

Poemeniinae: *Deuteroxorides, Podoschistus, Poemenia, Pseudorhyssa*

Rhyssinae: *Rhyssa, Rhyssella*

Stilbopinae: *Panteles, Stilbops*

Tersilochinae: *Allophroides, Aneuclis, Barycnemis, Diaparsis, Epistathmus, Gelanes, Heterocola, Phradis, Probles, Sathropterus, Spinolochus, Tersilochus*

Tryphoninae: *Acrotomus, Cladeutes, Cosmoconus, Cteniscus, Ctenochira, Cycasis, Dyspetes, Eclytus, Eridolius, Erromenus, Excavarus, Exenterus, Exyston, Grypocentrus, Hercus, Idiogramma, Kristotomus, Monoblastus, Neleges, Neliopisthus, Netelia, Oedemopsis, Orthomiscus, Otoblastus, Phytodietus, Polyblastus, Smicroplectrus, Sphinctus, Thymaris, Tryphon*

Xoridinae: *Ischnoceros, Odontocolon, Xorides*

SUPERFAMILY MYMAROMMATOIDEA

The only family in this group has sometimes been included in the Chalcidoidea.

Family Mymarommatidae (1 genus, 1 species)
There are only 14 species known of this family worldwide, with just one species, *Palaeomymar anomalum*, recorded from Britain. They are extremely small, less than 1 mm long and can be confused with the Mymaridae, but they can be separated using a good family key such as that in Gauld and Bolton (1988); see also Debauche (1948). The biology of this group is unknown, but from their small size they are assumed to be egg-parasites.

British genus: *Palaeomymar*

SUPERFAMILY PLATYGASTROIDEA

The two families in this group were formerly included in the Proctotrupoidea in the belief that they showed some affinity with the Diapriidae, but most authors now accept that they belong in a separate superfamily. However, there are still doubts about the monophyly of the Scelionidae. The two families can be separated with the family key of Gauld and Bolton (1988). Königsmann (1978) discussed the higher classification of this group.

Family Platygastridae (15 genera, *c*. 157 species)
Although this is a large family, the taxonomy is very little studied and there is considerable doubt about how many species are genuinely recorded from Britain. They are all very small species, usually between 1–3 mm long.

The known platygastrids are endoparasitoids, and most attack Diptera, especially the gall-forming Cecidomyiidae. Some are parasitoids of mealybugs (Pseudococcidae) and at least one species attacks whiteflies (Aleyrodidae). Other species are known from beetle eggs and some are reported from hymenopterous galls, though these may well be parasitoids of cecidomyiid flies that are inquilines in the galls. Some species are of potential value in the biological control of cecidomyids and whiteflies, both of which can be crop pests.

Identification: with no recent works on the British species, and very little published on the European fauna, identification is very difficult and it will be necessary to consult the specialist literature listed in Noyes et al. (1999). Several works like Kozlov (1987c), Skuhravá et al. (1984) and Masner and Huggert (1989) will be found useful, as will the key to whitefly parasites by Gerling (1990). Vlug (1995) provided a world catalogue.

British genera: *Acerotella, Allotropa, Amblyaspis, Euxestonotus, Inostemma, Iphitrachelus, Isocybus, Isostasius, Leptacis, Metaclisis, Piestopleura, Platygaster, Platystasius, Synopeas, Trichacis*

Family Scelionidae (14 genera, *c*. 102 species)
Rather like the Platygastridae, this is another large family with many British species that is in urgent need of revision. They are all small undistinguished wasps whose biology is little studied. The known species are egg parasites of a wide range of invertebrate hosts, though most seem to be host-specific, either at species level or within a narrow range of host species. There are apparently biological differences between the three subfamilies, although records are scarce. Most Telenominae seem to be associated with Lepidoptera or Hemiptera, and some have been used as biological control agents. The Scelioninae show the widest variation in host preferences, with records from the eggs of Orthoptera, Hemiptera and spiders. Although the Teleasinae is a large subfamily it is little studied, and the few known species are associated with beetle eggs. One peculiarity of this family is that several species have been recorded as being carried phoretically by potential hosts, especially on the hairy bodies of Lepidoptera. When the host lays its eggs the scelionid wasp is in place to parasitize them immediately.

Identification: as with the Platygastridae there are no recent works on the British species, and very little has been published on the European fauna, so it will be necessary to consult the specialist literature listed in Noyes et al. (1999) for identification. Some generally useful works are Kozlov and Kononova (1983, 1990), Huggert (1979) and Masner (1976, 1980).

British subfamilies and genera:
Scelioninae: *Baeus, Gryon, Opisthacantha, Plesiobaeus, Scelio, Sparasion, Thoron*
Teleasinae: *Teleas, Trimorus, Trisacantha, Xenomerus*
Telenominae: *Allophanurus, Telenomus, Trissolcus*

SUPERFAMILY PROCTOTRUPOIDEA

The exact relationships between the three families in this group are not clear as they are each somewhat distinct, both morphologically and biologically. They can be separated using the family key in Gauld and Bolton (1988); Masner (1993) is also useful, and nomenclature can be checked against the world catalogue by Johnson (1992).

Family Diapriidae (38 genera, *c*. 300 species)
With around 300 British species this large family is little studied, and the biology of most species is not known. Adult diapriids are small brown or black wasps, often found in damp, shaded areas; many species are common, and the existing keys to the British species work reasonably well, so it surprising that so little is known about them. The known associations are with the Diptera, though some attack ants or beetles; most seem to be endoparasitoids on pupae.

There are three subfamilies in Britain: the Belytinae, Ismarinae and Diapriinae, and they are briefly described separately. Identification of the British fauna to subfamily is covered by Nixon (1980), with *Ismarus* included in the Belytinae.

Members of the Belytinae are usually found in damp wooded situations, and most are known to attack Mycetophilidae and Sciaridae (Diptera). One species, *Synacra paupera*, has been used to control the greenhouse pest *Bradysia difformis* (= *paupera*) (Sciaridae).

The Diapriinae is a large subfamily whose members are found in wide variety of habitats, often in the soil, though some are in the intertidal zone. Most species are pupal endoparasitoids of higher Diptera, inside the puparium. Many are gregarious endoparasitoids, and 30–50 individuals can emerge from a single fly puparium. Several species are associated with ants, though the exact relationship is not always clear. One wingless species, *Platymischus dilatatus*, can be found in large numbers in the intertidal zone where it attacks the sepsid fly *Orygma luctuosa* that breeds in rotting seaweed. A few species are of some economic importance as they attack Dipteran pests like the carrot fly, *Psila rosae* (Psilidae) and the frit fly, *Oscinella frit* (Chloropidae).

Members of the small subfamily Ismarinae seem to be widespread though not commonly found. They are hyperparasitoids that develop within dryinid larvae parasitizing Cicadellidae (Hemiptera).

Identification: the handbooks by Nixon (1957) on the Belytinae (including the Ismarinae), and Nixon (1980) on the Diapriinae are still useful, but many extra references need to be read in order to be up to date; see Noyes et al. (1999). The works by Hellén (1963, 1964) and Kozlov (1987a) are also worth consulting.

British subfamilies and genera:

Belytinae: *Acanopsilus, Acanosema, Acanthopsilus, Aclista, Acropiesta, Anommatium, Aprestes, Belyta, Cinetus, Diphora, Eumiota, Macrohynnis, Miota, Oxylabis, Pamis, Pantoclis, Pantolyta, Paroxylabis, Polypeza, Psilomma, Rhynchopsilus, Synacra, Synbelyta, Zygota*

Diapriinae: *Aneurhynchus, Basalys, Diapria, Entomacis, Idiotypa, Labolips, Monelata, Paramesius, Platymischus, Psilus, Spilomicrus, Tetramopria, Trichopria*

Ismarinae: *Ismarus*

Family Heloridae (1 genus, 3 species)
The Heloridae are often regarded as a relict group; the family contains the single, morphologically isolated genus *Helorus*, with just three species in Britain, though only seven species have been described worldwide. They are small shiny black wasps that are not often seen in the wild. They are all endoparasitoids of the larvae of Chrysopidae (Neuroptera) and may be quite common if chrysopids are reared in captivity. The helorid eggs are injected into the chrysopid larva and the newly hatched larvae do not start to develop until the host spins its cocoon in which to pupate. The helorid then grows rapidly and pupates within the host cocoon. There can be several generations in a year, but development depends on the host; some chrysopids overwinter as a diapausing larva, and then the helorid remains undeveloped for several months.

Identification: it is necessary to consult the specialized literature listed in Noyes et al. (1999) such as Pschorn-Walcher (1971).

British genus: *Helorus*

Family Proctotrupidae (11 genera, 40 species)
This is a moderate sized family in Britain; most species are small to medium sized and shiny black in colour. Like some Diapriidae they can be very common in damp woodland habitats, where they parasitize the larvae of Coleoptera, or sometimes Mycetophilidae (Diptera); these host larvae are either hidden under bark, leaf-litter or in fungi. There is a list of host associations of each genus in Gauld and Bolton (1988). Females oviposit into the host larva, and some species are gregarious endoparasitoids. As in the Heloridae, the newly hatched larvae remain quiescent until the host prepares to pupate; the parasitoids then grow rapidly and eventually kill the host larva. In the gregarious species they emerge and pupate in characteristically neat rows with the ends of the pupae still attached to the host.

Identification: there are no recent works covering the British species, and it will be necessary to consult the specialized literature in Noyes et al. (1999) such as Townes and Townes (1981) and Kozlov (1987b).

British genera: *Brachyserphus, Codrus, Cryptoserphus, Disogmus, Exallonyx, Mischoserphus, Paracodrus, Parthenocodrus, Phaenoserphus, Proctotrupes, Tretoserphus*

SUPERFAMILY TRIGONALOIDEA

Family Trigonalidae (1 genus, 1 species)
There is just a single British species, *Pseudogonalos hahnii* (previously placed in *Trigonalis*), in this unusual family; this is also the only species known in Europe, though it is rarely seen. The family name has been spelled as Trigonalyidae, with the superfamily correspondingly spelled Trigonalyoidea.

The members of this family have an unusual biology, though it should be noted that the life history of the British species is not yet known. Most trigonalids have two successive hosts. The female wasp lays many very small minute eggs on leaves, and these do not hatch until they are eaten by the larva of Lepidopteran or a sawfly larva (which is a similar strategy to that used by some Tachinidae in the Diptera). Further development then seems to depend on one of two events occurring. If the primary host is parasitized by another endoparasitoid, such as an ichneumonid or braconid wasp, or a tachinid fly, then the trigonalid larva enters the parasitoid larva and develops as a hyperparasitoid. The second possibility is that development of the larva will continue if the primary host is taken by a predatory wasp of the family Vespidae. When the trigonalid larva is fed by a worker wasp to a larva then it develops as an endoparasitoid of the wasp larva.

Identification: family keys such as Gauld and Bolton (1988).

British genus: *Pseudogonalos*

SUBORDER SYMPHYTA

There is general agreement that this is the most primitive group of Hymenoptera, both in terms of their morphology and their phytophagous habit. However, this is not a monophyletic group as it is linked by plesiomorphic characters, though most authors still retain it for convenience. The sawflies and woodwasps are weak-flying and rather heavy-bodied insects that do not have the narrow 'waist' of the higher groups of Hymenoptera. The larvae of most groups are phytophagous, either as external feeders of plant tissue or else forming galls or mines. Woodwasp larvae bore into timber, and the Orussidae are parasites. Symphytan larvae have a close similarity to the caterpillars of Lepidoptera, but they usually have at least six pairs of abdominal prolegs, while Lepidoptera have five pairs at the most; sawfly larvae also lack the hooks or crochets seen on Lepidopteran prolegs. However, the internally feeding sawflies may have a modified morphology, with reduced prolegs. The free-living species often feed communally and rear up in unison when disturbed, as a defence mechanism (see Fig. 20. 78).

There are several works that can help identify the British Symphyta, especially the handbooks by Benson (1951, 1952, 1958) and Quinlan and Gauld (1981) though several of these are outdated. The work by Liston (1995, 1996) updates the nomenclature and classification, but there are also several specialist references that need to be consulted. A useful simplified guide to the British genera is by Wright (1990). A checklist of the British sawflies is on the following website: http://www.nhm.ac.uk/resources-rx/files/checklist-of-british-and-irish-sawflies-59046.pdf.

SUPERFAMILY CEPHOIDEA

Family Cephidae (5 genera, 12 species)
The members of this small family are slender-bodied compared with most other sawflies, with long antennae. Their larvae are stem-borers, with some genera living in woody plants or Rosaceae, and others in the stems of grasses and cereal crops. Damage to crops is sometimes caused, not by direct feeding, but because the larval activity causes the stem to fracture easily and the grain cannot therefore be harvested. The larvae have no prolegs and even the true legs are vestigial and unsegmented, typical adaptations of an internal feeder.

Identification: the handbook by Quinlan and Gauld (1981) generally superseded that of Benson (1951).

British genera: *Calameuta, Cephus, Hartigia, Janus, Trachelus*

SUPERFAMILY ORUSSOIDEA

Family Orussidae (1 genus, 1 species)
There is just one species recorded in Britain, *Orussus abietinus*, but as it has not been seen since the early 19th century it may well be extinct here. Although details of the life history of this family are not clear, it seems that most are parasitoids on wood-boring beetle larvae.

Identification: Benson (1951) or Quinlan and Gauld (1981), as well as the family key in Gauld and Bolton (1988).

British genus: *Orussus*

SUPERFAMILY PAMPHILIOIDEA

This group was previously known as the Megalodontoidea. Early works such as Benson (1951) suggested that the family Megalodontidae (now Megalodontesidae) also occurred in Britain, but there are no authenticated records and this family is now removed from the British list.

Family Pamphiliidae (4 genera, 19 species)
Adult pamphilids have a rather flattened appearance, with long thread-like antennae. Their larvae all feed on trees and do not have distinct abdominal prolegs.

Members of the subfamily Cephalciinae are all associated with pine trees; they spin silken webs underneath pine twigs, gradually biting off more needles and pulling them into the web. Some species can reach pest proportions in parts of Europe. The Pamphilinae feed in a similar way, but are confined to angiosperms such as Rosaceae, Betulaceae and Salicaceae. They spin silk in the same way as the cephalciines but often make a characteristic roll at the edge of the leaves. All pamphilid larvae spend their time on the underside of leaves or pine needles, walking upside down.

Identification: the handbook by Quinlan and Gauld (1981) generally superseded that of Benson (1951), but it is still necessary to consult some specialized literature as listed in Noyes et al. (1999).

British subfamilies and genera:
Cephalciinae: *Acantholyda, Cephalcia*
Pamphiliinae: *Neurotoma, Pamphilius*

SUPERFAMILY SIRICOIDEA

Family Siricidae (4 genera, 11 species)
Although eleven species of woodwasps have been recorded in Britain it is likely that only half of these reproduce here. Nonetheless they form a striking group of insects that are large and colourful. The most noticeable species, *Urocerus gigas*, known as the Horntail or Giant woodwasp (Fig. 20.76) causes

alarm every year as it is often assumed that the long stout ovipositor is a sting; in fact this species is completely harmless to people, despite its hornet-like appearance and large size of up to 40 mm in length. There are some common species of *Sirex*, also quite large at 15–30 mm long, in which the females are bluish-black in colour, while the males are mainly orange.

All the native species are in the subfamily Siricinae, and all are associated with conifers; the females use their stout ovipositors to insert eggs into the trees, where their larvae remain for two or three years before pupating just below the bark. The adults then emerge through characteristically large holes. Although woodwasps do not lay eggs in prepared timber, it is quite common for trees containing larvae to be felled and converted, and the larvae continue to develop though at a slower rate; this can lead to the emergence of the adult inside a building several years after the tree was cut down. Siricid larvae all require the presence of certain symbiotic fungi in their tunnels; the fungal spores are carried by the female woodwasp and injected into the tree when ovipositing.

The only member of the subfamily Tremicinae, *Tremex columba*, is a north American species that is occasionally introduced here; it lives in hardwood timber.

Identification: Quinlan and Gauld (1981).

British subfamilies and genera:
Siricinae: *Sirex, Urocerus, Xeris*
Tremicinae: *Tremex*

SUPERFAMILY TENTHREDINOIDEA

Family Argidae (3 genera, 15 species)
Although many sawfly families are most diverse in temperate regions, the Argidae are unusual in being mainly a tropical group; there are over 800 species described worldwide but only 15 in Britain. Adult argids are broad-bodied, heavy looking insects that fly rather slowly and are often seen feeding on umbelliferous flower-heads. Most of the British species are external feeders on woody plants or trees, although *Aprosthema melanura* is associated with legumes. *Arge ochropus* is known as the Rose sawfly (Figs. 20.77 & 20.78).

Identification: Quinlan and Gauld (1981).

British subfamilies and genera:
Arginae: *Arge*
Sterictiphorinae: *Aprosthema, Sterictiphora*

Fig. 20.76 *Uroceras gigas* (Siricidae) (Photo: Roger Key)

Fig. 20.77 *Arge ochropus* (Argidae) (Photo: Roger Key)

Fig. 20.79 *Trichiosoma* sp. (Cimbicidae) (Photo: Roger Key)

Fig. 20.78 *Arge ochropus* larvae (Argidae) (Photo: Roger Key)

Fig. 20.80 *Cimbex femoratus* larva (Cimbicidae) (Photo: Roger Key)

Family Blasticotomidae (1 genus, 1 species)
Worldwide this is a very small family with around 10 species known; just one, *Blasticotoma filiceti*, occurs in Britain. The larvae of this family are stem borers in ferns; there are very few British records, though from several different host species; most are from ferns in botanical gardens.

Identification: Quinlan and Gauld (1981).

British genus: *Blasticotoma*

Family Cimbicidae (4 genera, 14 species)
The cimbicid sawflies can be quite large and striking insects, up to 25 mm long, with short, slightly

clubbed antennae (Fig. 20.79); they fly rapidly, produce a clearly audible buzzing sound, and sometimes even attempting to bite when handled. Their larvae are external feeders, with eight pairs of abdominal prolegs.

The larger species are members of the subfamily Cimbicinae, and their larvae feed on woody angiosperms, especially trees. A few species have common names based on their host plant, including *Trichiosoma tibiale*, the Hawthorn sawfly and *Cimbex femoratus*, the Birch sawfly. They are usually solitary feeders, which curl up when disturbed (Fig. 20.80). Pupation occurs in a cocoon, which may be underground or attached to the host plant. Members of the subfamily Abiinae are smaller and generally less common; their larvae feed on various climbing and herbaceous plants.

Identification: Quinlan and Gauld (1981) key all the British species, though some genera such as *Cimbex* and *Trichiosoma* can be difficult to determine to species level.

British subfamilies and genera:
Abiinae: *Abia, Zaraea*
Cimbicinae: *Cimbex, Trichiosoma*

Family Diprionidae (5 genera, 9 species)
The members of this small family of sawflies are all associated with coniferous trees; they are large-bodied, slow-flying insects which have serrated antennae in the females, and plumose antennae in the males. Their larvae are external feeders with eight pairs of abdominal prolegs.

Members of the subfamily Diprioninae feed on pine and spruce trees, and the commonest species is *Diprion pini*, the Pine sawfly (Fig. 20.81).

The only British member of the Monocteninae, *Monoctenus juniperi*, is known only from a few Scottish sites, feeding on juniper.

Identification: Quinlan and Gauld (1981).

British subfamilies and genera:
Diprioninae: *Diprion, Gilpinia, Microdiprion, Neodiprion*
Monocteninae: *Monoctenus*

Family Tenthredinidae (c. 75 genera, c. 400 species)
This is by far the largest family of sawflies, containing around 80% of the British species of Symphyta.

In such a large family it is inevitable that there is much variation in morphology and life histories, and the main differences are given in the brief sub-family descriptions below. Although some groups are leaf-miners or gall-causers most tenthredinid larvae are external leaf-feeders with 6–8 pairs of abdominal prolegs, though these may be absent in the mining species.

The subfamilial classification used by Benson (1952, 1958) is now out of date and the sytem used here conforms with that described in Gauld and Bolton (1988), which should be consulted for further details.

There are nearly 50 species in the subfamily Allantinae, and they were formerly included in the Blennocampinae by Benson (1952). Several species are brightly coloured, especially the larger black and yellow ones. The larvae of allantines are all external feeders on a wide variety of host plants, although Rosaceae are the most commonly used. Some species like *Athalia rosae*, the Turnip sawfly, are occasional pests. Even though several groups formerly in the Blennocampinae have been placed elsewhere, this still remains a somewhat heterogeneous group. The larvae of most species are external feeders on the foliage of various angiosperms, particularly Rosaceae. The two genera in the Dolerinae were formerly placed in the Selandriinae. The British species are usually found in damp habitats, where the larvae feed mainly on grasses, rushes, sedges and horsetails (Equisetaceae). The members of the Heterarthrinae were formerly included in the Blennocampinae. Several are leaf-miners, but the larvae of *Caliroa* can skeletonize orchard trees; the common *C. cerasi* can be a serious pest.

On a world scale the subfamily Nematinae is large and diverse, and it forms a large proportion of the symphytan fauna in northern latitudes. It is the largest subfamily in Britain, containing about 200 species. Several genera are large and contain difficult species complexes. Most nematines are found on trees, including Rosaceae, and their larvae are sometimes external feeders on leaves, though many are leaf rollers, miners or gall-formers. *Nematus ribesii*, the Gooseberry sawfly, is a common species in this group, while *Pontania proxima* causes galls on willow trees. Selandriine larvae are mainly feeders on ferns, though a number are associated with the monocotyledonous families of grasses, sedges and rushes; a few feed on herbaceous plants. Most are external feeders but the larva of *Heptamelus ochroleucus* bores through the stems of ferns and may be a minor pest in gardens, especially in Scotland.

Fig. 20.81 *Diprion pini* **larva (Diprionidae) (Photo: Roger Key)**

Many of the large and colourful British sawflies are to be found in the Tenthredininae, which includes around 60 species. Their larvae are mainly external feeders on the leaves of a wide range of herbaceous plants in many families, but many larvae are nocturnal and stay concealed during the day. *Tenthredo scrophulariae* (Figs. 20.82 & 20.83) is a typical species in this group, as is *Macrophya montana* (Fig. 20.84).

Identification: two of the handbooks by Benson can still be helpful in this family, though the nomenclature and subfamilial classification have changed considerably; it is essential to consult specialized references listed in Noyes et al. (1999). Benson (1952) can be used for the Allantinae, Blennocampinae, Dolerinae, Heterarthrinae and Tenthredininae, while Benson (1958) is useful for the Nematinae.

British subfamilies and genera:

Allantinae: *Allantus, Ametastegia, Apethymus, Athalia, Empria, Eriocampa, Harpiphorus, Monostegia, Monsoma, Taxonus*

Blennocampinae: *Ardis, Blennocampa, Cladardis, Claremontia, Eutomostethus, Halidamia, Monophadnoides, Monophadnus, Paracharactus, Pareophora, Periclista, Phymatocera, Rhadinoceraea, Stethomostus, Tomostethus*

Dolerinae: *Dolerus, Loderus*

Heterarthrinae: *Caliroa, Endelomyia, Fenella, Fenusa, Fenusella, Heterarthrus, Metallus, Parna, Profenusa, Scolioneura*

Nematinae: *Amauronematus, Anoplonyx, Cladius, Craesus, Dineura, Endophytus, Euura, Hemichroa, Hoplocampa, Mesoneura, Nematinus, Nematus, Pachynematus, Phyllocolpa, Platycampus, Pontania, Pristicampus, Pristiphora, Pseudodineura*

Selandriinae: *Aneugmenus, Birka, Brachythops, Dulophanes, Heptamelus, Pseudohemitaxonus, Selandria, Stromboceros, Strongylogaster*

Tenthredininae: *Aglaostigma, Macrophya, Pachyprotasis, Perineura, Rhogogaster, Sciapteryx, Tenthredo, Tenthredopsis*

SUPERFAMILY XIPHRYDIOIDEA

The Xiphydriidae were previously included in the superfamily Siricoidea.

Family Xiphydriidae (1 genus, 3 species)
There are just three species in the genus *Xiphydria* in Britain; they have a rather spherical head on a long 'neck' (Fig. 20.85). The larvae are all borers in deciduous trees and hence the group is often termed

Fig. 20.82 *Tenthredo scrophulariae* mating pair (Tenthredinidae) (Photo: Roger Key)

Fig. 20.83 *Tenthredo scrophulariae* larva (Tenthredinidae) (Photo: Roger Key)

Fig. 20.84 *Macrophya montana* (Tenthredinidae) (Photo: Roger Key)

Fig. 20.85 *Xiphydria camelus* (Xiphydriidae) (Photo: Roger Key)

woodwasps like the Siricidae; as in the latter family the larvae need symbiotic fungi in their tunnels to survive.

Identification: Quinlan and Gauld (1981) is useful, but another species, *Xiphydria longicollis*, has since been added to the British list; see Noyes et al. (1999) or Wright (1990) for details.

British genus: *Xiphydria*

SUPERFAMILY XYELOIDEA

Family Xyelidae (1 genus, 2 species)
There are just two British species in this family, which is generally accepted as being the most primitive in the Symphyta. Both are small members of the genus *Xyela*, though only one is commonly found. Their larvae feed on developing male pine cones, which become distorted or lose their pollen prematurely.

Identification: the handbook by Quinlan and Gauld (1981) superseded that of Benson (1951).

British genus: *Xyela*

REFERENCES

ALEKSEEV, V.N. 1987a. Superfamily Ceraphronoidea. Ceraphronidae. In: MEDVEDEV, G.S. (ed.) *Keys to the insects of the European part of the USSR* III, Part 2. Amerind Publishing, New Delhi, India, pp. 1240–57.

ALEKSEEV, V.N. 1987b. Superfamily Ceraphronoidea. Megaspilidae. In: MEDVEDEV, G.S. (ed.) *Keys to the insects of the European part of the USSR* III, Part 2. Amerind Publishing, New Delhi, India, pp. 1216–57.

ASKEW, R.R. 1968. Hymenoptera 2. Chalcidoidea Section (b). *Handbooks for the identification of British insects* 8(2b): 39 pp.

ASKEW, R.R. 1971. *Parasitic insects*. Heinemann, London.

ASKEW, R.R. 1984. The biology of gall wasps. In: ANANTHAKRISHNAN, T.N. (ed.) *Biology of gall insects*. Arnold, London, pp. 223–71.

AUBERT, J.F. 1978. *Les Ichneumonides ouest-paléarctiques et leurs hôtes*. 2. (Banchinae et suppl. aux Pimplinae). É.D.I.F.A.T., Paris, 318 pp.

BENSON, R.B. 1951. Hymenoptera Symphyta. Section A. *Handbooks for the identification of British insects* 6(2a): 49 pp.

BENSON, R.B. 1952. Hymenoptera Symphyta. Section B. *Handbooks for the identification of British insects* 6(2b): 51–137.

BENSON, R.B. 1958. Hymenoptera Symphyta. Section C. *Handbooks for the identification of British insects* 6(2c): 139–252.

BETTS, C. 1986. *The hymenopterist's handbook*. Amateur Entomologists' Society, London.

BOLTON, B. 1994. *Identification guide to the ant genera of the world*. Harvard University Press, Cambridge, MA.

BOLTON, B. 1995. *A new general catalogue of the ants of the world*. Harvard University Press, Cambridge, MA.

BOLTON, B. & COLLINGWOOD, C.A. 1975. Hymenoptera: Formicidae. *Handbooks for the identification of British insects* 6(3c): 34 pp.

BOUCEK, Z. 1952. The first revision of the European species of the family Chalcididae (Hymenoptera). *Sborník Entomologického Oddeleni Národního Musea v Praze* 27 (Suppl. 1): 108 pp.

BOUCEK, Z. 1988. *Australasian Chalcidoidea (Hymenoptera). A biosystematic revision of genera of fourteen families, with a reclassification of species.* CAB International, Wallingford.

BOUCEK, Z. & ASKEW, R.R. 1968. Palaearctic Eulophidae sine Tetrastichinae. In: DELUCCHI, V. & REMAUDIÈRE, G. (eds.) *Index of entomophagous insects* 3. Le François, Paris, 260 pp.

BOUCEK, Z. & RASPLUS, J.-Y. 1991. *Illustrated key to west-palaearctic genera of Pteromalidae (Hymenoptera: Chalcidoidea)*. Institut National de la Recherche Agronomique, Paris.

CRANE, E. 2009. *Apis* species (honey bees). In: RESH, V.H. & CARDÉ, R.T. (eds.) *Encyclopedia of insects* (2nd edn.). Academic Press/Elsevier, San Diego & London, pp. 31–2.

DAY, M.C. 1988. Spider wasps (Hymenoptera: Pompilidae). *Handbooks for the identification of British Insects* 6(4): 60 pp.

DEBAUCHE, H.R. 1948. Étude sur les Mymarommidae et les Mymaridae de la Belgique (Hym., Chalcidoidea). *Mémoires du Musée Royal d'Histoire Naturelle de Belgique* 108: 248 pp.

EADY, R.D. & QUINLAN, J. 1963. Hymenoptera Cynipoidea. Key to families and subfamilies and Cynipinae (including galls). *Handbooks for the identification of British insects* 8(1a): 81 pp.

EDWARDS, M. & JENNER, M. 2009. *Field guide to the bumblebees of Great Britain and Ireland* (revised edn.). Ocelli Publishing, Eastbourne.

EDWARDS, R. 1980. *Social wasps*. Rentokil, East Grinstead.

FALK, S. 1991. A review of the scarce and threatened bees, wasps and ants in Great Britain. *Research and Survey in Nature Conservation* 35. Nature Conservancy Council, Peterborough.

FERGUSSON, N.D.M. 1986. Charipidae, Ibaliidae & Figitidae (Hymenoptera: Cynipoidea). *Handbooks for the identification of British insects* 8(1c): 55 pp.

FERRIÈRE, C. 1965. *Hymenoptera Aphelinidae d'Europe et du bassin Mediterranean.* Masson et Cie, Paris.

FERRIÈRE, C. & KERRICH, G.J. 1958. Hymenoptera 2. Chalcidoidea. Section (a) Agaontidae, Leucospidae, Chalcididae, Eucharitidae, Perilampidae, Cleonymidae and Thysanidae. *Handbooks for the identification of British insects* 8(2a): 40 pp.

FITTON, M.G., SHAW, M.R. & GAULD, I.D. 1988. Pimpline ichneumon-flies. *Handbooks for the identification of British insects* 7(1): 110 pp.

FRANKS, N.R. 2009. Ants. In: RESH, V.H. & CARDÉ, R.T. (eds.) *Encyclopedia of insects* (2nd edn.). Academic Press / Elsevier, San Diego & London, pp. 24–7.

GAULD, I.D. 1991. The Ichneumonidae of Costa Rica, 1. *Memoirs of the American Entomological Institute* 47: 589 pp.

GAULD, I.D. & BOLTON, B. (eds.) 1988. *The Hymenoptera.* British Museum (Natural History) and Oxford University Press [Reprinted 1996, with minor alterations and additions].

GAULD, I.D. & MITCHELL, P.A. 1977. Ichneumonidae, subfamilies Orthopelmatinae and Anomaloninae. *Handbooks for the identification of British insects* 7(2b): 32 pp.

GERLING, D. 1990. Natural enemies of whiteflies: predators and parasitoids. In: GERLING, D. (ed.) *Whiteflies: their bionomics, pest status and management.* Intercept, Andover, pp. 147–85.

GIBSON, G.A.P. 1989. Phylogeny and classification of Eupelmidae, with a revision of the world genera of Calosotinae and Metapelmatinae. *Memoirs of the Entomological Society of Canada* 149: 121 pp.

GIBSON, G.A.P. 1995. Parasitic wasps of the subfamily Eupelminae: classification and revision of world genera (Hymenoptera: Chalcidoidea: Eupelmidae). *Memoirs on Entomology, International* 5: 421 pp.

GODFRAY, H.C.J. 1994. *Parasitoids: behavioural and evolutionary ecology.* Princeton University Press, Princeton, NJ.

GRAHAM, M.W.R. DE V. 1991. A reclassification of the European Tetrastichinae (Hymenoptera: Eulophidae): revision of the remaining genera. *Memoirs of the American Entomological Institute* 49: 322 pp.

GRIMALDI, D. & ENGEL, M.S. 2005. *Evolution of the insects.* Cambridge University Press, Cambridge.

HANSSON, C. 1985. Taxonomy and biology of the palaearctic species of Chrysocharis Forster, 1856 (Hymenoptera: Eulophidae). *Entomologica Scandinavica* (Suppl.) 26: 130 pp.

HELLÉN, W. 1963. Die Diapriinen Finnlands (Hymenoptera: Proctotrupoidea). *Fauna Fennica* 14: 35 pp.

HELLÉN, W. 1964. Die Ismarinen und Belytinen Finnlands (Hymenoptera: Proctotrupoidea). *Fauna Fennica* 18: 68 pp.

HILPERT, H. 1992. Zur Systematik der Gattung Ichneumon Linnaeus, 1758 in der Westpalaearktis (Hymenoptera, Ichneumonidae, Ichneumoninae). *Entomofauna* (Suppl.) 6: 389 pp.

HÖLLDOBLER, B. & WILSON, E.O. 1990. *The ants.* Harvard University Press, Cambridge, MA.

HUGGERT, L. 1979. Revision of the west palaearctic species of the genus Idris Forster s.l. (Hymenoptera, Proctotrupoidea: Scelionidae). *Entomologica Scandinavica* (Suppl.) 12: 60 pp.

JOHNSON, N. 1992. Catalog of world Proctotrupoidea excluding Platygastridae. *Memoirs of the American Entomological Institute* 51: 825 pp.

KASPARYAN, D.R. 1973. Ichneumonidae (subfamily Tryphoninae) tribe Tryphonini. *Fauna SSSR* (New Series) 106: 320 pp.

KASPARYAN, D.R. (ed.) 1981. Hymenoptera, Ichneumonidae. *Keys to insects of the European parts of the USSR* 3(3): 688 pp.

KASPARYAN, D.R. 1990. Ichneumonidae: subfamily Tryphoninae, tribe Exenterini, subfamily Adelognathinae. *Fauna SSSR* (New Series) 141: 340 pp.

KIMSEY, L.S. & BOHART, R.M. 1990. *The chrysidid wasps of the world.* Oxford University Press, Oxford.

KÖNIGSMANN, E. 1978. Das phylogenetische System der Hymenoptera. Teil 3: Terebrantes (Unterordnung Apocrita). *Deutsches Entomologische Zeitschrift* 25: 55 pp.

KOSTER, A. 1986. Het genus Hylaeus in Nederland (Hymenoptera, Colletidae). *Zoologische Bijdragen* no. 36.

KOZLOV, M.A. 1987a. Superfamily Proctotrupoidea (Proctotrupoids). Diapriidae. In: MEDVEDEV, G.S. (ed.) *Keys to the insects of the European part of the USSR* III, Part 2. Amerind Publishing, New Delhi, India, pp. 1000–110.

KOZLOV, M.A. 1987b. Superfamily Proctotrupoidea (Proctotrupoids). Proctotrupidae. In: MEDVEDEV, G.S. (ed.) *Keys to the insects of the European part of the USSR* III, Part 2. Amerind Publishing, New Delhi, India, pp. 991–1000.

KOZLOV, M.A. 1987c. Superfamily Proctotrupoidea (Proctotrupoids). Platygastridae. In: MEDVEDEV, G.S. (ed.) *Keys to the insects of the European part of the USSR* III, Part 2. Amerind Publishing, New Delhi, India, pp. 1110–212.

KOZLOV, M.A. & KONONOVA, S.V. 1983. Telenominae of the USSR (Hymenoptera, Scelionidae, Telenominae). *Opredeliteli po Faune SSSR* 136: 329 pp.

KOZLOV, M.A. & KONONOVA, S.V. 1990. Stselionidy fauny SSSR: (Hymenoptera, Scelionidae, Scelioninae) [Scelioninae of the fauna of the USSR: (Hymenoptera, Scelionidae, Scelioninae)]. *Opredeliteli po Faune SSSR* 161: 344 pp.

LISTON, A.D. 1995. *Compendium of European sawflies.* Chalastos Press, Gottfrieding, 190 pp.

LISTON, A.D. 1996. *Compendium of European sawflies: Supplement.* Chalastos Press, Gottfrieding, 16 pp.

MASNER, L. 1976. Revisionary notes and keys to world genera of Scelionidae (Hymenoptera: Proctotrupoidea). *Memoirs of the Entomological Society of Canada* 97: 87 pp.

MASNER, L. 1980. Key to genera of Scelionidae of the holarctic region with descriptions of new genera and species. *Memoirs of the Entomological Society of Canada* 113: 54 pp.

MASNER, L. 1993. Superfamily Proctotrupoidea. In: GOULET, H. & HUBER, J.T. (eds.) *Hymenoptera of the world: an identification guide to families.* Agriculture Canada, Ottawa: pp. 537–57.

MASNER, L. & HUGGERT, L. 1989. World review and keys to genera of the subfamily Inostemmatinae with reassignment of the taxa to the Platygastrinae and Sceliotrachelinae (Hym., Platygastridae). *Memoirs of the Entomological Society of Canada* 147: 214 pp.

MORGAN, D. 1984. Hymenoptera: Chrysididae. *Handbooks for the identification of British insects* 6(5): 37 pp.

MORLEY, C. 1903–15. *Ichneumonologia Britannica* 1–5. Keys, Plymouth & Brown, London.

MUESEBECK, C.F.W. 1979. Proctotrupoidea and Ceraphronoidea. In: KROMBEIN, K.V., HURD, P.D., SMITH, D.R. & BURKS, B.D. (eds.) *Catalog of Hymenoptera in America North of Mexico* 1: 1121–95.

NIKOL'SKAYA, M. 1952. Chalcids of the fauna of the USSR (Chalcidoidea). *Opredeliteli po Faune SSSR, Izdavaemie Zoologicheskim Institutom Akademii Nauk SSR* 44: 575 pp. Akademiya Nauk SSSR, Moscow and Leningrad [In Russian; English translation: 1963: Israeli Program for Scientific Translations, Jerusalem: 593 pp.].

NIXON, G.E. J. 1957. Hymenoptera, Proctotrupoidea, Diapriidae subfamily Belytinae. *Handbooks for the identification of British insects* 8(3dii): 107 pp.

NIXON, G.E. J. 1980. Diapriidae (Diapriinae) Hymenoptera, Proctotrupoidea. *Handbooks for the identification of British insects* 8(3di): 55 pp.

NOYES, J.S., FITTON, M.G., QUICKE, D.L.J., NOTTON, D.G., ELSE, G.R., FERGUSSON, N.D.M., BOLTON, B., LEWIS, S. & TAREL, L.C. 1999. Hymenoptera: the bees, wasps and ants. In: BARNARD, P.C. (ed.) *Identifying British insects and arachnids: an annotated bibliography of key works.* Cambridge University Press, Cambridge, pp. 196–319.

O'TOOLE, C. & RAW, A. 1991. *Bees of the world.* Blandford Press, Poole.

OLMI, M. 1984. A revision of the Dryinidae (Hymenoptera). *Memoirs of the American Entomological Institute* 37(2): 1913 pp.

PECK, O., BOUCEK, Z. & HOFFER, A. 1964. Keys to the Chalcidoidea of Czechoslovakia (Insecta: Hymenoptera). *Memoirs of the Entomological Society of Canada* 34: 120 pp.

PERKINS, J.F. 1959. Ichneumonidae, key to subfamilies and Ichneumoninae – I. *Handbooks for the identification of British insects* 7(2ai): 116 pp.

PERKINS, J.F. 1960. Ichneumonidae, subfamilies Ichneumoninae II, Alomyinae, Agriotypinae and Lycorininae. *Handbooks for the identification of British insects* 7(2aii): 117–213.

PERKINS, J.F. 1976. Hymenoptera: Bethyloidea (excluding Chrysididae). *Handbooks for the identification of British insects* 6(3a): 38 pp.

PSCHORN-WALCHER, H. 1971. Heloridae et Proctotrupidae. *Insecta Helvetica* 4, Hymenoptera. Fotorotar, Zurich, Switzerland, 64 pp.

QUICKE, D.L.J. 1997. *Parasitic wasps.* Chapman & Hall, London.

QUICKE, D.L.J. 2009. Hymenoptera (Ants, bees, wasps). In: RESH, V.H. & CARDÉ, R.T. (eds.) *Encyclopedia of insects* (2nd edn.). Academic Press/Elsevier, San Diego & London, pp. 473–84.

QUINLAN, J. 1978. Hymenoptera Cynipoidea Eucoilidae. *Handbooks for the identification of British insects* 8(1b): 58 pp.

QUINLAN, J. & GAULD, I.D. 1981. Symphyta (except Tenthredinidae) *Handbooks for the identification of British insects* 6(2a): 67 pp.

REDFERN, M. & ASKEW, R.R. 1998. *Plant galls* (revised edn.). Naturalists' Handbooks no. 17. Richmond Publishing, Slough, 99 pp.

REDFERN, M. & SHIRLEY, P. 2002. British plant galls: identification of galls on plants and fungi. Field Studies Council, OP270.

RICHARDS, O.W. 1977. Hymenoptera. Introduction and key to families (2nd edn.). *Handbooks for the identification of British insects* 6(1): 100 pp.

RICHARDS, O.W. 1980. Scolioidea, Vespoidea and Sphecoidea (Hymenoptera, Aculeata). *Handbooks for the identification of British insects* 6(3b): 118 pp.

SANBORNE, M. 1984. A revision of the world species of Sinophorus (Ichneumonidae). *Memoirs of the American Entomological Institute* 38: 403 pp.

SAUNDERS, E. 1896. *The Hymenoptera Aculeata of the British Islands.* Reeve, London [available to download from the website of the Bees, Wasps and Ants Recording Society].

SCHAUFF, M.E. 1984. The holarctic genera of Mymaridae (Hymenoptera: Chalcidoidea). *Memoirs of the Entomological Society of Washington* 12: 67 pp.

SHARKEY, M.J. 1993. Family Braconidae. In: GOULET, H. & HUBER, J. (eds.) *Hymenoptera of the world: an identification guide to families.* Agriculture Canada, Research Branch, Ottawa.

SHAW, M.R. 1997. *Rearing parasitic Hymenoptera.* Amateur Entomologists' Society, London.

SHAW, M.R. & ASKEW, R.R. 1976. Parasites. In: HEATH, J. (ed.) *The moths and butterflies of Great Britain and Ireland.* Vol. 1. Blackwell Scientific Publications and Curwen Press, Oxford, pp. 24–56.

SHAW, M.R. & HUDDLESTON, T. 1991. Classification and biology of braconid wasps (Hymenoptera: Braconidae). *Handbooks for the identification of British insects* 7(11): 126 pp.

SHORT, J.R.T. 1978. The final larval instars of the Ichneumonidae. *Memoirs of the American Entomological Institute* 25: 508 pp.

SKINNER, G.J. 1987. *Ants of the British Isles.* Shire Natural History, Princes Risborough.

SKINNER, G.J. & ALLEN, G.W. 1996. *Ants.* Naturalists' Handbooks no. 24. Richmond Publishing, Slough, 83 pp.

SKUHRAVÁ, M., SKUHRAVÝ, V. & BREWER, J.W. 1984. Biology of gall midges. In: ANANTHAKRISHNAN, T.N. (ed.) *The biology of gall insects.* Arnold, London, pp. 169–222.

SPRADBERY, J.P. 1973. *Wasps.* Sidgwick & Jackson, London.

STARÝ, P. 1966. *Aphid parasites of Czechoslovakia.* Junk, Den Haag, 242 pp.

STEP, E. 1932. *Bees, wasps, ants and allied insects of the British Isles.* Warne, London.

SUGONJAEV, E.S. 1964. Palaearctic species of the genus Blastothrix Mayr (Hymenoptera, Chalcidoidea) with remarks on their biology and economic importance. Part I. *Entomologicheskoe Obozrenie* 43(2): 368–90 [In Russian; English translation: *Entomological Review, Washington* 43: 189–98].

SUGONJAEV, E.S. 1965. Palaearctic species of the genus Blastothrix Mayr (Hymenoptera, Chalcidoidea) with remarks on their biology and economic importance. Part II. *Entomologicheskoe Obozrenie* 44: 395–410 [In Russian; English translation: *Entomological Review, Washington* 44: 225–33].

TAEGER, A. 1989. *Die Orgilus-Arten der Paläarktis (Hymenoptera, Braconidae).* Akademie der Landwirtschaftswissenschaften der Deutschen Demokratischen Republik, Berlin.

TOWNES, H. 1969. The genera of Ichneumonidae 1. *Memoirs of the American Entomological Institute* 11: 300 pp.

TOWNES, H. 1970a. The genera of Ichneumonidae 2. *Memoirs of the American Entomological Institute* 12: 537 pp.

TOWNES, H. 1970b. The genera of Ichneumonidae 3. *Memoirs of the American Entomological Institute* 13: 307 pp.

TOWNES, H. 1971. The genera of Ichneumonidae 4. *Memoirs of the American Entomological Institute* 17: 372 pp.

TOWNES, H. 1983. Revisions of twenty genera of Gelini (Ichneumonidae). *Memoirs of the American Entomological Institute* 35: 281 pp.

TOWNES, H. & TOWNES, M. 1981. A revision of the Serphidae (Hymenoptera). *Memoirs of the American Entomological Institute* 32: 541 pp.

TRJAPITZIN, V.A. 1978. Hymenoptera II. Chalcidoidea 7. Encyrtidae. In: Medvedev, G.S. (ed.) *Opredeliteli Nasekomykh Evropeyskoy Chasti SSR* 3: 236–328 [In Russian. English translation by United States Department of Agriculture and published 1987, Amerind Publishing Co. Pvt. Ltd., New Delhi].

TRJAPITZIN, V.A. 1989. Parasitic Hymenoptera of the Fam. Encyrtidae of Palaearctic. *Opredeliteli po Faune SSSR Izdavaemie Zoologicheskim Institutom AN SSSR* 158. Leningrad, Nauka, 489 pp. [In Russian].

VLUG, H.J. 1995. Catalogue of the Platygastridae (Platygastroidea) of the world. *Hymenopterorum Catalogus (nova editio)* 19: 168 pp.

WESTRICH, P. 1989. *Die Wildbienen Baden-Württembergs.* Ulmer, Stuttgart.

WILLIAMS, R. 2006. *Oak-galls in Britain.* Privately published, 2 vols., 453 pp.

WILLMER, P. 1985. *Bees, ants and wasps: a key to genera of the British aculeates.* Field Studies Council (AIDGAP) OP7, 28 pp.

WRIGHT, A. 1990. *British sawflies (Hymenoptera: Symphyta): a key to adults of the genera occurring in Britain.* Field Studies Council (AIDGAP) no. 203. 60 pp.

YEO, P.F. & CORBET, S.A. 1995. *Solitary wasps* (2nd edn.). Naturalists' Handbooks no. 3. Richmond Publishing, Slough.

YU, D.S. 1993. *TAXA. A biosystematic data management system. Insecta Hymenoptera Ichneumonidae, 1900–1990* [A database package for PC-compatible computers]. Yu, Lethbridge, Canada.

YU, D.S. & HORSTMANN, K. 1997. Catalogue of world Ichneumonidae (Hymenoptera). *Memoirs of the American Entomological Institute* 58: 1558 pp.

ZAHRADNÍK, J. 1998. *Bees and wasps.* Blitz Editions, Leicester.

WEBSITES

http://www.bwars.com/
The Bees, Wasps and Ants Recording Society is the important starting point for any work on British aculeates; there is much useful information with lists of works for identification.

http://hymenopterists.org/
The International Society of Hymenopterists.

http://www.brc.ac.uk/DBIF/homepage.aspx
A general site listing known foodplants of British insects.

http://www.leafmines.co.uk/index.htm
A general site covering the leaf-mining insects.

21 Order Lepidoptera: the butterflies and moths

c. 2570 species in 72 families

The close relationship between the Lepidoptera and Trichoptera is often cited as one of the classic sister-group relationships at ordinal level in all the insects. The flattened scales on the wings of butterflies and moths are clearly derived from the normal hairs found on caddisfly (and many other) wings; indeed such scales occur sporadically in several other insect groups, even in the Archaeognatha, for example. The most primitive Lepidoptera have retained functional mandibles, but otherwise the two orders have very different mouthparts and the higher Lepidoptera have developed a tubular proboscis or haustellum with which to imbibe liquids. It is in the life histories that the clearest difference between the two groups is seen; Trichoptera larvae are all aquatic and hardly ever feed on plant tissue, whereas Lepidopteran caterpillars are essentially terrestrial and have become one of the most important groups of phytophagous insects. When the two groups diverged, some time in the Jurassic, one can only imagine that their common ancestor lived in damp habitats, perhaps feeding on liverworts and maybe fungi. Such a mode of life could become terrestrial and phytophagous, or else aquatic and detritivorous and eventually predatory, without too much difficulty. The larvae of both groups also produce silk from oral glands, which in the caterpillar is used to spin a pupal cocoon or sometimes a feeding web, and in the Trichoptera forms the basis of the larval case or a feeding net, and again creates a pupal shelter. Whatever the true origins of the Lepidopteran ancestor, the monophyly of the group is not in doubt (Grimaldi & Engel, 2005).

The Lepidoptera have always been a favourite group to study; many are conspicuous and colourful both as adults and larvae and their morphology, taxonomy and life histories have been intensively investigated for centuries, by amateurs and professionals alike. Despite this familiarity with the British fauna, there are always a few species added to the British list each year, and even the well-known butterflies sometimes produce a surprise, as mentioned below. Inevitably the butterflies and larger moths have attracted more interest, especially to the amateur naturalist, and this led to a strange artificial system of classification in which the three main groups were the butterflies (sometimes given the spurious scientific name of Rhopalocera), the Macrolepidoptera and the Microlepidoptera. The Rhopalocera contains the true butterflies in the superfamily Papilionoidea, plus the skippers, which are now placed in a separate superfamily, the Hesperioidea. The Macrolepidoptera included the larger moths such as the Noctuoidea and Geometroidea, plus several others; the Microlepidoptera included the small moths that were long regarded as taxonomically difficult. This division began to break down as the phylogenies of the various groups were unravelled, and it became clear that some superfamilies like the Hepialoidea were closely related to 'micro' groups like Tineoidea, but were treated as honorary 'macros'. Although this division is no longer used in scientific literature it still persist in common usage and even the latest identification guides to British moths include all the former 'macros' even though some of these are actually smaller in size than some 'micros', which are ignored.

The Royal Entomological Society Book of British Insects, First Edition. Peter C. Barnard.
© 2011 Royal Entomological Society. Published 2011 by Blackwell Publishing Ltd.

Part of the reason for this persistence of an outdated system lies in the undoubted fact that most butterflies and large moths can be identified by a simple comparison with illustrations in a book, and the most sophisticated piece of optical equipment needed is likely to be just a hand lens. However, even in these 'macro' groups it is becoming increasingly clear that the separation of some species relies on examination of the genitalia, which usually necessitates a stereo microscope. This has long been the case for many small species in the 'micros', so at last the distinction between the two traditional groupings is disappearing.

Although there are many popular works on various groups of British Lepidoptera, the definitive series for identification is the *Moths and Butterflies of Great Britain and Ireland* ('MBGBI', which is usually pronounced MOGBI by those in the know). Like so many similar series it is still incomplete, and already some of the earlier volumes need revision. Nonetheless, the eight volumes published so far, out of the twelve projected, are essential for any serious study of the fauna, particularly for the small families where no other guide exists (Emmet, 1996a; Emmet & Heath, 1989, 1991; Emmet & Langmaid, 2002a, 2002b; Heath, 1976a; Heath & Emmet, 1979, 1983, 1985). This means that the main gaps are the Geometroidea, Tortricoidea, Pterophoroidea and Pyraloidea, most of which are reasonably well covered elsewhere, as noted below. There are similar European initiatives, including the *Microlepidoptera of Europe*, *Noctuidae Europaeae* and the most recent one *Palaearctic Macrolepidoptera*, of which just one volume on the Notodontidae has been published to date (Schintlmeister, 2008). These series are mentioned below when relevant to the British fauna.

Deciding which book to use for identifying the larger moths often comes down to an individual preference for photos of living specimens (Manley, 2009), photos of set specimens (Skinner, 2009), or paintings of set specimens (Waring & Townsend, 2009); the keen worker will want to compare all three. A useful book for illustrating a wide range of European 'micros' is Parenti (2000), though it is not comprehensive; it is also hard to find copies. For identifying larvae of the larger species, the traditional works were Buckler (1886–1901), updated by Haggett (1981), or Stokoe and Stovin (1948). Among the more recent publications the book by Carter and Hargreaves (1986) uses paintings, while Porter's (1997) book contains photos of the caterpillars of a good range of families; fortunately it has just been reprinted as the original was hard to find. Butterfly books are listed under the section Papilionoidea below. The classic

work for identifying the smaller moths was Meyrick (1928) and although superseded for most families some workers still like to compare the detailed descriptions therein. For general works on Lepidoptera, Scoble (1995) gives a good summary of the morphology and biology of the whole group; beginners to the study of British Lepidoptera will find the practical handbooks by Sokoloff (1980) and Dickson (1992) helpful, together with the well-illustrated guide by Leverton (2001). The classic review of the biology of moths was by Ford (1972), and the work was completely revised by Majerus (2002); moth natural history was also reviewed by Young (1997).

There is no published checklist of the British Lepidoptera that contains full synonymy; Kloet and Hincks (1972) is out of date, and the list by Bradley (2000) is perhaps best treated as a label list as its attempts to reconcile an outdated numbering system with modern taxonomy are often less than successful. A more complete European list was by Karsholt and Razowski (1996), but there are several on-line lists that are kept more or less up to date.

There is a long tradition of publishing local lists and guides to the Lepidoptera of large or small geographical regions in Britain, and these are often an important source of information. Chalmers-Hunt's (1989) catalogue of such lists was updated in the chapter by Parsons et al. (1999).

There are around 8500 species in 87 families in Europe, with at least 160,000 species in 120 families worldwide. A brief world overview of the Lepidoptera is provided by Powell (2009).

SPECIES OF CONSERVATION CONCERN

The following 38 species of Lepidoptera are the subject of various degrees of legal protection, with details given under each family: *Pareulype berberata*, *Siona lineata*, *Thalera fimbrialis* and *Thetidia smaragdaria* (Geometridae); *Carterocephalus palaemon*, *Erynnis tages*, *Hesperia comma* and *Thymelicus action* (Hesperiidae); *Aricia artaxerxes*, *Celastrina argiolus*, *Cupido minimus*, *Hamearis lucina*, *Lycaena dispar*, *Lysandra bellargus*, *Lysandra coridon*, *Maculinea arion*, *Neozephyrus quercus*, *Plebejus argus*, *Satyrium pruni*, *Satyrium w-album* and *Thecla betulae* (Lycaenidae); *Acosmetia caliginosa* and *Gortyna borelii* subsp. *lunata* (Noctuidae); *Aglais polychloros*, *Apatura iris*, *Argynnis adippe*, *Boloria euphrosyne*, *Coenonympha tullia*, *Danaus plexippus*, *Erebia epiphron*, *Euphydryas aurinia*, *Melitaea athalia* and *Melitaea cinxia* (Nymphalidae); *Papilio machaon* (Papilionidae); *Gonepteryx rhamni*

HIGHER CLASSIFICATION OF BRITISH LEPIDOPTERA

Suborder Glossata
    Superfamily Alucitoidea
        Family Alucitidae (1 genus, 1 species)
    Superfamily Bombycoidea
        Family Endromidae (1 genus, 1 species)
        Family Saturniidae (1 genus, 2 species)
        Family Sphingidae (14 genera, 18 species)
    Superfamily Choreutoidea
        Family Choreutidae (4 genera, 6 species)
    Superfamily Cossoidea
        Family Cossidae (3 genera, 3 species)
    Superfamily Drepanoidea
        Family Drepanidae (5 genera, 7 species)
        Family Thyatiridae (8 genera, 9 species)
    Superfamily Epermenioidea
        Family Epermeniidae (2 genera, 8 species)
    Superfamily Eriocranioidea
        Family Eriocraniidae (1 genus, 8 species)
    Superfamily Gelechioidea
        Family Agonoxenidae (4 genera, 6 species)
        Family Amphisbatidae (4 genera, 6 species)
        Family Autostichidae (2 genera, 4 species)
        Family Batrachedridae (1 genus, 3 species)
        Family Blastobasidae (1 genus, 4 species)
        Family Chimabachidae (2 genera, 3 species)
        Family Coleophoridae (4 genera, 110 species)
        Family Cosmopterigidae (7 genera, 17 species)
        Family Depressariidae (6 genera, 51 species)
        Family Elachistidae (6 genera, 49 species)
        Family Ethmiidae (1 genus, 6 species)
        Family Gelechiidae (53 genera, 170 species)
        Family Momphidae (1 genus, 15 species)
        Family Oecophoridae (17 genera, 25 species)
        Family Scythrididae (1 genus, 12 species)
        Family Stathmopodidae (1 genus, 1 species)
    Superfamily Geometroidea
        Family Geometridae (137 genera, 312 species)
    Superfamily Gracillarioidea
        Family Bucculatricidae (1 genus, 13 species)
        Family Douglasiidae (1 genus, 2 species)
        Family Gracillariidae (14 genera, 97 species)
        Family Roeslerstammiidae (1 genus, 2 species]
    Superfamily Hepialoidea
        Family Hepialidae (1 genus, 5 species)
    Superfamily Hesperioidea
        Family Hesperiidae (6 genera, 8 species)
    Superfamily Incurvarioidea
        Family Adelidae (3 genera, 15 species)
        Family Heliozelidae (2 genera, 5 species)
        Family Incurvariidae (2 genera, 5 species)
        Family Prodoxidae (1 genus, 7 species)

    Superfamily Lasiocampoidea
        Family Lasiocampidae (10 genera, 12 species)
    Superfamily Nepticuloidea
        Family Nepticulidae (5 genera, 106 species)
        Family Opostegidae (2 genera, 4 species)
    Superfamily Noctuoidea
        Family Arctiidae (21 genera, 33 species)
        Family Ctenuchidae (1 genus, 1 species)
        Family Lymantriidae (8 genera, 11 species)
        Family Noctuidae (168 genera, 415 species)
        Family Nolidae (2 genera, 6 species)
        Family Notodontidae (16 genera, 27 species)
        Family Thaumetopoeidae (1 genus, 2 species)
    Superfamily Papilionoidea
        Family Lycaenidae (16 genera, 20 species)
        Family Nymphalidae (23 genera, 31 species)
        Family Papilionidae (1 genus, 1 species)
        Family Pieridae (7 genera, 11 species)
        Family Riodinidae (1 genus, 1 species)
    Superfamily Pterophoroidea
        Family Pterophoridae (18 genera, 43 species)
    Superfamily Pyraloidea
        Family Crambidae (63 genera, 129 species)
        Family Pyralidae (48 genera, 90 species)
    Superfamily Schreckensteinioidea
        Family Schreckensteiniidae (1 genus, 1 species)
    Superfamily Sesioidea
        Family Sesiidae (6 genera, 14 species)
    Superfamily Tineoidea
        Family Psychidae (15 genera, 21 species)
        Family Tineidae (29 genera, 62 species)
    Superfamily Tischerioidea
        Family Tischeriidae (2 genera, 6 species)
    Superfamily Tortricoidea
        Family Tortricidae (c. 90 genera, 390 species)
    Superfamily Yponomeutoidea
        Family Acrolepiidae (3 genera, 6 species)
        Family Bedelliidae (1 genus, 1 species)
        Family Glyphipterygidae (2 genera, 8 species)
        Family Heliodinidae (1 genus, 1 species)
        Family Lyonetiidae (2 genera, 10 species)
        Family Plutellidae (3 genera, 7 species)
        Family Yponomeutidae (13 genera, 54 species)
        Family Ypsolophidae (2 genera, 16 species)
    Superfamily Zygaenoidea
        Family Limacodidae (2 genera, 2 species)
        Family Zygaenidae (3 genera, 10 species)
Suborder Zeugloptera
    Superfamily Micropterigoidea
        Family Micropterigidae (1 genus, 5 species)

and *Leptidea sinapis* (Pieridae); *Pyropteron chrysidiformis* (Sesiidae); *Zygaena viciae* (Zygaenidae).

There are 176 species on the UKBAP list, as follows: *Nematopogon magna* and *Nemophora fasciella* (Adelidae); *Arctia caja, Coscinia cribraria* subsp. *bivittata, Spilosoma lubricipeda, Spilosoma luteum* and *Tyria jacobaeae* (Arctiidae); *Coleophora hydrolapathella, Coleophora tricolor, Coleophora vibicella* and *Coleophora wockeella* (Coleophoridae); *Cossus cossus* (Cossidae); *Agrotera nemoralis, Anania funebris* and *Pyrausta sanguinalis* (Crambidae); *Agonopterix atomella* and *Agonopterix capreolella* (Depressariidae); *Watsonalla binaria* (Drepanidae); *Epermenia insecurella* (Epermeniidae); *Syncopacma albipalpella* and *Syncopacma suecicella* (Gelechiidae); *Aleucis distinctata, Aplasta ononaria, Aspitates gilvaria* subsp. *gilvaria, Chesias legatella, Chesias rufata, Chiasmia clathrata, Cyclophora pendularia, Cyclophora porata, Ecliptopera silaceata, Ennomos erosaria, Ennomos fuscantaria, Ennomos quercinaria, Entephria caesiata, Epione vespertaria, Epirrhoe galiata, Eulithis mellinata, Eupithecia extensaria* subsp. *occidua, Eustroma reticulatum, Hemistola chrysoprasaria, Idaea dilutaria, Idaea ochrata* subsp. *cantiata, Lithostege griseata, Lycia hirtaria, Lycia zonaria* subsp. *britannica, Macaria carbonaria, Macaria wauaria, Melanthia procellata, Minoa murinata, Orthonama vittata, Pareulype berberata, Pelurga comitata, Perizoma albulata* subsp. *albulata, Rheumaptera hastata, Scopula marginepunctata, Scotopteryx bipunctaria, Scotopteryx chenopodiata, Siona lineata, Thalera fimbrialis, Timandra comae, Trichopteryx polycommata, Xanthorhoe decoloraria* and *Xanthorhoe ferrugata* (Geometridae); *Phyllonorycter sagitella* and *Phyllonorycter scabiosella* (Gracillariidae); *Hepialus humuli* (Hepialidae); *Carterocephalus palaemon, Erynnis tages, Pyrgus malvae* and *Thymelicus acteon* (Hesperiidae); *Malacosoma neustria* and *Trichiura crataegi* (Lasiocampidae); *Aricia artaxerxes, Cupido minimus, Maculinea arion, Plebejus argus, Satyrium w-album* and *Thecla betulae* (Lycaenidae); *Orgyia recens* (Lymantriidae); *Stigmella zelleriella* (Nepticulidae); *Acosmetia caliginosa, Acronicta psi, Acronicta rumicis, Agrochola helvola, Agrochola litura, Agrochola lychnidis, Allophyes oxyacanthae, Amphipoea oculea, Amphipyra tragopoginis, Anarta cordigera, Apamea anceps, Apamea remissa, Aporophyla lutulenta, Archanara neurica, Asteroscopus sphinx, Atethmia centrago, Athetis pallustris, Blepharita adusta, Brachylomia viminalis, Caradrina morpheus, Catocala promissa, Catocala sponsa, Celaena haworthii, Celaena leucostigma, Chortodes brevilinea, Chortodes extrema, Cosmia diffinis, Dasypolia templi, Diarsia rubi, Dicycla oo, Diloba caeruleocephala, Eugnorisma glareosa, Euxoa nigricans, Euxoa tritici, Graphiphora augur, Hadena albimacula, Heliophobus reticulata, Heliothis*

maritima, Hoplodrina blanda, Hydraecia micacea, Hydraecia osseola subsp. *hucherardi, Jodia croceago, Luperina nickerlii* subsp. *leechi, Melanchra persicariae, Melanchra pisi, Mesoligia literosa, Mythimna comma, Noctua orbona, Oria musculosa, Orthosia gracilis, Paracolax tristalis, Pechipogo strigilata, Polia bombycina, Protolampra sobrina, Rhizedra lutosa, Shargacucullia lychnitis, Stilbia anomala, Tholera cespitis, Tholera decimalis, Trisateles emortualis, Tyta luctuosa, Xanthia gilvago, Xanthia icteritia, Xestia agathina, Xestia alpicola* subsp. *alpina, Xestia ashworthii, Xestia castanea* and *Xylena exsoleta* (Noctuidae); *Argynnis adippe, Boloria euphrosyne, Boloria selene, Coenonympha pamphilus, Coenonympha tullia, Erebia epiphron, Euphydryas aurinia, Hipparchia semele, Lasiommata megera, Limenitis camilla, Melitaea athalia* and *Melitaea cinxia* (Nymphalidae); *Aplota palpella* (Oecophoridae); *Leptidea sinapis* (Pieridae); *Lampronia capitella* (Prodoxidae); *Sciota hostilis* (Pyralidae); *Hamearis lucina* (Riodinidae); *Scythris siccella* (Scythrididae); *Pyropteron chrysidiformis* (Sesiidae); *Hemaris tityus* (Sphingidae); *Cymatophorima diluta* (Thyatiridae); *Eudarcia richardsoni* and *Nemapogon picarella* (Tineidae); *Celypha woodiana* and *Grapholita pallifrontana* (Tortricidae); *Adscita statices, Zygaena loti* subsp. *scotica* and *Zygaena viciae* subsp. *argyllensis* (Zygaenidae).

The following species were removed from the UKBAP list in 2007 for various reasons: *Hydrelia sylvata* and *Thetidia smaragdaria* subsp. *maritima* (Geometridae); *Hesperia comma* (Hesperiidae); *Phyllodesma ilicifolia* (Lasiocampidae); *Lycaena dispar* and *Lysandra bellargus* (Lycaenidae); *Hypena rostralis, Lygephila craccae, Mythimna turca, Schrankia taenialis* and *Xestia rhomboidea* (Noctuidae).

Some key publications on conservation issues in the Lepidoptera are Feltwell (1995), Harding and Green (1991), Parsons (1993, 1996), Pollard and Yates (1993) and Pullin (1995).

The Families of British Lepidoptera

SUBORDER GLOSSATA

SUPERFAMILY ALUCITOIDEA

Family Alucitidae (1 genus, 1 species)
The only British species in this family is *Alucita hexadactyla*, the Many-plumed moth (Fig. 21.1), easily distinguished by having each wing divided into six feathery 'plumes', which correspond to the principal wing-veins. It is a very common species,

Fig. 21.1 *Alucita hexadactyla*, Many-plumed moth (Alucitidae) (Photo: Roger Key)

Fig. 21.2 *Saturnia pavonia*, Emperor moth larva (Saturniidae) (Photo: Colin Rew)

found in most months of the year, and its larva feeds on honeysuckle.

Identification: Beirne (1952); Manley (2009).

British genus: *Alucita*

SUPERFAMILY BOMBYCOIDEA

Family Endromidae (1 genus, 1 species)
The only British species in this family is *Endromis versicolora*, the Kentish glory, which, despite its common name, is now found only in Scotland. It is a day-flying moth whose larvae feed on young birch leaves.

Identification: Young (1991); Waring and Townsend (2009); Manley (2009); Skinner (2009).

British genus: *Endromis*

Family Saturniidae (1 genus, 2 species)
There are just two British species in this family, *Saturnia pavonia*, the Emperor moth and *S. pyri*, the Great peacock moth, though the latter is only an occasional migrant from southern Europe. The larva of *S. pavonia* is very striking (Fig. 21.2); it feeds on various woody plants and is quite common throughout much of Britain.

Identification: Goater (1991b); Waring and Townsend (2009); Manley (2009); Skinner (2009).

British genus: *Saturnia*

Family Sphingidae (14 genera, 18 species)
The small family of hawk-moths are well-known to many people because of the combination of their large size and striking coloration. There are many

Fig. 21.3 *Smerinthus ocellata*, Eyed hawk-moth (Sphingidae) (Photo: Roger Key)

common species though several are regular migrants rather than residents such as the day-flying *Macroglossum stellatarum*, the Hummingbird hawk-moth, which hovers in front of nectar-rich flowers with its long proboscis extended deep into the flower. Some species warn off predatory birds using flashes of colour or patterns, such as the eye-spots of *Smerinthus ocellata*, the Eyed hawk-moth (Fig. 21.3). Many sphingid larvae are just as impressive as the adult moths (Fig. 21.4); some have eye-spots as well as a prominent 'tail' (Fig. 21.5) while others are excellent at camouflage (Fig. 21.6). Hawk-moth caterpillars feed on a variety of shrubs and trees, and many are specific to certain host plants.

Hemaris tityus (Narrow-bordered bee hawk-moth) is on the UKBAP list.

Fig. 21.4 *Smerinthus ocellata*, Eyed hawk-moth larva (Sphingidae) (Photo: Roger Key)

Fig. 21.5 *Hippotion celerio*, Silver-striped hawk-moth larva (Sphingidae) (Photo: Peter Barnard)

Fig. 21.6 *Laothoe populi*, Poplar hawk-moth larva on willow leaf (Sphingidae) (Photo: Peter Barnard)

Identification: Gilchrist (1979); Waring and Townsend (2009); Manley (2009); Skinner (2009).

British subfamilies and genera:
  Macroglossinae: *Daphnis, Deilephila, Hemaris, Hippotion, Hyles, Macroglossum, Proserpinus*
  Smerinthinae: *Laothoe, Mimas, Smerinthus*
  Sphinginae: *Acherontia, Agrius, Hyloicus, Sphinx*

SUPERFAMILY CHOREUTOIDEA

Family Choreutidae (4 genera, 6 species)
This is traditionally a family of 'micro' moths that were previously included in the Glyphipterygidae; they are day-flying moths that have very broad wings, somewhat resembling Tortricidae, though most species rest with their wings held slightly apart. *Anthophila fabriciana* is a very common species whose larva feeds on nettles. The other species in the family feed on a variety of plants.

Identification: Pelham-Clinton (1985c); Manley (2009) covers the more common species.

British genera: *Anthophila, Choreutis, Prochoreutis, Tebenna*

SUPERFAMILY COSSOIDEA

Family Cossidae (3 genera, 3 species)
The three British species in this family are all large and distinctively coloured. *Phragmataecia castaneae* is the Reed leopard, a rare inhabitant of East Anglia where its larva feeds on reeds; *Zeuzera pyrina* is the Leopard, a black and white spotted moth that is quite widespread and whose larva feeds on various trees and woody shrubs; and *Cossus cossus* is the Goat moth that gets its name from the goat-like smell of the larva, which feeds for up to four years in tree trunks.
  Cossus cossus (Goat moth) is on the UKBAP list.

Identification: Skinner (1985a); Waring and Townsend (2009); Manley (2009); Skinner (2009).

British subfamilies and genera:
  Cossinae: *Cossus*
  Zeuzerinae: *Phragmataecia, Zeuzera*

SUPERFAMILY DREPANOIDEA

Family Drepanidae (5 genera, 7 species)
Most members of this family are known as hook-tips from the shape of the tip of the fore wing (Fig. 21.7), but the exception is *Cilix glaucata*, the Chinese character moth (Fig. 21.8), which is a strikingly white and translucent moth. All are nocturnal fliers and their larvae, which have a characteristic hump-backed appearance, feed on various trees.

Watsonalla binaria (Oak hook-tip) is on the UKBAP list.

Fig. 21.7 *Watsonalla binaria*, Oak hook-tip moth (Drepanidae) (Photo: Roger Key)

Identification: Goater (1991c); Waring and Townsend (2009); Manley (2009); Skinner (2009).

British genera: *Cilix, Drepana, Falcaria, Sabra, Watsonalla*

Family Thyatiridae (8 genera, 9 species)
Some authors treat this small group as a subfamily of the Drepanidae, although superficially they resemble some Noctuidae; most have very distinctive markings and are easy to identify. There are several common species whose larvae feed on bramble, such as *Thyatira batis*, the Peach blossom, and *Habrosyne pyritoides*, the Buff arches moth (Fig. 21.9). Many other species feed on various trees.

Cymatophorima diluta (Oak lutestring) is on the UKBAP list.

Identification: Goater (1991d); Waring and Townsend (2009); Manley (2009); Skinner (2009).

British genera: *Achlya, Cymatophorina, Habrosyne, Ochropacha, Polyploca, Tethea, Tetheella, Thyatira*

SUPERFAMILY EPERMENIOIDEA

Family Epermeniidae (2 genera, 8 species)
Some authors have placed this obscure family in the Yponomeutoidea, but its relationships with other families are yet to be resolved. The wings of all species are narrow and pointed, and they often have tufts of raised scales that give a toothed effect

Fig. 21.8 *Cilix glaucata*, Chinese character moth (Drepanidae) (Photo: Roger Key)

Fig. 21.9 *Habrosyne pyritoides*, Buff arches moth (Thyatiridae) (Photo: Roger Key)

Fig. 21.10 *Epermenia chaerophyllella* larva on hogweed (Epermeniidae) (Photo: Roger Key)

to the profile of the moth when at rest. The larvae feed in silk webbing on a variety of plants, often umbellifers (Fig. 21.10).

Epermenia insecurella (Chalk-hill lance-wing) is on the UKBAP list.

Identification: Godfray and Sterling (1996); Manley (2009) shows two species of *Epermenia*.

British genera: *Epermenia, Phaulernis*

SUPERFAMILY ERIOCRANIOIDEA

Family Eriocraniidae (1 genus, 8 species)
All the British species of this small family are currently placed in the genus *Eriocrania*, though some earlier authors divided this genus into smaller groups. They are generally considered as primitive moths, some of which have vestigial mandibles, though the adults do not feed. They are all metallic purple or gold-coloured and day-flying, with a fore wing length of 5–6 mm. The larvae are leaf-miners that feed on birch and a few other trees.

Identification: Heath (1976c); Manley (2009) illustrates five of the eight British species.

British genus: *Eriocrania*

SUPERFAMILY GELECHIOIDEA

There are many small families in this group, most of which were split off from larger ones, but there is still much work to be done in elucidating the relationships of several of the families.

Family Agonoxenidae (4 genera, 6 species)
The members of this small family were placed in the Momphidae by some previous authors, though others have grouped them with the Oecophoridae or Cosmopterigidae. They are small moths with a fore wing length of 5–6 mm, but often strikingly patterned. Most of the known larvae bore into trees, either under the bark of the trunk or into smaller twigs. One species, *Blastodacna atra*, is known as the Apple-pith moth because it can be a pest of fruit trees, while *B. hellerella* is a common inhabitant of hawthorn berries.

Identification: Koster (2002b); Manley (2009) shows two species, *Blastodacna hellerella* and *Dystebenna stephensi* (under Cosmopterigidae).

British genera: *Blastodacna, Chrysoclista, Dystebenna, Spuleria*

Family Amphisbatidae (4 genera, 6 species)
Some authors place these as a subgroup of the Oecophoridae, they are small species with a fore wing length of 5–7 mm, but often brightly patterned. At least two species are quite common in Britain, *Amphisbatis incongruella* and *Pseudatemelia josephinae*; some members of this family build a portable larval case made from leaves or plant stems, and they feed on a variety of low-growing plants.

Identification: Harper et al. (2002) within the Oecophoridae. Manley (2009) illustrates *Hypercallia citrinalis* and *Telechrysis tripuncta* (under Oecophoridae).

British genera: *Amphisbatis, Hypercallia, Pseudatemelia, Telechrysis*

Family Autostichidae (2 genera, 4 species)
This family has also been known as the Symmocidae, and they were originally included within the Gelechiidae. *Symmoca signatella* has whitish wings with brown markings, but the three species of *Oegoconia* have brown and yellow wings, their fore wing length is around 6–8 mm. The biology of this family is little studied, but the known larvae seem to feed on decaying plant material.

Identification: Bland (2002).

British genera: *Oegoconia, Symmoca*

Family Batrachedridae (1 genus, 3 species)
The members of the single British genus *Batrachedra* were previously placed in the Momphidae or the Coleophoridae. They are small, rather nondescript species with whitish or yellowish wings, with a fore wing length of 4–7 mm. From European records it is known that the larva of one species lives in catkins or buds of poplar or willow trees, while another

Fig. 21.11 *Blastobasis decolorella* (Blastobasidae) (Photo: Roger Key)

lives in the needles of Norway spruce, but their biology in Britain is virtually unknown.

Identification: Koster (2002a).

British genus: *Batrachedra*

Family Blastobasidae (1 genus, 4 species)
The relationships of this family to the Oecophoridae, Momphidae and Coleophoridae are yet to be clarified. Although traditionally regarded as 'micro' moths, the fore wing can reach 10 mm in the single British genus *Blastobasis*. One of the commonest species is *B. decolorella* (Fig. 21.11), which can vary greatly in the degree of wing patterning. The biology of this family is little known, but some larvae seem to feed on plant material or general detritus.

Identification: Dickson (2002); Manley (2009) illustrates two species.

British genera: *Blastobasis*

Family Chimabachidae (2 genera, 3 species)
These were formerly treated as a subfamily of the Oecophoridae. They are unusual in showing sexual dimorphism in wing shape; the male wings are broad and oval, while those of the female are short and narrower, with a pointed apex. They are quite large moths for this group of families, with a fore wing length up to 14 mm in the males and the wings are greyish or brownish. All three species are quite widespread in Britain, and their larvae spin leaves together on various trees and shrubs.

Identification: Harper et al. (2002) within the Oecophoridae. Manley (2009) illustrates all three British species (under Oecophoridae).

British genera: *Dasystoma, Diurnea*

Family Coleophoridae (4 genera, 110 species)
This is a large family of rather small moths, which often resemble each other quite closely and have a fore wing length of 5–8 mm. Larvae of *Augasma aeratella* cause galls on *Polygonum aviculare*, but all the other species construct a portable case out of silk. There are several types of cases, some of which incorporate pieces of plant material and it is sometimes possible to identify a larva from a combination of the case structure and its host plant; such a key is provided by Emmet et al. (1996). Coleophorids feed on a wide variety of plants, and there are many common species, but the group is not as popular as it might be amongst amateur naturalists, probably because identification of adults requires microscopic examination of the genitalia.

Coleophora hydrolapathella (Water-dock case-bearer), *Coleophora tricolor* (Basil-thyme case-bearer), *Coleophora vibicella* (Large gold case-bearer) and *Coleophora wockeella* (Betony case-bearer) are on the UKBAP list.

Identification: Emmet et al. (1996); Manley (2009) shows a few common species.

British genera: *Augasma, Coleophora, Goniodoma, Metriotes*

Family Cosmopterigidae (7 genera, 17 species)
This family name was previously spelled Cosmopterygidae, based on the misspelled generic name *Cosmopteryx*. The status of this family is still in some doubt, as the three subfamilies are rather different in morphology although the larvae are all leaf-miners. Most species have narrow wings, often with distinctive colour patterns, and with a fore wing length of 4–8 mm, although *Limnaecia phragmitella* reaches 10 mm. The latter species is quite common throughout Britain on *Typha* species, but many other members of the family are very local in distribution.

Identification: Koster (2002d); Manley (2009) shows two species, *Limnaecia phragmitella* and *Cosmopterix pulchrimella*.

British subfamilies and genera:
Antequerinae: *Euclemensia, Limnaecia, Pancalia*
Chrysopeleiinae: *Sorhagenia*
Cosmopteriginae: *Anatrachyntis, Cosmopterix, Pyroderces*

Family Depressariidae (6 genera, 51 species)
This group was previously treated as a subfamily of the Oecophoridae. Some species are moderately large with a fore wing length up to 12 mm; most have rather broad wings but with rather subdued wing markings, which make species identification difficult in some genera, particularly the large genus *Agonopterix*. Most larvae in this group spin the

leaves of their food plants together while feeding, and they use a wide variety of plant groups. There is a key to larvae, based on food plant, in Harper et al. (2002).

The following two species are on the UKBAP list: *Agonopterix atomella* (Greenweed flat-body moth), *Agonopterix capreolella* (Fuscous flat-body moth).

Identification: Harper et al. (2002) within the Oecophoridae. Manley (2009) illustrates several species of *Agonopteryx, Depressaria, Luquetia* and *Semioscopis* (under Oecophoridae).

British genera: *Agonopterix, Depressaria, Exaeretia, Levipalpus, Luquetia, Semioscopis*

Family Elachistidae (6 genera, 49 species)
This family was often treated as a repository for species that did not seem to fit well anywhere else; many of these have been removed to their rightful place but the family remains a moderately large one, and there is undoubtedly much work still to do on the British fauna. They are small moths, with a fore wing length of around 4–6 mm; many are difficult to separate on wing pattern and even the genitalic differences are sometimes rather slight. All the known larvae are leaf-miners on various plant families, with several recorded from grasses.

Identification: Bland (1996); Manley (2009) illustrates four species of *Elachista*. Some of the included genera are regarded as synonyms by some authors.

British genera: *Biselachista, Cosmiotes, Elachista, Mendesia, Perittia, Stephensia*

Family Ethmiidae (1 genus, 6 species)
This is a rather distinct though small family. All the British species are moderately sized, with a fore wing length up to 13 mm and all except one are conspicuously patterned with black and white (Fig. 21.12); they can superficially resemble Yponomeutidae. The exception is *Ethmia pyrausta*, a greyish species known only from a few Scottish sites. Most of the British species are quite local, and the larvae feed on Boraginaceae or Ranunculaceae.

Identification: Sattler (2002); Manley (2009) shows three species.

British genus: *Ethmia*

Family Gelechiidae (53 genera, 170 species)
This large family has a volume of *Moths and Butterflies of Great Britain and Ireland* all to itself (Emmet & Langmaid, 2002b). Although most species are quite small, with fore wing lengths from about 4–8 mm, they often show quite distinctive colour patterns

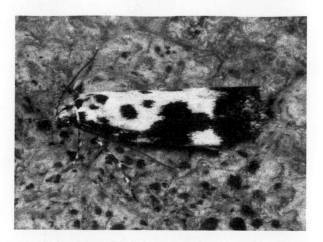

Fig. 21.12 *Ethmia quadrillella* (Ethmiidae) (Photo: Roger Key)

Fig. 21.13 *Helcystogramma rufescens* larva (Gelechiidae) (Photo: Roger Key)

that can help identification. The adult moths often rest with their body held up at an angle to the substrate, and the antennae are held over the closed wings in a gentle curve. The larvae feed on a wide variety of food plants, ranging from lichens and mosses to gymnosperms and all kinds of flowering plants, although most species are quite specific in their choice of food plant. Gelechiid larvae are always concealed from view in some way; a few create galls, some are miners in stems, roots, flowers or seeds, but even the leaf-feeding species spin a silk web, and some larvae are strongly patterned (Fig. 21.13). Although the genitalia in this group are rather complex and sometimes asymmetrical, with some unique structures that are difficult to homologize, they are generally easy to use for species identification. Thus, there is scope for the amateur naturalist to contribute to the knowledge of this

Fig. 21.14 *Nothris congressariella* (Gelechiidae) (Photo: Roger Key)

group, as there is still much to be discovered about their biology.

Nothris congressariella (Fig. 21.14) is one of the scarcer species, known only from a few sites in south-west England and some islands; its distribution is limited by its food plant, *Scrophularia scorodonia*.

Syncopacma albipalpella (Slate sober moth) and *Syncopacma suecicella* (Western sober moth) are on the UKBAP list.

Identification: Emmet and Langmaid (2002b); Manley (2009) shows some common species.

British subfamilies and genera:

Dichomeridinae: *Acompsia, Brachmia, Dichomeris, Helcystogramma*

Gelechiinae: *Altenia, Anacampsis, Anarsia, Apodia, Aproaerema, Argolamprotes, Aristotelia, Aroga, Athrips, Bryotropha, Carpatolechia, Caryocolum, Chionodes, Chrysoesthia, Coleotechnites, Eulamprotes, Exoteleia, Gelechia, Gnorimoschema, Hypatima, Isophrictis, Mesophleps, Metzneria, Mirificarma, Monochroa, Neofaculta, Neofriseria, Nothris, Parachronistis, Phthorimaea, Prolita, Psamathocrita, Pseudotelphusa, Psoricoptera, Ptocheuusa, Recurvaria, Scrobipalpa, Scrobipalpula, Sophronia, Stenolechia, Syncopacma, Teleiodes, Teleiopsis, Xenolechia, Xystophora*

Pexicopiinae: *Pexicopia, Platyedra, Sitotroga, Thiotricha*

Family Momphidae (1 genus, 15 species)

All the British species are in the genus *Mompha*; this family is more restricted than previously. All are small species with a fore wing length of 3–8 mm, and some have distinct orange markings on brown and white wings. The larvae mine the stems or

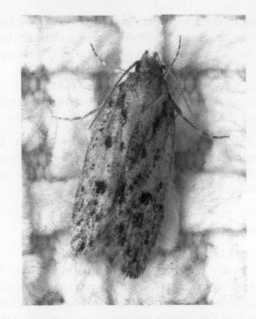

Fig. 21.15 *Hofmannophila pseudospretella*, Brown house moth (Oecophoridae) (Photo: Peter Barnard)

leaves of their food plants, which are mostly in the Onagraceae, especially *Epilobium*.

Identification: Koster (2002c); Manley (2009) shows four species.

British genus: *Mompha*

Family Oecophoridae (17 genera, 25 species)

This family is now much more restricted than in the treatments by previous authors such as Harper et al. (2002); nonetheless it still contains some very common and striking species. Two are common pests in houses and have appropriate vernacular names: *Hofmannophila pseudospretella*, the Brown house moth (Fig. 21.15), and *Endrosis sarcitrella*, the White-shouldered house moth. Their larvae can feed on a variety of general detritus and they commonly breed in the material that builds up in the corners of rooms, under floorboards and so on; in heated houses they can have several generations per year. Other oecophorid larvae seem to feed on fungi in dead wood, dried vegetation and similar habitats. As with the Gelechiidae there is much still to be discovered about the biology of this family, and there is scope for the amateur naturalist to contribute here.

Aplota palpella (Scarce brown streak) is on the UKBAP list.

Identification: Harper et al. (2002); Manley (2009) shows some common species.

278

British genera: *Alabonia, Aplota, Batia, Bisigna, Borkhausenia, Carcina, Crassa, Dasycera, Denisia, Endrosis, Epicallima, Esperia, Hofmannophila, Oecophora, Pleurota, Schiffermuelleria, Tachystola*

Family Scythrididae (1 genus, 12 species)

The British members of this small family are all in the genus *Scythris*. Although day-flying, they are rarely seen by the casual observer, partly because they rarely fly far but instead hop through vegetation, staying close to the ground. They are also quite small, with a fore wing length from 4–7 mm, and only a few species have distinctive markings. Their larvae all spin silk webs on their food plants, which include *Helianthemum* and heathers. Most species are very local in distribution, and there are few British records.

Scythris siccella (Least owlet) is on the UKBAP list.

Identification: Bengtsson (2002); Manley (2009) shows one species.

British genus: *Scythris*

Family Stathmopodidae (1 genus, 1 species)

This family contains the single British species *Stathmopoda pedella*, which was formerly placed in the Oecophoridae. It is easily recognized by its distinctive resting posture, standing on its front and mid legs, and holding its hind legs out at right-angles to the body. Both the wings and the prominent hind legs are black (or dark brown) and yellow, so the overall effect may be a warning coloration. The larvae feed on the fruits of alder, but it is a rather local species, mainly confined to south-east England.

Identification: Harper et al. (2002) in Oecophoridae. Manley (2009) illustrates *S. pedella*.

British genus: *Stathmopoda*

SUPERFAMILY GEOMETROIDEA

Family Geometridae (137 genera, 312 species)

This is a very large and well-known family that includes some widely distributed and easily recognized species. There are six subfamilies currently recognized, which form quite distinct groups, briefly described below. In general geometrid moths have narrow bodies and large triangular wings, almost like those of butterflies, but they hold the wings flat on the substrate when at rest; most are nocturnal and are not very strong fliers, hence very few arrive as migrants. The larvae of this family are distinctive; they have lost the first three pairs of abdominal prolegs, leaving just the true legs at the

Fig. 21.16 *Odontopera bidentata*, Scalloped hazel larva (Geometridae) (Photo: Roger Key)

front and prolegs at the back, and they move by looping their bodies, bringing the rear of the body up to the front and then moving the front forwards again (Fig. 21.16). This looping motion gives them the appearance of 'pacing out' and measuring distances and they give the family its name Geometridae, which roughly translates as earth-measurer; in the USA they are called inchworms.

The subfamily Alsophilinae includes just one British species, *Alsophila aescularia*, the March moth, which is a common species on the wing in early spring; its larva feeds on various trees and shrubs. The Archiearinae contains only two species in the genus *Archiearis*, sometimes known as the orange underwings, but this is a very distinct group; they are day-flying moths that can be seen flying strongly in spring sunshine, and their larvae feed on birch and aspen.

The Ennominae is a large group of around 95 species that includes moths known generally as thorns, umbers and beauties, together with many others such as *Biston betularia*, the Peppered moth, used in several experiments on natural selection. The subfamily Geometrinae includes 10 British species, which are generally called the emeralds from their pale green colour (Fig. 21.17).

The Larentiinae has over 160 British species: around 110 are known as carpets from the richly patterned wings (Fig. 21.18), together with more than 50 species known as pugs in *Eupithecia* and a few related genera; there are several other small genera in this group (Fig. 21.19). The subfamily Sterrhinae includes around 40 species, some known as mochas and others as waves, from the wavy lines that make up the main wing-patterns (Fig. 21.20).

Fig. 21.17 *Geometra papilionaria*, Large emerald moth (Geometridae) (Photo: Roger Key)

Fig. 21.19 *Odezia atrata*, Chimney sweeper moth (Geometridae) (Photo: Colin Rew)

Fig. 21.18 *Chloroclysta siterata*, Red-green carpet moth (Geometridae) (Photo: Colin Rew)

Fig. 21.20 *Idaea aversata*, Riband wave moth (Geometridae) (Photo: Colin Rew)

The following four species are listed on the Wildlife and Countryside Act 1981: *Pareulype berberata* (Barberry carpet), *Siona lineata* (Black-veined moth), *Thalera fimbrialis* (Sussex emerald) and *Thetidia smaragdaria* (Essex emerald moth).

The following species are on the UKBAP list: *Aleucis distinctata* (Sloe carpet), *Aplasta ononaria* (Rest harrow), *Aspitates gilvaria* subsp. *gilvaria* (Straw belle), *Chesias legatella* (The streak), *Chesias rufata* (Broom-tip), *Chiasmia clathrata* (Latticed heath), *Cyclophora pendularia* (Dingy mocha), *Cyclophora porata* (False mocha), *Ecliptopera silaceata* (Small phoenix), *Ennomos erosaria* (September thorn), *Ennomos fuscantaria* (Dusky thorn), *Ennomos*

quercinaria (August thorn), *Entephria caesiata* (Grey mountain carpet), *Epione vespertaria* (Dark bordered beauty), *Epirrhoe galiata* (Galium carpet), *Eulithis mellinata* (The spinach), *Eupithecia extensaria* subsp. *occidua* (Scarce pug), *Eustroma reticulatum* (Netted carpet), *Hemistola chrysoprasaria* (Small emerald), *Idaea dilutaria* (Silky wave), *Idaea ochrata* subsp. *cantiata* (Bright wave), *Lithostege griseata* (Grey carpet), *Lycia hirtaria* (Brindled beauty), *Lycia zonaria* subsp. *britannica* (Belted beauty), *Macaria carbonaria* (Netted mountain moth), *Macaria wauaria* (V-moth), *Melanthia procellata* (Pretty chalk carpet), *Minoa murinata* (Drab looper), *Orthonama vittata* (Oblique carpet), *Pareulype berberata* (Barberry carpet), *Pelurga comi-*

tata (Dark spinach), *Perizoma albulata* subsp. *albulata* (Grass rivulet), *Rheumaptera hastata* (Argent & Sable), *Scopula marginepunctata* (Mullein wave), *Scotopteryx bipunctaria* (Chalk carpet), *Scotopteryx chenopodiata* (Shaded broad-bar), *Siona lineata* (Black-veined moth), *Thalera fimbrialis* (Sussex emerald), *Timandra comae* (Blood-vein), *Trichopteryx polycommata* (Barred tooth-striped), *Xanthorhoe decoloraria* (Red carpet) and *Xanthorhoe ferrugata* (Dark-barred twin-spot carpet).

Identification: Waring and Townsend (2009); Manley (2009); Skinner (2009).

British subfamilies and genera:
Alsophilinae: *Alsophila*
Archiearinae: *Archiearis*
Ennominae: *Abraxas, Aethalura, Agriopis, Alcis, Aleucis, Angerona, Apeira, Apocheima, Aspitates, Biston, Bupalus, Cabera, Campaea, Cepphis, Charissa, Chiasmia, Cleora, Cleorodes, Colotois, Crocallis, Deileptenia, Dyscia, Ectropis, Ematurga, Ennomos, Epione, Erannis, Fagivorina, Glacies, Gnophos, Hylaea, Hypomecis, Isturgia, Itame, Ligdia, Lomaspilis, Lomographa, Lycia, Macaria, Menophra, Odontognophos, Odontopera, Opisthograptis, Ourapteryx, Pachycnemia, Paradarisa, Parectropis, Perconia, Peribatodes, Petrophora, Phigalia, Plagodis, Pseudopanthera, Selenia, Selidosema, Siona, Stegania, Tephronia, Theria*
Geometrinae: *Aplasta, Chlorissa, Comibaena, Geometra, Hemistola, Hemithea, Jodis, Pseudoterpna, Thalera, Thetidia*
Larentiinae: *Acasis, Anticlea, Anticollix, Aplocera, Asthena, Camptogramma, Carsia, Catarhoe, Chesias, Chloroclysta, Chloroclystis, Cidaria, Coenocalpe, Colostygia, Cosmorhoe, Costaconvexa, Discoloxia, Ecliptopera, Electrophaes, Entephria, Epirrhoe, Epirrita, Euchoeca, Eulithis, Euphyia, Eupithecia, Eustroma, Gymnoscelis, Horisme, Hydrelia, Hydriomena, Lampropteryx, Larentia, Lithostege, Lobophora, Melanthia, Mesoleuca, Minoa, Nebula, Odezia, Operophtera, Orthonama, Pareulype, Pasiphila, Pelurga, Perizoma, Phibalapteryx, Philereme, Plemyria, Pterapherapteryx, Rheumaptera, Scotopteryx, Spargania, Thera, Trichopteryx, Triphosa, Venusia, Xanthorhoe*
Sterrhinae: *Cyclophora, Idaea, Rhodometra, Scopula, Timandra*

SUPERFAMILY GRACILLARIOIDEA

Family Bucculatricidae (1 genus, 13 species)
This family was previously treated as a subfamily of the Lyonetiidae. They are very small species with a fore wing length of 3–4 mm, and all have rather similar mottled wing markings.

All larvae begin life as leaf-miners on various plants, but then moult underneath a flat silk cocoon and continue feeding on the leaf from the underside, cutting out 'windows' but leaving the upper epidermis of the leaf intact. On many substrates the pupal cocoon is surrounded by a palisade of vertical silken 'posts', which are highly distinctive. Many species are widespread in Britain, but only one or two are commonly found.

Identification: Emmet (1985b) under Lyonetiidae; Manley (2009) shows *Bucculatrix ulmella*, under Tineidae.

British genus: *Bucculatrix*

Family Douglasiidae (1 genus, 2 species)
The two species in this small family are both in the genus *Tinagma*; they are small greyish moths with a fore wing length of 3–4 mm. Their larvae are leaf- or stem-miners on *Echium* and both species are restricted to the south or east of England.

Identification: Agassiz (1985).

British genus: *Tinagma*

Family Gracillariidae (14 genera, 97 species)
Within this group the subfamily Phyllocnistinae is sometimes regarded as a distinct family.

Most larvae are found on various trees; the early instars are leaf-miners but only in the epidermis of the leaf where they drink the sap; a major change in the head and mouthpart morphology occurs after the second or third moult and the larvae then eat leaf parenchyma. One important difference in the Phyllocnistinae is that the larvae remain as sap-drinkers throughout their lives; the mouthparts only metamorphose before pupation, when they atrophy. There is a key to the larval mines, by food plant, in Emmet et al. (1985) and this is often easier than identifying the adults, which are rather similar to each other in appearance.

Phyllonorycter sagitella (Scarce aspen midget moth) and *Phyllonorycter scabiosella* (Surrey midget moth) are on the UKBAP list.

Identification: Emmet et al. (1985) for the Gracillariinae and Lithocolletinae; Emmet (1985c) for the Phyllocnistinae; Manley (2009) illustrates some common species.

British subfamilies and genera:
Gracillariinae: *Acrocercops, Aspilapteryx, Callisto, Caloptilia, Calybites, Dialectica, Leucospilapteryx, Parectopa, Parornix*
Lithocolletinae: *Cameraria, Phyllonorycter*
Phyllocnistinae: *Phyllocnistis*

Family Roeslerstammiidae (1 genus, 2 species)
The two British species in this family are also the only ones in Europe; both are in the genus *Roeslerstammia*, which was previously included in the Yponomeutidae. Both have distinctive coppery coloured wings, with a fore wing length of 5–6 mm; neither is common. The known larvae feed on *Tilia* leaves, mining in the first instar and then feeding externally.

Identification: Agassiz (1996).

British genus: *Roeslerstammia*

SUPERFAMILY HEPIALOIDEA

Family Hepialidae (1 genus, 5 species)
The swift-moths are generally accepted as being a primitive family of Lepidoptera and they are 'microlepidoptera' that are large enough to be treated as honorary 'macros', with a fore wing length of 15–20 mm. *Hepialus sylvina*, the Orange swift, is one of the brighter coloured species (Fig. 21.21).

They fly mainly in the evenings, have very short antennae and cannot feed; the males often perform special mating displays to attract females, releasing scents that humans can detect. In most species the females are larger than the males, and less strongly marked. Hepialid larvae all live underground, feeding on plant roots and often live for two years; they are all white in colour. Although all the British species are currently placed in *Hepialus*, some authors prefer to subdivide this genus.

Hepialus humuli (Ghost moth) is on the UKBAP list.

Identification: Heath (1976d); Waring and Townsend (2009); Manley (2009); Skinner (2009).

British genus: *Hepialus*

SUPERFAMILY HESPERIOIDEA

The skippers are traditionally linked informally with the Papilionoidea as 'butterflies'; for general identification guides see the list under Papilionoidea.

Family Hesperiidae (6 genera, 8 species)
The skippers are not directly related to the Papilionoidea but are nevertheless often regarded as 'primitive' butterflies; most guides to butterflies include them as a matter of course. Their clubbed antennae and triangular wings certainly resemble those of the true butterflies (Fig. 21.22) but several species rest with the fore and hind wings held apart (Fig. 21.23). They also have very broad bodies and heads. Hesperiids do not fly for long distances or high above the ground; in fact they are usually seen 'skipping' over rough grassland. Some species are still a common sight, but many have declined dramatically in recent years, just like many Papilionoidea. The larvae of most British species feed on various grasses, though there are some exceptions.

The following three species are listed on the Wildlife and Countryside Act 1981: *Carterocephalus palaemon* (Chequered skipper), *Hesperia comma* (Silver-spotted skipper) and *Thymelicus acteon* (Lulworth skipper). *Erynnis tages* (Dingy skipper) is on the The Wildlife (Northern Ireland) Order 1985. The following are on the UKBAP list: *Carterocephalus*

Fig. 21.21 *Hepialus sylvina*, Orange swift moth (Hepialidae) (Photo: Roger Key)

Fig. 21.22 *Ochlodes venata*, Large skippers mating (Hesperiidae) (Photo: Roger Key)

Fig. 21.23 *Thymelicus lineolus*, Essex skipper (Hesperiidae) (Photo: Peter Barnard)

Fig. 21.24 *Adela reaumurella* (Adelidae) (Photo: Roger Key)

palaemon (Chequered skipper) *Erynnis tages* (Dingy skipper), *Pyrgus malvae* (Grizzled skipper) and *Thymelicus acteon* (Lulworth skipper).

British subfamilies and genera:
Hesperiinae: *Hesperia, Ochlodes, Thymelicus*
Heteropterinae: *Carterocephalus*
Pyrginae: *Erynnis, Pyrgus*

SUPERFAMILY INCURVARIOIDEA

Family Adelidae (3 genera, 15 species)
Earlier authors placed this group with the Incurvariidae. The members of this family are noted for their extremely long antennae, especially in the males; their wings are metallic bronze or greenish coloured. The larvae feed on leaf-litter and they construct a portable case from fragments of leaves. Two of the most common species are *Adela reaumurella* (Fig. 21.24) and *Nemophora degeerella* (Fig. 21.25). The males of some species form mating swarms around the tops of trees and bushes and these can be mistaken for the long-horned caddis-flies (Leptoceridae), which behave similarly and also have very long antennae.

Nematopogon magna (Scarce long-horn moth) and *Nemophora fasciella* (Horehound long-horn moth) are on the UKBAP list.

Identification: Heath and Pelham-Clinton (1976) under Incurvariidae; Manley (2009) illustrates some common species.

British subfamilies and genera:
Adelinae: *Adela, Nemophora*
Nematopogoninae: *Nematopogon*

Fig. 21.25 *Nemophora degeerella* (Adelidae) (Photo: Roger Key)

Family Heliozelidae (2 genera, 5 species)
This is a little-known family of very small moths, with a fore wing length of 2–4 mm; the wings are often copper or bronze coloured. The larvae are leaf-miners on trees or woody shrubs, and at least two species are common throughout Britain.

Identification: Emmet (1976c).

British genera: *Antispila, Heliozela*

Family Incurvariidae (2 genera, 5 species)
These were formerly linked with the Adelidae, but the antennae are usually no longer than the fore wing; the male antennae in *Incurvaria* are strongly pectinate. The larvae feed on trees such as birch, hazel and hawthorn; they are initially leaf-miners, but then build portable cases from leaves, drop to the ground and continue feeding on dead leaves.

Identification: Heath and Pelham-Clinton (1976); Manley (2009) shows two species of *Incurvaria*.

British genera: *Incurvaria, Phylloporia*

Family Prodoxidae (1 genus, 7 species)
All the British species in this family are in the genus *Lampronia*, which was formerly included in the Incurvariidae. They superficially resemble *Incurvaria*, except that the male antennae are not pectinate. Most larvae also feed in a similar way, starting off as leaf-miners and later feeding from portable cases, but a few remain on the tree shoots and one even causes gall formation. Several species are common and widespread and some even reach pest proportions, such as *Lampronia rubiella*, the Raspberry moth, which can cause much damage to loganberries and raspberries.

Lampronia capitella (Currant shoot borer) is on the UKBAP list.

Identification: Heath and Pelham-Clinton (1976) under Incurvariidae.

British genus: *Lampronia*

SUPERFAMILY LASIOCAMPOIDEA

Family Lasiocampidae (10 genera, 12 species)
This is a very distinctive family of large moths; most species have a fore wing length of 15–28 mm in the male, and 22–38 mm in the female. The males have strongly feathered antennae with which they detect the pheromones secreted by females; both sexes are fat-bodied with a very hairy appearance, often strongly marked. The larvae are equally spectacular, often with a dense covering of hairs and colourful markings (Fig. 21.26); they feed on various trees.

There are many common and well-known species, including *Lasiocampa quercus*, the Oak eggar; *Macrothylacia rubi*, the Fox moth; and *Euthrix potatoria*, the Drinker moth (Fig. 21.27). *Malacosoma neustria*, the Lackey moth, is widespread in southern Britain, and has a spectacularly coloured larva (Fig. 21.28).

Malacosoma neustria (Lackey moth) and *Trichiura crataegi* (Pale eggar) are on the UKBAP list.

Identification: Goater (1991a); Waring and Townsend (2009); Manley (2009); Skinner (2009).

Fig. 21.27 *Euthrix potatoria*, Drinker moth (Lasiocampidae) (Photo: Roger Key)

Fig. 21.26 Lasiocampid larva (Photo: Colin Rew)

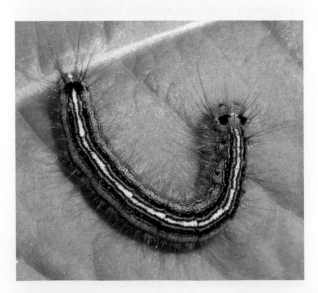

Fig. 21.28 *Malacosoma neustria*, Lackey moth larva (Lasiocampidae) (Photo: Peter Barnard)

British subfamilies and genera:

Lasiocampinae: *Eriogaster, Lasiocampa, Macrothylacia*
Malacosominae: *Malacosoma*
Pinarinae: *Dendrolimus, Gastropacha, Phyllodesma, Euthrix*
Poecilocampinae: *Poecilocampa, Trichiura*

SUPERFAMILY NEPTICULOIDEA

Family Nepticulidae (5 genera, 106 species)
This is a large family of very small-sized moths, with a fore wing length of just 2–3 mm. They have a rather similar appearance, partly because they have surprisingly large wing-scales, which, on such small wings, tends the limit the available patterns. The larvae are leaf-miners on trees or herbaceous shrubs, and there is a key to mines, based on food plants, in Emmet (1976a), which also contains much detailed information on their biology.

Stigmella zelleriella (Sandhill pigmy moth) is on the UKBAP list.

Identification: Emmet (1976a); Manley (2009) shows a few species.

British genera: *Bohemannia, Ectoedemia, Enteucha, Stigmella, Trifurcula*

Family Opostegidae (2 genera, 4 species)
The members of this small family were previously included in the Lyonetiidae. They are small species, with a fore wing length of 4–5 mm, and most have whitish wings. There are very few records of this group, and the biology is little studied, but there are possible larval records from *Rumex* and *Caltha*, though the mode of feeding is not clear.

Identification: Pelham-Clinton (1976).

British genera: *Opostega, Pseudopostega*

SUPERFAMILY NOCTUOIDEA

Family Arctiidae (21 genera, 33 species)
This family is sometimes treated as a subfamily of the Noctuidae. There are two distinct subfamilies with different morphologies and biologies. The Arctiinae include the tiger moths and ermine moths, and most are brightly coloured, a warning coloration to indicate that they are poisonous to predators (Fig. 21.29); the ermine moths are whitish or pale yellow. The subfamily Lithosiinae are much narrower-looking moths and can be quite small; they are commonly known as footmen, supposedly because their plain grey or yellow wings are wrapped round the body like the coat of a Victorian

Fig. 21.29 *Parasemia plantaginis*, Wood tiger moth (Arctiidae) (Photo: Roger Key)

Fig. 21.30 *Atolmis rubricollis*, Red-necked footman (Arctiidae) (Photo: Roger Key)

footman (Fig. 21.30). The larvae of Arctiinae are often densely hairy (Fig. 21.31) and some accumulate toxins from their food plant, which are retained in the adult stage, a common example being *Tyria jacobaeae*, the Cinnabar moth, which feeds on ragwort (Fig. 21.32).

Arctia caja (Garden tiger), *Coscinia cribraria* subsp. *bivittata* (Speckled footman), *Spilosoma lubricipeda* (White ermine), *Spilosoma luteum* (Buff ermine) and *Tyria jacobaeae* (Cinnabar moth) are on the UKBAP list.

Fig. 21.31 Arctiid larva (Photo: Colin Rew)

Fig. 21.33 *Euproctis similis*, Yellow-tail moth, young larva (Lymantriidae) (Photo: Roger Key)

Identification: Heath (1979b); Waring and Townsend (2009); Manley (2009); Skinner (2009).

British genus: *Amata*

Family Lymantriidae (8 genera, 11 species)
This family has previously been treated as a subfamily of the Noctuidae; they are sometimes known collectively as the tussock moths. They are quite large and hairy moths, but the two species of *Orgyia* are unusual in having wingless females, which remain in their pupal cocoon and attract flying males using pheromones. Most larvae feed on trees and the densely hairy caterpillars can sometimes reach pest proportions in urban streets, where every tree can be covered in caterpillars; their deciduous hairs are blown everywhere and can cause allergic reactions in local residents; *Euproctis chrysorrhoea*, the Browntail, is the usual culprit. The larva of *Eu. similis*, the Yellowtail, is similar (Fig. 21.33). One of the most bizarre larvae in this group is that of *Orgyia antiqua*, the Vapourer moth, which has arrays of different coloured tufts of hair (Fig. 21.34).

Orgyia recens (Scarce vapourer) is on the UKBAP list.

Identification: De Worms (1979b); Waring and Townsend (2009); Manley (2009); Skinner (2009).

British subfamilies and genera:
 Arctornithinae: *Arctornis*
 Leucominae: *Leucoma*
 Lymantriinae: *Lymantria*
 Nygmiinae: *Euproctis*
 Orgyinae: *Calliteara, Laelia, Orgyia*
 Incertae sedis: *Dicallomera*

Fig. 21.32 *Tyria jacobaeae*, Cinnabar moth larva (Arctiidae) (Photo: Roger Key)

Identification: De Worms (1979c); Waring and Townsend (2009); Manley (2009); Skinner (2009).

British subfamilies and genera:
 Arctiinae: *Arctia, Callimorpha, Coscinia, Diacrisia, Diaphora, Euplagia, Parasemia, Spilosoma, Tyria, Utetheisa*
 Lithosiinae: *Atolmis, Cybosia, Eilema, Lithosia, Miltochrista, Nudaria, Pelosia, Phragmatobia, Setina, Spiris, Thumatha*

Family Ctenuchidae (1 genus, 1 species)
The only British species in this family is *Amata phegea*, known as the Nine-spotted moth. It is sometimes placed with the Arctiidae or even the Noctuidae. It is a very distinctive species that has bluish-black wings with white spots, but it is an occasional immigrant from mainland Europe that does not breed here.

Fig. 21.34 *Orgyia antiqua*, Vapourer moth larva (Lymantriidae) (Photo: Colin Rew)

Fig. 21.36 *Acronicta* sp., Dagger moth (Noctuidae) (Photo: Colin Rew)

Fig. 21.35 *Lycophotia porphyrea*, True lover's knot moth (Noctuidae) (Photo: Roger Key)

Fig. 21.37 *Melanchra pisi*, Broom moth larva on bracken (Noctuidae) (Photo: Roger Key)

Family Noctuidae (168 genera, 415 species)

This is the largest family of Lepidoptera in Britain, and includes a wide variety of moths in several subfamilies; the Herminiinae have sometimes been treated as a distinct family, and the Dilobinae are sometimes included in the Notodontidae. The taxonomy of this family is so complex that not all authors agree on the arrangement of the subgroups. In general the noctuids are quite stout-bodied moths, frequently brownish in colour although nearly all have distinct wing-patterns that make identification of most species straightforward (Figs. 21.35 & 21.36). Most are nocturnal and fly readily to lights; they are powerful fliers and several species migrate long distances, often stopping at flowers to take nectar. Noctuid larvae are usually not hairy; some feed on leaves of trees, some on the leaves, stems or roots of herbaceous plants and grasses (Fig. 21.37). This is an extremely well-known group,

and the life histories of most species are known in detail; more information on the individual groups and species will be found in the identification books listed below.

Acosmetia caliginosa (Reddish buff) and *Gortyna borelii* subsp. *lunata* (Fisher's estuarine moth) are listed on the Wildlife and Countryside Act 1981.

The following are on the UKBAP list: *Acosmetia caliginosa* (Reddish buff), *Acronicta psi* (Grey dagger), *Acronicta rumicis* (Knot grass), *Agrochola helvola* (Flounced chestnut), *Agrochola litura* (Brown-spot pinion), *Agrochola lychnidis* (Beaded chestnut), *Allophyes oxyacanthae* (Green-brindled crescent), *Amphipoea oculea* (Ear moth), *Amphipyra tragopoginis*

(Mouse moth), *Anarta cordigera* (Small dark yellow underwing), *Apamea anceps* (Large nutmeg), *Apamea remissa* (Dusky brocade), *Aporophyla lutulenta* (Deep-brown dart), *Archanara neurica* (White-mantled wainscot), *Asteroscopus sphinx* (The sprawler), *Atethmia centrago* (Centre-barred sallow), *Athetis pallustris* (Marsh moth), *Blepharita adusta* (Dark brocade), *Brachylomia viminalis* (Minor shoulder-knot), *Caradrina morpheus* (Mottled rustic), *Catocala promissa* (Light crimson underwing), *Catocala sponsa* (Dark crimson underwing), *Celaena haworthii* (Haworth's minor), *Celaena leucostigma* (The crescent), *Chortodes brevilinea* (Fenn's wainscot), *Chortodes extrema* (The concolorous), *Cosmia diffinis* (White-spotted pinion), *Dasypolia templi* (Brindled ochre), *Diarsia rubi* (Small square-spot), *Dicycla oo* (Heart moth), *Diloba caeruleocephala* (Figure of eight), *Eugnorisma glareosa* (Autumnal rustic), *Euxoa nigricans* (Garden dart), *Euxoa tritici* (White-line dart), *Graphiphora augur* (Double dart), *Hadena albimacula* (White spot), *Heliophobus reticulata* (Bordered gothic), *Heliothis maritima* (Shoulder-striped clover), *Hoplodrina blanda* (Rustic), *Hydraecia micacea* (Rosy rustic), *Hydraecia osseola* subsp. *hucherardi* (Marsh mallow moth), *Jodia croceago* (Orange upperwing), *Luperina nickerlii* subsp. *leechi* (Sandhill rustic – Cornish subsp.), *Melanchra persicariae* (Dot moth), *Melanchra pisi* (Broom moth), *Mesoligia literosa* (Rosy minor), *Mythimna comma* (Shoulder-striped wainscot), *Noctua orbona* (Lunar yellow underwing), *Oria musculosa* (Brighton wainscot), *Orthosia gracilis* (Powdered quaker), *Paracolax tristalis* (Clay fan-foot), *Pechipogo strigilata* (Common fan-foot), *Polia bombycina* (Pale shining brown), *Protolampra sobrina* (Cousin German), *Rhizedra lutosa* (Large wainscot), *Shargacucullia lychnitis* (Striped lychnis), *Stilbia anomala* (The anomalous), *Tholera cespitis* (Hedge rustic), *Tholera decimalis* (Feathered gothic), *Trisateles emortualis* (Olive crescent), *Tyta luctuosa* (Four-spotted moth), *Xanthia gilvago* (Dusky-lemon sallow), *Xanthia icteritia* (Sallow), *Xestia agathina* (Heath rustic), *Xestia alpicola* subsp. *alpina* (Northern dart), *Xestia ashworthii* (Ashworth's rustic), *Xestia castanea* (Neglected rustic) and *Xylena exsoleta* (Sword-grass).

Identification: Bretherton et al. (1979) and Heath and Emmet (1983); Waring and Townsend (2009); Manley (2009); Skinner (2009).

British subfamilies and genera:
Acontiinae: *Acontia, Aedia*
Acronictinae: *Acronicta, Craniophora, Moma, Simyra*
Amphipyrinae: *Amphipyra*

Aventiinae: *Laspeyria*
Bolitobiinae: *Parascotia*
Bryophilinae: *Cryphia*
Calpinae: *Anomis, Scoliopteryx*
Catocalinae: *Callistege, Catephia, Catocala, Dysgonia, Euclidia, Lygephila, Minucia, Tathorhynchus*
Condicinae: *Acosmetia*
Cuculliinae: *Cucullia, Shargacucullia*
Dilobinae: *Diloba*
Eriopinae: *Callopistria*
Eublemminae: *Eublemma*
Eustrotiinae: *Deltote*
Hadeninae: *Anarta, Brithys, Cerapteryx, Egira, Hada, Hecatera, Heliophobus, Lacanobia, Mamestra, Melanchra, Mythimna, Orthosia, Pachetra, Panolis, Papestra, Polia, Sideridis, Tholera*
Heliothinae: *Helicoverpa, Heliothis, Periphanes, Pyrrhia*
Herminiinae: *Herminia, Macrochilo, Paracolax, Pechipogo, Zanclognatha*
Hypeninae: *Hypena*
Hypenodinae: *Hypenodes, Schrankia*
Metoponiinae: *Panemeria, Tyta*
Noctuinae: *Actebia, Agrotis, Anaplectoides, Axylia, Cerastis, Coenophila, Diarsia, Eugnorisma, Eurois, Euxoa, Graphiphora, Lycophotia, Naenia, Noctua, Ochropleura, Peridroma, Protolampra, Rhyacia, Spaelotis, Standfussiana, Xestia*
Nolinae: *Bena, Earias, Nycteola, Pseudoips*
Oncocnemidinae: *Calophasia, Stilbia, Xylocampa*
Pantheinae: *Colocasia*
Phytometrinae: *Colobochyla, Phytometra, Trisateles*
Plusiinae: *Abrostola, Autographa, Chrysodeixis, Cornutiplusia, Ctenoplusia, Diachrysia, Euchalcia, Macdunnoughia, Megalographa, Plusia, Polychrysia, Syngrapha, Thysanoplusia, Trichoplusia*
Psaphidinae: *Allophyes, Asteroscopus, Meganephria*
Rivulinae: *Rivula*
Xyleninae: *Actinotia, Agrochola, Amphipoea, Antitype, Apamea, Aporophyla, Archanara, Arenostola, Atethmia, Athetis, Blepharita, Brachylomia, Calamia, Caradrina, Celaena, Charanyca, Chilodes, Chortodes, Coenobia, Conistra, Cosmia, Dasypolia, Dicycla, Dryobota, Dryobotodes, Dypterygia, Elaphria, Eremobia, Euplexia, Eupsilia, Gortyna, Hoplodrina, Hydraecia, Hyppa, Ipimorpha, Jodia, Leucochlaena, Lithomoia, Lithophane, Luperina, Mesapamea, Mesoligia, Mormo, Nonagria, Oligia, Omphaloscelis, Oria, Parastichtis, Phlogophora, Photedes, Polymixis, Rhizedra, Sedina, Spodoptera, Thalpophila, Trachea, Trigonophora, Xylena*

Family Nolidae (2 genera, 6 species)
Although six species in this family are on the British list, only four are considered as resident. All are

relatively small species with a fore wing length of 8–11 mm and they are basically whiteish or grey with various darker markings. Their larvae feed on various trees and shrubs, but *Nola cucullatella*, the Short-cloaked moth, is probably the only species that could be called common.

Identification: Revell (1979); Waring and Townsend (2009); Manley (2009); Skinner (2009).

British genera: *Meganola, Nola*

Family Notodontidae (16 genera, 27 species)

This family includes the prominent and kitten moths; all are stout moths with hairy bodies and complex wing-patterns. The antennae are relatively long and often feathered in the males. Although only about 21 species can be considered as resident, many of these are quite common and well-known species. *Phalera bucephala*, the Buff-tip, is a good example of a moth that seems to have a rather conspicuous patterning but which actually blends well into its background of birch twigs (Fig. 21.38). The larvae in this group are often very striking, with bright colours and strange appendages, a common example being *Cerura vinula*, the Puss moth (Fig. 21.39). Most larvae feed on various trees and shrubs.

Identification: De Worms (1979a); Waring and Townsend (2009); Manley (2009); Skinner (2009).

British subfamilies and genera:

Heterocampinae: *Harpyia, Peridea, Stauropus*
Notodontinae: *Cerura, Drymonia, Furcula, Gluphisia, Leucodonta, Notodonta, Odontosia, Pheosia, Pterostoma, Ptilodon, Ptilophora*

Phalerinae: *Phalera*
Pygaerinae: *Clostera*

Family Thaumetopoeidae (1 genus, 2 species)

Some authors have included this group as a subfamily of the Notodontidae. The two species on the British list are *Thaumetopoea pityocampa*, the Pine processionary and *Th. processionea*, the Oak processionary; neither is currently resident but both are forest pests in mainland Europe so their presence here could present a risk. The larvae are hairy and live in communal webs.

Identification: Heath (1979a); Waring and Townsend (2009); Manley (2009); Skinner (2009).

British genus: *Thaumetopoea*

SUPERFAMILY PAPILIONOIDEA

All the books on butterflies treat the families together, and they also include the Hesperioidea. There are of course numerous books on British butterflies and deciding which one(s) to use is often a matter of personal choice. One important point to note is that it was only in 2001 that the presence of *Leptidea reali* was recognized (in Ireland) so any book published before then will not distinguish it from *L. sinapis* (Pieridae).

Among the better books for identification are Higgins and Riley (1970), Howarth (1984), Thomas (1986), Riley (2007) and Thomas and Lewington

Fig. 21.38 *Phalera bucephala*, Buff-tip moth among birch twigs (Notodontidae) (Photo: Roger Key)

Fig. 21.39 *Cerura vinula*, Puss moth larva (Notodontidae) (Photo: Roger Key)

(2010). There are also many books dealing with the ecology and conservation of butterflies, and these include Dennis (1977, 1992), Dennis and Shreeve (1996), Asher et al. (2001) and Fox et al. (2006). Other useful works are May (2003) on food plants; Whalley (1980) on watching rather than collecting specimens, and the classic work by Ford (1945).

Family Lycaenidae (16 genera, 20 species)

Although often loosely called the blues, this family includes the coppers (Fig. 21.40) and hairstreaks, together with some that do not fit any of these categories such as *Aricia agestis*, the Brown argus (Figs. 21.41 & 21.42). Several of the blues are commonly seen, especially *Celastrina argiolus*, the Holly blue,

which is a common sight in urban gardens; its caterpillar feeds on holly in the spring, while the second brood are often on ivy (Fig. 21.43).

The following 11 species are listed on the Wildlife and Countryside Act 1981: *Aricia artaxerxes* (Northern brown argus), *Cupido minimus* (Small blue), *Lycaena dispar* (Large copper), *Lysandra bellargus* (Adonis blue), *Lysandra coridon* (Chalk-hill blue), *Maculinea arion* (Large blue butterfly), *Plebejus argus* (Silver-studded blue), *Satyrium pruni* (Black hairstreak), *Satyrium w-album* (White-letter hairstreak) and *Thecla betulae* (Brown hairstreak).

Celastrina argiolus (Holly blue) and *Neozephyrus quercus* (Purple hairstreak) are listed on The Wildlife (Northern Ireland) Order 1985.

Fig. 21.40 *Lycaena phlaeas*, Small copper butterfly (Lycaenidae)(Photo: Colin Rew)

Fig. 21.42 *Aricia agestis*, underside (Lycaenidae) (Photo: Roger Key)

Fig. 21.41 *Aricia agestis*, Brown argus butterfly (Lycaenidae) (Photo: Roger Key)

Fig. 21.43 *Celastrina argiolus*, Holly blue larva (Lycaenidae) (Photo: Roger Key)

The following are on the UKBAP list: *Aricia artaxerxes* (Northern brown argus), *Cupido minimus* (Small blue), *Maculinea arion* (Large blue butterfly), *Plebejus argus* (Silver-studded blue), *Satyrium w-album* (White-letter hairstreak), *Thecla betulae* (Brown hairstreak).

Some genera are treated as subgenera only. *Hamearis* (Riodinidae) is sometimes included in this family.

British subfamilies and genera:

Lycaeninae: *Lycaena*

Polyommatinae: *Cacyreus, Celastrina, Cupido, Everes, Lampides, Lysandra, Phengaris (= Maculinea), Plebejus, Polyommatus*

Theclinae: *Callophrys, Favonius* (incl. *Neozephyrus*), *Satyrium, Thecla*

Incertae sedis: *Aricia, Cyaniris*

Family Nymphalidae (23 genera, 31 species)

Some subfamilies in this group were previously treated as distinct families, especially the Danainae and Satyrinae. They include many common and familiar species such as fritillaries (Fig. 21.44), *Polygonium c-album*, the Comma (Fig. 21.45) and *Pyronia tithonus*, the Gatekeeper (Figs. 21.46 & 21.47). Several members of this family have very different patterns on the underside of the wings, usually cryptic (Fig. 21.48). The larvae feed on various plants and most are easily recognized to species (Fig. 21.49). Like most butterflies the pupae are not enclosed in a cocoon (Fig. 21.50) and the clearly visible emergence of the adult butterfly is one of the classic sights that always enthrals the observer (Figs. 21.51 & 21.52).

Fig. 21.45 *Polygonia c-album*, Comma butterfly (Nymphalidae) (Photo: Roger Key)

Fig. 21.46 *Pyronia tithonus*, Gatekeeper butterfly (Nymphalidae) (Photo: Peter Barnard)

Fig. 21.44 *Argynnis aglaja*, Dark green fritillaries mating (Nymphalidae) (Photo: Roger Key)

Fig. 21.47 *Pyronia tithonus* mating (Nymphalidae) (Photo: Colin Rew)

Fig. 21.48 *Aglais urticae*, Small tortoiseshell butterfly, underside (Nymphalidae) (Photo: Colin Rew)

Fig. 21.50 *Polygonia c-album* pupa (Nymphalidae) (Photo: Roger Key)

Fig. 21.49 *Polygonia c-album*, Comma butterfly larva (Nymphalidae) (Photo: Roger Key)

Fig. 21.51 *Polygonia c-album* emerging (Nymphalidae) (Photo: Roger Key)

The following nine species are listed on the Wildlife and Countryside Act 1981: *Aglais polychloros* (Large tortoiseshell), *Apatura iris* (Purple emperor), *Argynnis adippe* (High brown fritillary), *Boloria euphrosyne* (Pearl-bordered fritillary), *Coenonympha tullia* (Large heath), *Erebia epiphron* (Mountain ringlet), *Euphydryas aurinia* (Marsh fritillary), *Melitaea athalia* (Heath fritillary) and *Melitaea cinxia* (Glanville fritillary).

Danaus plexippus (Milkweed) is listed on the Bonn Convention.

The following are on the UKBAP list: *Argynnis adippe* (High brown fritillary), *Boloria euphrosyne* (Pearl-bordered fritillary), *Boloria selene* (Small

pearl-bordered fritillary), *Coenonympha pamphilus* (Small heath), *Coenonympha tullia* (Large heath), *Erebia epiphron* (Mountain ringlet), *Euphydryas aurinia* (Marsh fritillary), *Hipparchia semele* (Grayling), *Lasiommata megera* (Wall), *Limenitis camilla* (White admiral), *Melitaea athalia* (Heath fritillary) and *Melitaea cinxia* (Glanville fritillary).

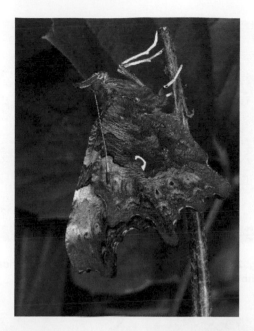

Fig. 21.52 *Polygonia c-album* underside (Nymphalidae) (Photo: Roger Key)

Fig. 21.54 *Pieris napi*, Green-veined white butterfly (Pieridae) (Photo: Peter Barnard)

Fig. 21.53 *Papilio machaon*, Swallowtail butterfly (Papilionidae) (Photo: Roger Key)

Fig. 21.55 *Pieris brassicae*, Large white butterfly, young larvae (Pieridae) (Photo: Roger Key)

British subfamilies and genera:
  Apaturinae: *Apatura*
  Danainae: *Danaus*
  Heliconiinae: *Argynnis, Boloria, Issoria*
  Limenitidinae: *Limenitis*
  Melitaeinae: *Euphydryas, Melitaea*
  Nymphalinae: *Aglais, Araschnia, Inachis, Nymphalis, Polygonia, Vanessa*
  Satyrinae: *Aphantopus, Coenonympha, Erebia, Hipparchia, Lasiommata, Maniola, Melanargia, Pararge, Pyronia*

Family Papilionidae (1 genus, 1 species)
The only British species in this family is *Papilio machaon*, the Swallowtail (Fig. 21.53) familiar to eve-ryone yet very local to the fens of East Anglia and rarely seen in the wild.

Papilio machaon (Swallowtail) is listed on the Wildlife and Countryside Act 1981.

British genus: *Papilio*

Family Pieridae (7 genera, 11 species)
This family includes the familiar white butterflies that feed on brassicas, as well as species like *Pieris napi*, the Green-veined white (Fig. 21.54). Both *P. rapae*, the Small white and *P. brassicae*, the Large white can become pests on brassica plants in gardens; their larvae feed gregariously and often change in appearance as they grow (Figs. 21.55 &

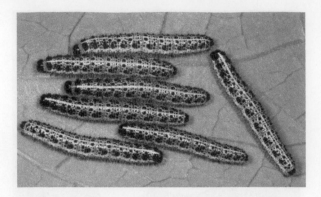

Fig. 21.56 *Pieris brassicae*, mature larvae (Pieridae) (Photo: Roger Key)

Fig. 21.58 *Amblyptilia acanthadactyla* (Pterophoridae) (Photo: Roger Key)

Fig. 21.57 *Gonepteryx rhamni*, Brimstone butterfly (Pieridae) (Photo: Roger Key)

Fig. 21.59 *Oxyptilus parvidactylus* (Pterophoridae) (Photo: Roger Key)

21.56). Other common species in this family include *Gonepteryx rhamni*, the Brimstone (Fig. 21.57). *Leptidea reali*, Réal's wood white, was found in Ireland in 2001, though it does not occur in the UK.

Leptidea sinapis (Wood white) is listed on the Wildlife and Countryside Act 1981 and is also on the UKBAP list. *Gonepteryx rhamni* (Brimstone) is listed on the The Wildlife (Northern Ireland) Order 1985.

British subfamilies and genera:
   Coliadinae: *Colias, Gonepteryx*
   Dismorphiinae: *Leptidea*
   Pierinae: *Anthocharis, Aporia, Pieris, Pontia*

Family Riodinidae (1 genus, 1 species)
The single British species, *Hamearis lucina*, the Duke of Burgundy fritillary, is sometimes included in the Lycaenidae. It superficially resembles a small fritil-

lary (of the family Nymphalidae) and its larvae feed on *Primula*, though it is becoming an endangered species in this country.

Hamearis lucina (Duke of Burgundy fritillary) is listed on the Wildlife and Countryside Act 1981 and is on the UKBAP list.

British genus: *Hamearis*

SUPERFAMILY PTEROPHOROIDEA

Family Pterophoridae (18 genera, 43 species)
The plume-moths form a very distinctive family, which hold their narrow wings out straight when at rest, like a capital T (Figs. 21.58 & 21.59). Their larvae feed on various plant species, and several are very common throughout Britain.

Identification: Beirne (1952) can be used with care, though it is rather out of date; Manley (2009) illustrates many of the British species; also useful is the European review by Gielis (1996).

British subfamilies and genera:

Agdistinae: *Agdistis*

Pterophorinae: *Adaina, Amblyptilia, Buckleria, Capperia, Cnaemidophorus, Crombrugghia, Emmelina, Hellinsia, Marasmarcha, Merrifieldia, Oidaematophorus, Ovendenia, Oxyptilus, Platyptilia, Pselnophorus, Pterophorus, Stenoptilia*

SUPERFAMILY PYRALOIDEA

Family Crambidae (63 genera, 129 species)

These are sometimes treated as a subfamily of the Pyralidae; although many are quite small species, some of the larger ones are so familiar that they have been given common names, such as *Evergestis forficalis*, the Garden pebble; *Eurrhypara hortulata*, the Small magpie moth (Fig. 21.60; from its similarity to the Magpie moth, *Abraxas grossulariata*, in the Geometridae); *Pleuroptya ruralis*, the Mother-of-pearl moth; and *Elophila nymphaeata*, the Brown china-mark moth. The species of *Crambus* and *Agriphila* (Fig. 21.61) are generally known as grass moths (Fig. 21.62). Several species have aquatic larvae, including *Cataclysta lemnata*, the Small china-mark (Fig. 21.63), whose larva feeds on duckweed, and *Parapoynx* (often mis-spelled 'Paraponyx'). There are some very colourful moths in this family, such as *Pyrausta aurata* (Fig. 21.64).

Agrotera nemoralis (Beautiful pearl), *Anania funebris* (White-spotted sable moth) and *Pyrausta*

Fig. 21.61 *Agriphila tristella* (Crambidae) (Photo: Roger Key)

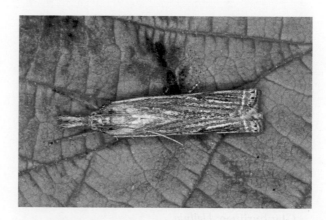

Fig. 21.62 Crambid (Photo: Colin Rew)

Fig. 21.60 *Eurrhypara hortulata*, Small magpie moth (Crambidae) (Photo: Colin Rew)

Fig. 21.63 *Cataclysta lemnata*, Small china-mark moth (Crambidae) (Photo: Peter Barnard)

Fig. 21.64 *Pyrausta aurata* (Crambidae) (Photo: Roger Key)

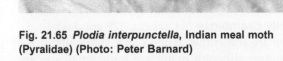

Fig. 21.65 *Plodia interpunctella*, Indian meal moth (Pyralidae) (Photo: Peter Barnard)

sanguinalis (Scarce crimson and gold) are on the UKBAP list.

Identification: Goater (1986).

British subfamilies and genera:

Acentropinae: *Acentria, Cataclysta, Elophila, Nymphula, Oligostigma, Parapoynx*

Crambinae: *Acigona, Agriphila, Ancylolomia, Calamotropha, Catoptria, Chilo, Chrysocrambus, Chrysoteuchia, Crambus, Euchromius, Friedlanderia, Pediasia, Platytes, Thisanotia*

Evergestinae: *Evergestis*

Glaphyriinae: *Hellula*

Odontiinae: *Cynaeda, Metaxmeste*

Pyraustinae: *Anania, Ebulea, Eurrhypara, Loxostege, Microstega, Nascia, Opsibotis, Ostrinia, Paracorsia, Paratalanta, Perinephela, Phlyctaenia, Psammotis, Pyrausta, Sclerocona, Sitochroa, Uresiphita*

Schoenobiinae: *Donacaula, Schoenobius*

Scopariinae: *Dipleurina, Eudonia, Scoparia*

Spilomelinae: *Agrotera, Antigastra, Conogethes, Diaphania, Diasemia, Diasemiopsis, Diplopseustis, Dolicharthria, Herpetogramma, Maruca, Mecyna, Nomophila, Palpita, Pleuroptya, Sceliodes, Spoladea, Udea*

Family Pyralidae (48 genera, 90 species)

This large family is often linked with the Crambidae; it includes several species of economic importance, some of which are cosmopolitan pests, such as *Plodia interpunctella*, the Indian meal moth (Fig. 21.65), and *Pyralis farinalis*, the Meal moth. Some species feed on the honeycombs of bee-hives, and these include *Aphomia sociella*, the Bee moth; *Achroia grisella*, the Lesser wax moth, and *Galleria mellonella*, the Wax moth (Figs. 21.66, 21.67 & 21.68).

Fig. 21.66 *Galleria mellonella*, Wax moth (Pyralidae) (Photo: Peter Barnard)

Fig. 21.67 *Galleria mellonella* larva (Pyralidae) (Photo: Peter Barnard)

Fig. 21.68 *Galleria mellonella* pupae (Pyralidae) (Photo: Peter Barnard)

Fig. 21.69 *Synanthedon tipuliformis*, Currant clearwing moth (Sesiidae) (Photo: Roger Key)

Fig. 21.70 *Synanthedon myopaeformis*, Red-belted clearwing moth (Sesiidae) (Photo: Roger Key)

Sciota hostilis (Scarce aspen knot-horn) is on the UKBAP list.

Identification: Goater (1986); Manley (2009) illustrates several species.

British subfamilies and genera:

Galleriinae: *Achroia, Aphomia, Corcyra, Galleria, Paralipsa*

Phycitinae: *Acrobasis, Ancylosis, Anerastia, Apomyelois, Assara, Cadra, Catastia, Cryptoblabes, Delplanqueia, Dioryctria, Eccopisa, Elegia, Ephestia, Epischnia, Etiella, Eurhodope, Euzophera, Gymnancyla, Homoeosoma, Hypochalcia, Matilella, Moitrelia, Myelois, Nephopterix, Nyctegretis, Oncocera, Ortholepis, Pempelia, Pempeliella, Phycita, Phycitodes, Pima, Plodia, Rhodophaea, Salebriopsis, Sciota, Selagia, Zophodia*

Pyralinae: *Aglossa, Endotricha, Hypsopygia, Pyralis, Synaphe*

SUPERFAMILY SCHRECKENSTEINIOIDEA

Family Schreckensteiniidae (1 genus, 1 species)
The only British species in this family is *Schreckensteinia festaliella*, which is also the only species in Europe. It is a small narrow-winged moth, with a fore wing length of 4–5mm; the larva feeds on species of *Rubus*, and is widespread throughout most of Britain.

Identification: Emmet (1996b).

British genus: *Schreckensteinia*

SUPERFAMILY SESIOIDEA

Family Sesiidae (6 genera, 14 species)
The members of this very distinct family are known as clear-wing moths. The large, transparent, scaleless patches on their wings, combined with their black and yellow or red coloration, give them a wasp-like appearance; some resemble various colourful Diptera (Figs. 21.69, 21.70 & 21.71). Even the more common species are rarely seen; they are day-flying but are most active just after emerging and when seeking a mate. The males are attracted by the scent of the females, and artificial versions of these pheromones are available commercially; this is

Fig. 21.71 *Synanthedon formicaeformis*, Red-tipped clearwing moth (Sesiidae) (Photo: Roger Key)

Fig. 21.72 *Moraphaga choragella* (Tineidae) (Photo: Roger Key)

often the best way to find some species. The larvae live in various trees and woody shrubs and can take two years to develop.

Pyropteron chrysidiformis (Fiery clearwing) is listed on the Wildlife and Countryside Act 1981, and is also on the UKBAP list.

Identification: Baker (1985); Waring and Townsend (2009); Manley (2009); Skinner (2009).

British genera: *Bembecia, Paranthrene, Pyropteron, Sesia, Synansphecia, Synanthedon*

SUPERFAMILY TINEOIDEA

Family Psychidae (15 genera, 21 species)
This is a strange family of moths; although some can have a fore wing length up to 15 mm they are regarded as 'micros' and therefore not included in most common moth books. Most adult psychids are rather nondescript though some females are wingless and reproduce parthenogenetically. The larvae are unusual in constructing portable cases, often quite elaborate in form, which closely resemble some kinds of caddisfly cases (Trichoptera). The basic structure of the case is of silk, but it is covered with sand, dried plant material or general dry debris, but its form is so characteristic in each species that it is possible to identify most of them by the case alone. Most larvae seem to feed on algae, lichens or mosses; some are found on grasses and some may feed on fungi.

Identification: Hättenschwiler (1985) includes a key to larval cases; Manley (2009) illustrates a few species.

British subfamilies and genera:
Epichnopteriginae: *Epichnopterix, Whittleia*
Naryciinae: *Dahlica, Diplodoma, Narycia*
Oiketicinae: *Acanthopsyche, Canephora, Pachythelia, Sterrhopterix*
Psychinae: *Bacotia, Luffia, Proutia, Psyche*
Taleporiinae: *Bankesia, Taleporia*

Family Tineidae (29 genera, 62 species)
This family contains several moths of economic importance as domestic pests, including *Tineola bisselliella*, the Common clothes moth; *Tinea pellionella*, the Case-bearing clothes moth; and *Trichophaga tapetzella*, the Tapestry moth. It is the subfamily Tineinae that includes these species whose larvae are able to digest keratin, so can feed on hair, wool, feathers and skin. In the wild they can live in birds' nests for example, but are clearly pre-adapted to living in human habitations. Most tineid larva feed on fungi and lichens, and most non-tineine adults have distinctive wing-patterns (Fig. 21.72).

The subfamily Hieroxestinae was previously treated as a distinct family.

Eudarcia richardsoni (Dorset tineid moth) and *Nemapogon picarella* (Pied tineid moth) are on the UKBAP list.

Identification: Pelham-Clinton (1985a, 1985b); Manley (2009) shows a few common species. For a world review of the biology of the Tineidae see Robinson (2009).

British subfamilies and genera:
Dryadaulinae: *Dryadaula*
Euplocaminae: *Euplocamus*
Hieroxestinae: *Oinophila, Opogona*

Meessiinae: *Eudarcia, Infurcitinea, Ischnoscia, Stenoptinea, Tenaga*

Myrmecozelinae: *Myrmecozela*

Nemapogoninae: *Archinemapogon, Nemapogon, Nemaxera, Triaxomasia, Triaxomera*

Perissomasticinae: *Ateliotum, Haplotinea*

Scardiinae: *Morophaga*

Setomorphinae: *Lindera, Setomorpha*

Teichobiinae: *Psychoides*

Tineinae: *Cephimallota, Cephitinea, Ceratophaga, Monopis, Niditinea, Tinea, Tineola, Trichophaga*

SUPERFAMILY TISCHERIOIDEA

Family Tischeriidae (2 genera, 6 species)

The members of this small family were formerly placed in the Lyonetiidae, but the group is now recognized as being sufficiently distinct to warrant its own superfamily. They are small moths, with a fore wing length of 3–5mm, and they are rather uniformly coloured with little patterning. The larvae produce blotch-mines, lined with silk, and feed on various trees and woody shrubs. Only *Tischeria marginea* seems to be common in Britain.

Identification: Emmet (1976b).

British genera: *Coptotriche, Tischeria*

SUPERFAMILY TORTRICOIDEA

Family Tortricidae (ca 90 genera, 390 species)

This large family includes several species that have been given common names based on the plants that they feed on, sometimes reaching pest proportions. These include *Rhyacionia buoliana*, the Pine shoot moth; *Tortrix viridana*, the Green oak tortrix; *Acleris comariana*, the Strawberry tortrix, and many more. Although there is much variation in this group, most tortricids have a characteristic way of resting with their wings held roof-like over the body (Fig. 21.73) though some have a more flattened appearance. Tortricid larvae can be of several colours (Fig. 21.74); they often spin leaves of their food plant into a covering around them, and this leaf-rolling is a common feature of many species. *Cydia pomonella* is the notorious Codlin moth that can cause extensive damage to apples, while *C. nigricana* causes damage to pea-pods (Fig. 21.75).

Celypha woodiana (Mistletoe marble) and *Grapholita pallifrontana* (Liquorice piercer) are on the UKBAP list.

Some earlier works recognized the family Cochylidae, but these are now included within the Tortricinae.

Fig. 21.73 *Agapeta hamana* (Tortricidae) (Photo: Roger Key)

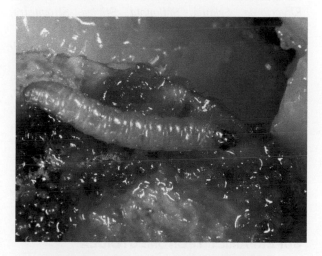

Fig. 21.74 *Grapholita funebrana* larva (Tortricidae) (Photo: Roger Key)

Fig. 21.75 *Cydia nigricana*, Pea moth larvae (Tortricidae) (Photo: Roger Key)

Identification: Bradley et al. (1973, 1979); Manley (2009) illustrates many species. It is also worth consulting the European review by Razowski (2002, 2003).

British subfamilies and genera:

Chlidanotinae: *Isotrias, Olindia*

Olethreutinae: *Acroclita, Ancylis, Apotomis, Argyroploce, Bactra, Celypha, Clavigesta, Crocidosema, Cydia, Dichrorampha, Enarmonia, Endothenia, Epiblema, Epinotia, Eriopsela, Eucosma, Eucosmomorpha, Eudemis, Gibberifera, Grapholita, Gypsonoma, Hedya, Lathronympha, Lobesia, Metendothenia, Notocelia, Olethreutes, Orthotaenia, Pammene, Pelochrista, Phaneta, Phiaris, Piniphila, Pristerognatha, Pseudococcyx, Pseudosciaphila, Retinia, Rhopobota, Rhyacionia, Selania, Spilonota, Stictea, Strophedra, Thiodia, Zeiraphera*

Tortricinae: *Acleris, Adoxophyes, Aethes, Agapeta, Aleimma, Aphelia, Archips, Argyrotaenia, Cacoecimorpha, Capua, Choristoneura, Clepsis, Cnephasia, Cochylidia, Cochylimorpha, Cochylis, Commophila, Ditula, Eana, Epagoge, Epichoristodes, Epiphyas, Eulia, Eupoecilia, Exapate, Falseuncaria, Gynnidomorpha, Hysterophora, Lozotaenia, Lozotaeniodes, Neosphaleroptera, Pandemis, Paramesia, Periclepsis, Phalonidia, Philedone, Philedonides, Phtheochroa, Pseudargyrotoza, Ptycholoma, Ptycholomoides, Sparganothis, Spatalistis, Syndemis, Tortricodes, Tortrix*

SUPERFAMILY YPONOMEUTOIDEA

Family Acrolepiidae (3 genera, 6 species)
This small family was previously treated as a subfamily of the Yponomeutidae. They are quite colourful small moths with a fore wing length of 4–7 mm, and can quite easily be separated by their wing-patterns. The larvae are leaf-miners on Asteraceae, Liliaceae or Solanaceae, though there are few records of many species.

Identification: Agassiz (1996) under Yponomeutidae; Manley (2009) illustrates two species of *Acrolepiopsis*.

British genera: *Acrolepia, Acrolepiopsis, Digitivalva*

Family Bedelliidae (1 genus, 1 species)
The only British species in this family is *Bedellia somnulentella*; it has narrow brownish wings with long fringes, with a fore wing length of 3–4 mm. There are not many records of this species, though it seems fairly widespread; the larva is a leaf-miner on various plants including bindweeds.

Identification: Emmet (1985b) under Lyonetiidae.

British genus: *Bedellia*

Family Glyphipterigidae (2 genera, 8 species)
The members of this small family are day-flying moths that can be recognized by their habit of raising and lowering their wings when resting. With a few exceptions the larvae feed in the seeds of grasses; several species are quite widespread in Britain.

The subfamily Orthotelinae was previously placed in the Yponomeutidae.

Identification: Pelham-Clinton (1985d) and Agassiz (1996) for *Orthotelia* (under Yponomeutidae); Manley (2009) shows two species of *Glyphipterix*.

British subfamilies and genera:

Glyphipteriginae: *Glyphipterix*
Orthoteliinae: *Orthotelia*

Family Heliodinidae (1 genus, 1 species)
The only British species in this family is *Heliodines roesella*; it has not been recorded since the 19th century and is probably extinct here. The larvae feed gregariously in spun leaves of Chenopodiaceae.

Identification: Emmet (1985d).

British genus: *Heliodines*

Family Lyonetiidae (2 genera, 10 species)
Most members of this small family have white wings marked with several yellowish brown streaks near the apex; the fore wing length is 2–4 mm. At least two species can form large populations on fruit trees, the best known being *Lyonetia clerkella*, the Apple leaf-miner (Fig. 21.76). The larvae form long, narrow and sinuous mines (Fig. 21.77); pupation takes place in a cocoon that is suspended on silk

Fig. 21.76 *Lyonetia clerkella*, **Apple leaf miner (Lyonetiidae) (Photo: Roger Key)**

Fig. 21.77 *Lyonetia clerkella* mines on apple leaf (Lyonetiidae) (Photo: Peter Barnard)

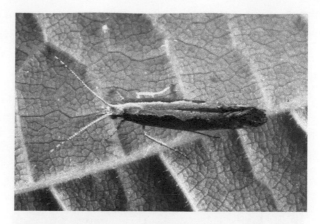

Fig. 21.79 *Plutella xylostella*, Diamond-back moth (Plutellidae) (Photo: Roger Key)

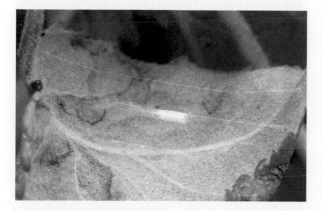

Fig. 21.78 *Lyoneta clerkella* cocoon on apple leaf (Lyonetiidae) (Photo: Peter Barnard)

Fig. 21.80 *Ypnomeuta evonymella* (Yponomeutidae) (Photo: Peter Barnard)

threads across a leaf, rather like a hammock (Fig. 21.78).

Identification: Emmet (1985b); Manley (2009) shows *Lyonetia clerkella*, under Tineidae.

British subfamilies and genera:
  Cemiostominae: *Leucoptera*
  Lyonetiinae: *Lyonetia*

Family Plutellidae (3 genera, 7 species)
Most members of this small family have wavy markings on the wings that look particularly prominent when the wings are held together at rest; the best-known is *Plutella xylostella*, the Diamond-back moth (Fig. 21.79). The larvae of most species feed in a web on crucifers and *P. xylostella* can be an occasional pest on brassica crops, especially when native populations are augmented by immigrants.

Identification: Agassiz (1996) under Yponomeutidae; Manley (2009) illustrates two species of *Plutella* under Yponomeutidae.

British genera: *Eidophasia, Plutella, Rhigognostis*

Family Yponomeutidae (13 genera, 54 species)
Some of the common species in this family can be quite large, with a fore wing length up to 12 mm; species of *Yponomeuta* are usually white with numerous black spots (Fig. 21.80) and they can be serious pests on certain trees. On occasions the communal webs spun by huge numbers of larvae can completely cover rows of trees (Fig. 21.81); the larvae are densely congregated within the web (Fig. 21.82) and pupation also occurs there (Fig. 21.83). Although the trees are completely defoliated they often recover with a second flush of leaves. There are still some taxonomic questions about species complexes in this family and it is not always possible to be sure of specific identifications.

Fig. 21.81 *Ypnonomeuta* **larval webs on cherry trees (Yponomeutidae) (Photo: Peter Barnard)**

Fig. 21.82 *Yponomeuta* **larvae in web (Yponomeutidae) (Photo: Peter Barnard)**

Fig. 21.83 *Yponomeuta* **larvae and pupae (Yponomeutidae) (Photo: Peter Barnard)**

Identification: Agassiz (1996); Manley (2009) shows some common species.

British subfamilies and genera:
  Argyresthiinae: *Argyresthia*
  Praydinae: *Atemelia*, *Prays*
  Scythropiinae: *Scythropia*
  Yponomeutinae: *Cedestis*, *Euhyponomeuta*, *Kessleria*, *Ocnerostoma*, *Paraswammerdamia*, *Pseudoswammerdamia*, *Swammerdamia*, *Yponomeuta*, *Zelleria*

Family Ypsolophidae (2 genera, 16 species)
The subfamily Ochsenheimeriinae was previously treated as a distinct family, while the Ypsolophinae were included in the Yponomeutidae. Many species of Ypsolophinae have a falcate wing-tip, and these 'hooks' are clearly visible when the moth is at rest. The larvae feed on various trees and shrubs; several species are quite common and widespread in Britain.

Identification: Emmet (1985a) for the Ochsenheimeriinae, and Agassiz (1996) for *Ypsolopha*; Manley (2009) illustrates a few species of *Ypsolopha*.

British subfamilies and genera:
  Ochsenheimeriinae: *Ochsenheimeria*
  Ypsolophinae: *Ypsolopha*

SUPERFAMILY ZYGAENOIDEA

Family Limacodidae (2 genera, 2 species)
There are just two British species in this family, *Apoda limacodes*, the Festoon; and *Heterogenea asella*, the Triangle. They are quite small species, with fore wing lengths of 7–9 mm in the males, but up to 12 mm in the females. The adult moths rest with their abdomens curled up between the wings. The larvae feed on trees and are rather slug-like; they give the family their name (from *Limax*, the Latin for a slug).

Identification: Skinner (1985b); Waring and Townsend (2009); Manley (2009); Skinner (2009).

British genera: *Apoda*, *Heterogenea*

Family Zygaenidae (3 genera, 10 species)
The two subfamilies are quite distinct: the Zygaeninae are the burnet moths belonging to the genus *Zygaena*, with their familiar red and black patterns (Fig. 21.84), while the Procridinae are the forester moths, which are metallic green (Fig. 21.85). They are all day-flying moths and although a few species are common, many are very restricted in distribution and are considered under threat. Zygaenid larvae feed on various low-growing

Fig. 21.84 *Zygaena trifolii*, Five-spot burnet moth (Zygaenidae) (Photo: Roger Key)

Fig. 21.86 *Micropterix calthella* (Micropterigidae) (Photo: Colin Rew)

Fig. 21.85 *Adscita geryon*, Cistus forester moth (Zygaenidae) (Photo: Roger Key)

plants such as trefoils and vetches, but each species has specific requirements.

Zygaena viciae (New Forest burnet) is listed on the Wildlife and Countryside Act 1981. *Adscita statices* (The forester), *Zygaena loti* subsp. *scotica* (Slender Scotch burnet) and *Zygaena viciae* subsp. *argyllensis* (New Forest burnet) are on the UKBAP list.

Identification: Tremewan (1985); Waring and Townsend (2009); Manley (2009); Skinner (2009).

British subfamilies and genera:
Procridinae: *Adscita*, *Jordanita*
Zygaeninae: *Zygaena*

SUBORDER ZEUGLOPTERA

SUPERFAMILY MICROPTERIGOIDEA

This suborder is considered the most primitive group of Lepidoptera; the adult moths have retained functional mandibles and feed on pollen.

Family Micropterigidae (1 genus, 5 species)
All five British species are placed in the genus *Micropterix*. These are small moths with pointed wings that are often metallic or purple; the fore wing length is 2–5 mm. Adults are often seen in the spring feeding on pollen (Fig. 21.86) and several species are quite common and widespread. The larvae have not all been described but they seem to feed on fungal hyphae in leaf-litter.

Identification: Heath (1976b); Manley (2009) illustrates three species.

British genus: *Micropterix*

REFERENCES

AGASSIZ, D.J.L. 1985. Douglasiidae. In: HEATH, J. & EMMET, A.M. (eds.) *The moths and butterflies of Great Britain and Ireland*. 2. Cossidae – Heliodinidae. Harley Books, Colchester, pp. 408–9.

AGASSIZ, D.J.L. 1996. Yponomeutidae (including Roeslerstammiidae). In: EMMET, A.M. (ed.) *The moths and butterflies of Great Britain and Ireland*. 3. Yponomeutidae – Elachistidae. Harley Books, Colchester, pp. 39–114.

ASHER, J., WARREN, M., FOX, R., HARDING, P., JEFFCOATE, G. & JEFFCOATE, S. 2001. *The millennium atlas of butterflies in Britain & Ireland*. Oxford University Press, Oxford.

BAKER, B.R. 1985. Sesiidae. In: HEATH, J. & EMMET, A.M. (eds.) *The moths and butterflies of Great Britain and Ireland.* 2. Cossidae – Heliodinidae. Harley Books, Colchester, pp. 369–88.

BEIRNE, B.P. 1952. *British pyralid and plume moths.* Wayside & Woodland Series, Warne, London [Also available on CD-ROM from Pisces Conservation].

BENGTSSON, B.Å. 2002. Scythrididae. In: EMMET, A.M. & LANGMAID, J.R. (eds.) *The moths and butterflies of Great Britain and Ireland.* 4(1). Oecophoridae – Scythrididae. Harley Books, Colchester, pp. 278–94.

BLAND, K.P. 1996. Elachistidae. In: EMMET, A.M. (ed.) *The moths and butterflies of Great Britain and Ireland.* 3. Yponomeutidae – Elachistidae. Harley Books, Colchester, pp. 339–410.

BLAND, K.P. 2002. Autostichidae. In: EMMET, A.M. & LANGMAID, J.R. (eds.) *The moths and butterflies of Great Britain and Ireland.* 4(1). Oecophoridae – Scythrididae. Harley Books, Colchester, pp. 188–95.

BRADLEY, J.D. 2000. *Checklist of Lepidoptera recorded from the British Isles.* Privately published.

BRADLEY, J.D., TREMEWAN, W.G. & SMITH, A. 1973. British tortricoid moths. Cochylidae and Tortricidae: Tortricinae. Ray Society, London [Also available on CD-ROM from Pisces Conservation].

BRADLEY, J.D., TREMEWAN, W.G. & SMITH, A. 1979. British tortricoid moths. Tortricidae: Olethreutinae. Ray Society, London [Also available on CD-ROM from Pisces Conservation].

BRETHERTON, R.F., GOATER, B. & LORIMER, R.I. 1979. Noctuidae: Noctuinae and Hadeninae. In: HEATH, J. & EMMET, A.M. (eds.) *The moths and butterflies of Great Britain and Ireland.* 9. Sphingidae–Noctuidae (Noctuinae and Hadeninae). Harley Books, Colchester, pp. 120–280.

BUCKLER, W. 1886–1901. *The larvae of the British butterflies and moths.* Ray Society, London. Volumes 1–9: 1, *The butterflies* (1886); 2, *The Sphinges or hawk-moths and part of the Bombyces* (1887); 3, *The concluding portion of the Bombyces* (1889); 4, *The first portion of the Noctuae* (1891); 5, *The second portion of the Noctuae* (1893); 6, *The third and concluding portion of the Noctuae* (1895); 7, *The first portion of the Geometrae* (1897); 8, *The concluding portion of the Geometrae* (1899); 9, *The Deltoides, Pyrales, Crambites, Tortrices, Tineae and Pterophori, concluding the work* (1901) [Also available on CD-ROM from Pisces Conservation].

CARTER, D.J. & HARGREAVES, B. 1986. *A field guide to caterpillars of butterflies and moths in Britain and Europe.* Collins, London [reprinted in 2001 as *Collins field guide to caterpillars of Britain and Europe*].

CHALMERS-HUNT, J.M. 1989. *Local lists of Lepidoptera or a bibliographical catalogue of local lists and regional accounts of the butterflies and moths of the British Isles.* Hedera Press, Uffington.

DENNIS, R.L.H. 1977. *The British butterflies: their origin and establishment.* Classey, Faringdon.

DENNIS, R.L.H. 1992. *The ecology of butterflies in Britain.* Oxford University Press, Oxford.

DENNIS, R.L.H. & SHREEVE, T.G. 1996. *Butterflies on British and Irish offshore islands: ecology and landscape.* Gem Publishing Company, Wallingford.

DE WORMS, C.G.M. 1979a. Notodontidae. In: HEATH, J. & EMMET, A.M. (eds) *The moths and butterflies of Great Britain and Ireland.* 9. Sphingidae–Noctuidae (Noctuinae and Hadeninae). Harley Books, Colchester, pp. 39–65.

DE WORMS, C.G.M. 1979b. Lymantriidae. In: HEATH, J. & EMMET, A.M. (eds) *The moths and butterflies of Great Britain and Ireland.* 9. Sphingidae–Noctuidae (Noctuinae and Hadeninae). Harley Books, Colchester, pp. 66–78.

DE WORMS, C.G.M. 1979c. Arctiidae. In: HEATH, J. & EMMET, A.M. (eds) *The moths and butterflies of Great Britain and Ireland.* 9. Sphingidae–Noctuidae (Noctuinae and Hadeninae). Harley Books, Colchester, pp. 78–111.

DICKSON, R. 1992. *A lepidopterist's handbook* (2nd edn.). Amateur Entomologist's Society, London.

DICKSON, R.J. 2002. Blastobasidae. In: EMMET, A.M. & LANGMAID, J.R. (eds.) *The moths and butterflies of Great Britain and Ireland.* 4(1). Oecophoridae – Scythrididae. Harley Books, Colchester, pp. 196–203.

EMMET, A.M. 1976a. Nepticulidae. In: HEATH, J. (ed.) *The moths and butterflies of Great Britain and Ireland.* 1. Micropterigidae – Heliozelidae. Curwen Press, London, pp. 171–267.

EMMET, A.M. 1976b. Tischeriidae. In: HEATH, J. (ed.) *The moths and butterflies of Great Britain and Ireland.* 1. Micropterigidae – Heliozelidae. Curwen Press, London, pp. 272–6.

EMMET, A.M. 1976c. Heliozelidae. In: HEATH, J. (ed.) *The moths and butterflies of Great Britain and Ireland.* 1. Micropterigidae – Heliozelidae. Curwen Press, London, pp. 300–6.

EMMET, A.M. 1985a. Ochsenheimeriidae. In: HEATH, J. & EMMET, A.M. (eds.) *The moths and butterflies of Great Britain and Ireland.* 2. Cossidae – Heliodinidae. Harley Books, Colchester, pp. 208–12.

EMMET, A.M. 1985b. Lyonetiidae. In: HEATH, J. & EMMET, A.M. (eds.) *The moths and butterflies of Great Britain and Ireland.* 2. Cossidae – Heliodinidae. Harley Books, Colchester, pp. 212–39.

EMMET, A.M. 1985c. Phyllocnistidae. In: HEATH, J. & EMMET, A.M. (eds.) *The moths and butterflies of Great Britain and Ireland.* 2. Cossidae – Heliodinidae. Harley Books, Colchester, pp. 363–8.

EMMET, A.M. 1985d. Heliodinidae. In: HEATH, J. & EMMET, A.M. (eds.) *The moths and butterflies of Great Britain and Ireland.* 2. Cossidae – Heliodinidae. Harley Books, Colchester, pp. 410–11.

EMMET, A.M. (ed.) 1996a. *The moths and butterflies of Great Britain and Ireland.* 3. Yponomeutidae – Elachistidae. Harley Books, Colchester.

EMMET, A.M. 1996b. Schreckensteiniidae. In: EMMET, A.M. (ed.) *The moths and butterflies of Great Britain and Ireland.* 3. Yponomeutidae – Elachistidae. Harley Books, Colchester, pp. 123–5.

EMMET, A.M. & HEATH, J. (eds.) 1989. *The moths and butterflies of Great Britain and Ireland.* 7(1). Hesperiidae – Nymphalidae. Harley Books, Colchester.

EMMET, A.M. & HEATH, J. (eds.) 1991. *The moths and butterflies of Great Britain and Ireland.* 7(2). Lasiocampidae – Thyatiridae. Harley Books, Colchester.

EMMET, A.M. & LANGMAID, J.R. (eds.) 2002a. *The moths and butterflies of Great Britain and Ireland.* 4(1). Oecophoridae – Scythrididae. Harley Books, Colchester.

EMMET, A.M. & LANGMAID, J.R. (eds.) 2002b. *The moths and butterflies of Great Britain and Ireland.* 4(2). Gelechiidae. Harley Books, Colchester.

EMMET, A.M., LANGMAID, J.R., BLAND, K.P., CORLEY, M.F.V. & RAZOWSKI, J. 1996. Coleophoridae. In: EMMET, A.M. (ed.) *The moths and butterflies of Great Britain and Ireland.* 3. Yponomeutidae – Elachistidae. Harley Books, Colchester, pp. 126–338.

EMMET, A.M., WATKINSON, I.A. & WILSON, M.R. 1985. Gracillariidae. In: HEATH, J. & EMMET, A.M. (eds.) *The moths and butterflies of Great Britain and Ireland.* 2. Cossidae – Heliodinidae. Harley Books, Colchester, pp. 244–363.

FELTWELL, J. 1995. *The conservation of butterflies in Britain, past and present.* Wildlife Matters, Battle.

FORD, E.B. 1945. *Butterflies.* Collins New Naturalist, London.

FORD, E.B. 1972. *Moths* (3rd edn.). Collins New Naturalist, London.

FOX, R., ASHER, J., BRERETON, T., ROY, D. & WARREN, M. 2006. *The state of butterflies in Britain and Ireland.* Pisces Publications, Oxford.

GIELIS, C. 1996. Pterophoridae. In: HUEMER, P., KARSHOLT, O. & LYNEBORG, L. (eds.) *Microlepidoptera of Europe* 1: 222 pp.

GILCHRIST, W.L.R.E. 1979. Sphingidae. In: HEATH, J. & EMMET, A.M. (eds.) *The moths and butterflies of Great Britain and Ireland.* 9. Sphingidae – Noctuidae (Noctuinae and Hadeninae). Harley Books, Colchester, pp. 20–39.

GOATER, B. 1986. *British pyralid moths – a guide to their identification.* Harley Books, Colchester.

GOATER, B. 1991a. Lasiocampidae. In: EMMET, A.M. & HEATH, J. (eds.) *The moths and butterflies of Great Britain and Ireland.* 7(2). Lasiocampidae – Thyatiridae. Harley Books, Colchester, pp. 306–23.

GOATER, B. 1991b. Saturniidae. In: EMMET, A.M. & HEATH, J. (eds.) *The moths and butterflies of Great Britain and Ireland.* 7(2). Lasiocampidae – Thyatiridae. Harley Books, Colchester, pp. 324–6.

GOATER, B. 1991c. Drepanidae. In: EMMET, A.M. & HEATH, J. (eds.) *The moths and butterflies of Great Britain and Ireland.* 7(2). Lasiocampidae – Thyatiridae. Harley Books, Colchester, pp. 331–9.

GOATER, B. 1991d. Thyatiridae. In: EMMET, A.M. & HEATH, J. (eds.) *The moths and butterflies of Great Britain and Ireland.* 7(2). Lasiocampidae – Thyatiridae. Harley Books, Colchester, pp. 340–54.

GODFRAY, H.C.J. & STERLING, P.H. 1996. Epermeniidae. In: EMMET, A.M. (ed.) *The moths and butterflies of Great Britain and Ireland.* 3. Yponomeutidae – Elachistidae. Harley Books, Colchester, pp. 115–23.

GRIMALDI, D. & ENGEL, M.S. 2005. *Evolution of the insects.* Cambridge University Press, Cambridge.

HAGGETT, G.M. 1981. *Larvae of the British Lepidoptera not figured by Buckler.* British Entomological & Natural History Society, London.

HARDING, P.T. & GREEN, S.V. 1991. *Recent surveys and research on butterflies in Britain and Ireland: a species index and bibliography.* Biological Records Centre, Abbots Ripton.

HARPER, M.W., LANGMAID, J.R. & EMMET, A.M. 2002. Oecophoridae. In: EMMET, A.M. & LANGMAID, J.R. (eds.) *The moths and butterflies of Great Britain and Ireland.* 4(1). Oecophoridae – Scythrididae. Harley Books, Colchester, pp. 43–177.

HÄTTENSCHWILER, P. 1985. Psychidae. In: HEATH, J. & EMMET, A.M. (eds.) *The moths and butterflies of Great Britain and Ireland.* 2. Cossidae – Heliodinidae. Harley Books, Colchester, pp. 128–51.

HEATH, J. (ed.) 1976a. *The moths and butterflies of Great Britain and Ireland.* 1. Micropterigidae – Heliozelidae. Curwen Press, London.

HEATH, J. 1976b. Micropterigidae. In: HEATH, J. (ed.) *The moths and butterflies of Great Britain and Ireland.* 1. Micropterigidae – Heliozelidae. Curwen Press, London, pp. 150–5.

HEATH, J. 1976c. Eriocraniidae. In: HEATH, J. (ed.) *The moths and butterflies of Great Britain and Ireland.* 1. Micropterigidae – Heliozelidae. Curwen Press, London, pp. 156–65.

HEATH, J. 1976d. Hepialidae. In: HEATH, J. (ed.) *The moths and butterflies of Great Britain and Ireland.* 1. Micropterigidae – Heliozelidae. Curwen Press, London, pp. 166–70.

HEATH, J. 1979a. Thaumetopoeidae. In: HEATH, J. & EMMET, A.M. (eds.) *The moths and butterflies of Great Britain and Ireland.* 9. Sphingidae – Noctuidae (Noctuinae and Hadeninae). Harley Books, Colchester, pp. 65–6.

HEATH, J. 1979b. Ctenuchidae. In: HEATH, J. & EMMET, A.M. (eds.) *The moths and butterflies of Great Britain and Ireland.* 9. Sphingidae – Noctuidae (Noctuinae and Hadeninae). Harley Books, Colchester, pp. 111–12.

HEATH, J. & EMMET, A.M. (eds.) 1979. *The moths and butterflies of Great Britain and Ireland.* 9. Sphingidae – Noctuidae (Noctuinae and Hadeninae). Harley Books, Colchester.

HEATH, J. & EMMET, A.M. (eds.) 1983. *The moths and butterflies of Great Britain and Ireland.* 10. Noctuidae (Cuculliinae to Hypeninae) and Agaristidae. Harley Books, Colchester.

HEATH, J. & EMMET, A.M. (eds.) 1985. *The moths and butterflies of Great Britain and Ireland.* 2. Cossidae – Heliodinidae. Harley Books, Colchester.

HEATH, J. & PELHAM-CLINTON, E.C. 1976. Incurvariidae. In: HEATH, J. (ed.) *The moths and butterflies of Great Britain and Ireland.* 1. Micropterigidae – Heliozelidae. Curwen Press, London, pp. 277–300.

HIGGINS, L.G. & RILEY, N.D. 1970. *A field guide to the butterflies of Britain and Europe.* Collins, London.

HOWARTH, T.G. 1984. *Colour identification guide to butterflies of the British Isles* (revised edn.). Viking, Harmondsworth.

KARSHOLT, O. & RAZOWSKI, J. (eds.) 1996. *The Lepidoptera of Europe: a distributional checklist.* Apollo Books, Stenstrup.

KLOET, G.S. & HINCKS, W.D. 1972. A check-list of British Insects. 2. Lepidoptera (2nd edn.). *Handbooks for the identification of British insects* 11(2): 153 pp.

KOSTER, J.C. 2002a. Batrachedridae. In: EMMET, A.M. & LANGMAID, J.R. (eds.) *The moths and butterflies of Great Britain and Ireland*. 4(1). Oecophoridae – Scythrididae. Harley Books, Colchester, pp. 204–10.

KOSTER, J.C. 2002b. Agonoxenidae. In: EMMET, A.M. & LANGMAID, J.R. (eds.) *The moths and butterflies of Great Britain and Ireland*. 4(1). Oecophoridae – Scythrididae. Harley Books, Colchester, pp. 211–23.

KOSTER, J.C. 2002c. Momphidae. In: EMMET, A.M. & LANGMAID, J.R. (eds.) *The moths and butterflies of Great Britain and Ireland*. 4(1). Oecophoridae – Scythrididae. Harley Books, Colchester, pp. 224–54.

KOSTER, J.C. 2002d. Cosmopterigidae. In: EMMET, A.M. & LANGMAID, J.R. (eds.) *The moths and butterflies of Great Britain and Ireland*. 4(1). Oecophoridae – Scythrididae. Harley Books, Colchester, pp. 255–78.

LEVERTON, R. 2001. *Enjoying moths*. Poyser Natural History, London.

MAJERUS, M. 2002. *Moths*. HarperCollins, New Naturalist, London.

MANLEY, C. 2009. *British moths and butterflies: a photographic guide* (revised edn.). A & C Black, London, 352 pp.

MAY, P.R. 2003. *Larval foodplants of the butterflies of Great Britain and Ireland*. Amateur Entomologists' Society, London.

MEYRICK, E. 1928. *A revised handbook of British Lepidoptera*. Watkins & Doncaster, London.

PARENTI, U. 2000. *A guide to the microlepidoptera of Europe*. Museo Regionale di Scienze Naturali, Turin.

PARSONS, M.S. 1993. *A review of the scarce and threatened pyralid moths of Great Britain*. Joint Nature Conservation Committee, Peterborough (UK Nature Conservation no. 11, 97 pp.).

PARSONS, M.S. 1996. *A review of the scarce and threatened ethmiine, stathmopodine and gelechiid moths of Great Britain*. Joint Nature Conservation Committee, Peterborough (UK Nature Conservation no. 16, 130 pp.).

PARSONS, M.S., ROBINSON, G.S., HONEY, M.R. & CARTER, D.J. 1999. Lepidoptera: the moths and butterflies. In: BARNARD, P.C. (ed.) *Identifying British insects and arachnids: an annotated bibliography of key works*. Cambridge University Press, Cambridge, pp. 145–70.

PELHAM-CLINTON, E.C. 1976. Opostegidae. In: HEATH, J. (ed.) *The moths and butterflies of Great Britain and Ireland*. 1. Micropterigidae – Heliozelidae. Curwen Press, London, pp. 268–71.

PELHAM-CLINTON, E.C. 1985a. Tineidae. In: HEATH, J. & EMMET, A.M. (eds.) *The moths and butterflies of Great Britain and Ireland*. 2. Cossidae – Heliodinidae. Harley Books, Colchester, pp. 152–207.

PELHAM-CLINTON, E.C. 1985b. Hieroxestidae. In: HEATH, J. & EMMET, A.M. (eds.) *The moths and butterflies of Great Britain and Ireland*. 2. Cossidae – Heliodinidae. Harley Books, Colchester, pp. 240–3.

PELHAM-CLINTON, E.C. 1985c. Choreutidae. In: HEATH, J. & EMMET, A.M. (eds.) *The moths and butterflies of Great Britain and Ireland*. 2. Cossidae – Heliodinidae. Harley Books, Colchester, pp. 389–99.

PELHAM-CLINTON, E.C. 1985d. Glyphipterygidae. In: HEATH, J. & EMMET, A.M. (eds.) *The moths and butterflies of Great Britain and Ireland*. 2. Cossidae – Heliodinidae. Harley Books, Colchester, pp. 400–7.

POLLARD, E. & YATES, T.J. 1993. *Monitoring butterflies for ecology and conservation*. Chapman & Hall. London.

PORTER, J. 1997. *The colour identification guide to caterpillars of the British Isles*. Viking, Harmondsworth [Reprinted 2010].

POWELL, J.A. 2009. Lepidoptera (moths, butterflies). In: RESH, V.H. & CARDÉ, R.T. (eds.) *Encyclopedia of insects* (2nd edn.). Academic Press/Elsevier, San Diego & London, pp. 559–87.

PULLIN, A.S. 1995. *Ecology and conservation of Butterflies*. Chapman & Hall, London.

RAZOWSKI, J. 2002. *Tortricidae of Europe. Vol. 1: Tortricinae and Chlidanotinae*. Frantisek Slamka, Bratislava.

RAZOWSKI, J. 2003. Tortricidae of Europe. Vol. 2: Olethreutinae. Frantisek Slamka, Bratislava.

REVELL, R.J. 1979. Nolidae. In: HEATH, J. & EMMET, A.M. (eds.) *The moths and butterflies of Great Britain and Ireland*. 9. Sphingidae – Noctuidae (Noctuinae and Hadeninae). Harley Books, Colchester, pp. 112–20.

RILEY, A.M. 2007. *British and Irish butterflies: the complete identification, field and site guide to the species, subspecies and forms*. Brambleby Books, Luton, 352 pp.

ROBINSON, G.S. 2009. *Biology, distribution and diversity of tineid moths*. Southdene Sdn Bhd, Kuala Lumpur, and Natural History Museum, London.

SATTLER, K. 2002. Ethmiidae. In: EMMET, A.M. & LANGMAID, J.R. (eds.) *The moths and butterflies of Great Britain and Ireland*. 4(1). Oecophoridae – Scythrididae. Harley Books, Colchester, pp. 178–87.

SCHINTLMEISTER, A. 2008. Notodontidae. *Palaearctic Macrolepidoptera*, vol. 1. Apollo Books, Stenstrup.

SCOBLE, M.J. 1995. *The Lepidoptera: form, function and diversity*. Oxford University Press, Oxford.

SKINNER, B. 1985a. Cossidae. In: HEATH, J. & EMMET, A.M. (eds.) *The moths and butterflies of Great Britain and Ireland*. 2. Cossidae – Heliodinidae. Harley Books, Colchester, pp. 69–74.

SKINNER, B. 1985b. Limacodidae. In: HEATH, J. & EMMET, A.M. (eds.) *The moths and butterflies of Great Britain and Ireland*. 2. Cossidae – Heliodinidae. Harley Books, Colchester, pp. 124–7.

SKINNER, B. 2009. *Colour identification guide to moths of the British Isles* (3rd edn.). Apollo Books, Stenstrup.

SOKOLOFF, P. 1980. *Practical hints for collecting and studying microlepidoptera*. Amateur Entomologist's Society, London.

STOKOE, W.J. & STOVIN, G.H.T. 1948. *The caterpillars of British moths*. 2 vols. Warne, London.

THOMAS, J.A. 1986. *Butterflies of the British Isles*. Hamlyn, London.

THOMAS, J. & LEWINGTON, R. 2010. *The butterflies of Britain and Ireland*. British Wildlife Publishing, Milton on Stour.

TREMEWAN, W.H. 1985. Zygaenidae. In: HEATH, J. & EMMET, A.M. (eds.) *The moths and butterflies of Great Britain and Ireland*. 2. Cossidae – Heliodinidae. Harley Books, Colchester, pp. 74–123.

WARING, P. & TOWNSEND, M. 2009. *Field guide to the moths of Great Britain and Ireland* (2nd edn.). British Wildlife Publishing, Milton on Stour.

WHALLEY, P.E.S. 1980. *Butterfly watching*. Severn House Publishers, London.

YOUNG, M.R. 1991. Endromidae. In: EMMET, A.M. & HEATH, J. (eds.) *The moths and butterflies of Great Britain and Ireland*. 7(2). Lasiocampidae – Thyatiridae. Harley Books, Colchester, pp. 327–30.

YOUNG, M. 1997. *The natural history of moths*. Poyser, London.

WEBSITES

http://www.ukbutterflies.co.uk/index.php
Covers all the British butterflies.

http://ukmoths.org.uk/
A well-maintained site with photos of many British species from all families.

http://www.wildguideuk.com/lepidoptera.htm
This site is useful for photos of the smaller moths.

http://www.nhm.ac.uk/research-curation/research/projects/hostplants/
A world database of Lepidoptera and their host plants.

http://www.leafmines.co.uk/index.htm
The British leaf-miners of all insect groups.

22 Order Mecoptera: the scorpionflies

4 species in 2 families

Traditionally the Mecoptera have been regarded as closely related to the Diptera and, although this is still accepted, there is increasing evidence that at least one family, the Boreidae, has a closer relationship with the Siphonaptera. If this is true, then it might have implications for the monophyly of the Mecoptera, or else it might mean that the Siphonaptera would be reduced to a subgroup; see Grimaldi and Engel (2005) for a discussion of these issues.

For the time being, the Mecoptera are an easily recognized group, characterized by the elongate ventrally pointing rostrum or 'beak', which has biting and chewing mouthparts at its tip. Both the adult and larval stages are scavengers on a variety of dead animals, usually insects.

Identification of this group is easily carried out with the key by Plant (1997), and notes on the distribution of the British species are given in the atlas by Plant (1994).

There are 23 species of Mecoptera in three families in Europe (with many subspecies), and around 600 species in nine families worldwide. A brief world overview of the Mecoptera is given by Byers (2009).

HIGHER CLASSIFICATION OF BRITISH MECOPTERA

Family Boreidae (1 genus, 1 species)
Family Panorpidae (1 genus, 3 species)

SPECIES OF CONSERVATION CONCERN

No species of Mecoptera have any legal protection and none is on the UKBAP list.

The Families of British Mecoptera

Family Boreidae (1 genus, 1 species)
The only British species in this family is *Boreus hyemalis*, often known as the snow flea. It is a small insect, around 5 mm long, and is not often seen though it is probably much more common than current records suggest. Both sexes are flightless; in the females the wings are reduced to tiny scales, but in the males they are modified into spine-like structures. Females have a long ovipositor (Fig. 22.1). The adults and larvae live amongst moss on moors and heathland where they scavenge on small dead invertebrates, and can be captured in pitfall traps in such areas; otherwise they are hard to find. Adult *Boreus* are unusual among British insects in being present over the winter months, from around October to April; they can be active even when snow is on the ground, and this is when they are most likely to be seen by the casual observer (Fig. 22.2).

British genus: *Boreus*

Family Panorpidae (1 genus, 3 species)
The members of this family have the common name of scorpion flies (though the name is often applied to the Mecoptera as a whole); this is because the enlarged genital capsule of the male is held recurved over the body, rather like a scorpion's sting

The Royal Entomological Society Book of British Insects, First Edition. Peter C. Barnard.
© 2011 Royal Entomological Society. Published 2011 by Blackwell Publishing Ltd.

Fig. 22.1 *Boreus hyemalis* on moss (Boreidae) (Photo: Roger Key)

Fig. 22.4 *Panorpa* female (Panorpidae) (Photo: Roger Key)

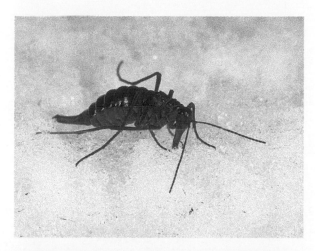

Fig. 22.2 *Boreus hyemalis* on snow (Boreidae) (Photo: Roger Key)

(Fig. 22.3). The female's abdomen ends in a straight point (Fig. 22.4). All three British species are found in rough grassland and woodland margins, often among brambles; the adults are not strong fliers and are usually caught by beating or sweeping suitable vegetation, though they often simply fold their wings and drop to the ground when disturbed. On current evidence it seems that *Panorpa communis* and *P. germanica* are widespread in Britain, and that *P. cognata* is more restricted in distribution, but more records are still needed.

It used to be thought that the distinctive wing-markings in this family were sufficient to identify species (Fraser, 1959), but it is now known that these are not reliable, particularly because they seem to vary in intensity in different parts of the country. Modern keys such as Plant (1997) use genitalic characters exclusively.

British genus: *Panorpa*

REFERENCES

Byers, G.W. 2009. Mecoptera. In: Resh, V.H. & Cardé, R.T. (eds.) *Encyclopedia of insects* (2nd edn.). Academic Press/Elsevier, San Diego & London, pp. 611–14.

Fraser, F.C. 1959. Mecoptera, Megaloptera and Neuroptera. *Handbooks for the identification of British insects* 1(12, 13): 40 pp.

Grimaldi, D. & Engel, M.S. 2005. *Evolution of the insects*. Cambridge University Press, Cambridge.

Plant, C.W. 1994. *Provisional atlas of the lacewings and allied insects (Neuroptera, Megaloptera, Raphidioptera and Mecoptera) of Britain and Ireland*. Biological Records Centre, Huntingdon.

Plant, C.W. 1997. *A key to the adults of British lacewings and their allies (Neuroptera, Megaloptera, Raphidioptera and*

Fig. 22.3 *Panorpa* male (Panorpidae) (Photo: Roger Key)

Mecoptera). Field Studies Council (AIDGAP) no. 245. 90 pp.

WEBSITES

http://nbi.cucaera.co.uk/
A useful overview of the British species, with distribution records.

http://www.brc.ac.uk/recording_schemes.asp
The home page of the Biological Records Centre contains details of the Neuropterida Recording Scheme, which includes the Mecoptera.

http://researcharchive.calacademy.org/research/ entomology/Entomology_Resources/mecoptera
A world checklist of Mecoptera.

23

Order Megaloptera: the alderflies

3 species in 1 family

In early texts the Megaloptera were included within the Neuroptera, together with the Raphidioptera, but they have long been recognized as a distinct order. The monophyly of this group of three orders, sometimes termed the Neuropterida, is not in question, but the exact relationships between them are yet to be fully resolved; Grimaldi and Engel (2005) discuss the current evidence.

Adult Megaloptera have broad flattened heads with biting mouthparts, and two pairs of very similar wings; all the larvae are freshwater predators. The three British species can be readily identified using Plant's (1997) or Elliott's (2009) keys, and distributional notes are in Plant's (1994) atlas.

There are 10 species in 1 family in Europe, with around 330 species in 2 families worldwide. A brief world overview of the Megaloptera is given by Anderson (2009).

HIGHER CLASSIFICATION OF BRITISH MEGALOPTERA

All three British species are in the family Sialidae.

SPECIES OF CONSERVATION CONCERN

No species of Megaloptera have any legal protection and none is on the UKBAP list.

Family Sialidae (1 genus, 3 species)

These are wide-bodied insects with broad heads, and the body and wings are dark (Fig. 23.1). They fly rather slowly though readily, in sunshine, and often rest on trees overhanging the water; this is presumably where their common name of alderflies has its origin, though there is no direct link with alder trees. They are also commonly seen on reed-banks at the side of freshwater ponds and streams. Adults are often found with pollen on their heads and antennae, suggesting that they may visit flowers to feed (see Fig. 23.1). The larvae live in the silt at the bottom of suitable water-bodies, where they prey on other small invertebrates, the younger instars feeding on micro-organisms and crustaceans, and the larger ones on oligochaete worms and chironomid larvae. They have ten larval instars and the lifecycle is usually two years long. The larvae have a single, long terminal filament, with lateral filaments on the abdominal segments; like the adult they have a broad, flattened head with biting mouthparts. Fully grown larvae leave the water in the spring to pupate in damp soil nearby.

Most earlier texts on the British alderflies mention only *Sialis lutaria* and *S. fuliginosa*, but a third species (*S. nigripes*) was recognized in 1977; re-examination of museum specimens showed that it had been overlooked for over a century. This was because previous keys used features of the wing venation, which are now known to be unreliable.

British genus: *Sialis*

The Royal Entomological Society Book of British Insects, First Edition. Peter C. Barnard.
© 2011 Royal Entomological Society. Published 2011 by Blackwell Publishing Ltd.

Fig. 23.1 *Sialis lutaria* (Sialidae) (Photo: Roger Key)

REFERENCES

ANDERSON, N.H. 2009. Megaloptera. In: Resh, V.H. & Cardé, R.T. (eds.) *Encyclopedia of insects* (2nd edn.). Academic Press/Elsevier, San Diego & London, pp. 620–3.

ELLIOTT, J.M. 2009. Freshwater Megaloptera and Neuroptera of Britain and Ireland. *Scientific Publications of the Freshwater Biological Association* 65: 71 pp.

GRIMALDI, D. & ENGEL, M.S. 2005. *Evolution of the insects.* Cambridge University Press, Cambridge.

PLANT, C.W. 1994. *Provisional atlas of the lacewings and allied insects (Neuroptera, Megaloptera, Raphidioptera and Mecoptera) of Britain and Ireland.* Biological Records Centre, Huntingdon.

PLANT, C.W. 1997. *A key to the adults of British lacewings and their allies (Neuroptera, Megaloptera, Raphidioptera and Mecoptera).* Field Studies Council (AIDGAP) no. 245. 90 pp.

WEBSITES

http://nbi.cucaera.co.uk/

This site has useful summaries of known distributions of the three British species.

http://insects.tamu.edu/research/neuropterida/neuroweb.html

Information on world Neuropterida; its content is gradually being migrated to the following site.

http://lacewing.tamu.edu/

24 Order Neuroptera: the lacewings

69 species in 6 families

Since the time of Linnaeus the Neuroptera were often treated as a convenient receptacle for the diverse groups of endopterygote insects that did not fit anywhere else; the modern placement of the group and its relations with the Megaloptera and Raphidioptera are discussed by Grimaldi and Engel (2005). Now that they are clearly defined as a monophyletic order within the Neuropterida, they still show a wide diversity of size, habits and lifecycles. To avoid confusion with the name Neuroptera in its older broad sense, some authors employ the name Planipennia for the modern restricted usage. Even within the limited British fauna there is a considerable range, including the tiny Coniopterygidae, or wax flies, which resemble Hemipteran whiteflies; the giant lacewing *Osmylus*, which has large strongly patterned wings and a semi-aquatic larva; the Sisyridae, or sponge-flies, which have aquatic larvae that feed on freshwater sponges; the large ant-lion, which could be mistaken for a dragonfly, whose larva builds a conical pit in sandy soil; and both the brown lacewings (Hemerobiidae) and green lacewings (Chrysopidae), which are familiar visitors to gardens. All the British Neuroptera have predatory larvae, and many adults are also predators on small insects; the Chrysopidae are particularly regarded as beneficial insects in the garden because of the large numbers of aphids they consume.

There has long been considerable amateur interest in this fairly small but attractive group of insects, initially inspired by Killington's (1936–7) monograph, though this is now very out of date. For the same reason Fraser's (1959) handbook is no longer useful, because it contained many errors even at the time of publication. However, Plant's (1997) key will work for adults of all groups, with just two species added to the British list since its publication: *Peyerimhoffina gracilis* (Chrysopidae) and *Hemerobius handschini* (Hemerobiidae). Elliott's (2009) book is recommended for the freshwater families, not least because it also covers the known larvae, and distributional data on all the British species is included in Plant (1994).

There are 290 species of Neuroptera in 12 families in Europe, with nearly 6000 species in 17 families worldwide. The standard work on the European fauna is Aspöck et al. (1980); a useful world overview of the Neuroptera is given by Tauber et al. (2009).

THE HIGHER CLASSIFICATION OF BRITISH NEUROPTERA

Suborder Hemerobiiformia
    Family Chrysopidae (8 genera, 20 species)
    Family Coniopterygidae (6 genera, 13 species)
    Family Hemerobiidae (7 genera, 31 species)
    Family Osmylidae (1 genus, 1 species)
    Family Sisyridae (1 genus, 3 species)
Suborder Myrmeleontiformia
    Family Myrmeleontidae (1 genus, 1 species)

SPECIES OF CONSERVATION CONCERN

No species of Neuroptera have any legal protection; one is on the UKBAP list: *Megalomus hirtus* (Hemerobiidae).

The Royal Entomological Society Book of British Insects, First Edition. Peter C. Barnard.
© 2011 Royal Entomological Society. Published 2011 by Blackwell Publishing Ltd.

The Families of British Neuroptera

SUBORDER HEMEROBIIFORMIA

Family Chrysopidae (8 genera, 20 species)
The members of this family are generally known as green lacewings or, less commonly, golden-eyes, though the latter name is certainly appropriate (Fig. 24.1). The taxonomy of this group has been complicated in recent years by the discovery that the cosmopolitan 'common' green lacewing, *Chrysoperla carnea*, is actually a complex of species, which are difficult to separate morphologically. Currently it is considered that the two British species in the complex are *C. carnea* and *C. lucasina*, though this may change in the future. Not only is *C. carnea* the commonest species in Britain, but it also frequently enters buildings to overwinter, gradually turning brownish until it flies off in the spring. There are two main groups of green lacewings; the pale, yellowish-green and unmarked species (Fig. 24.2)

and the darker bluish-green species with black markings (Fig. 24.3). The two species of *Nothochrysa* are quite different in being brown with an orange head and should not be confused with *Osmylus* (Osmylidae). Chrysopid larvae often have very distinctive markings (Figs. 24.4 & 24.5) and they are

Fig. 24.3 Bluish-green species of Chrysopidae (Photo: Roger Key)

Fig. 24.1 Golden eyes of Chrysopidae (Photo: Roger Key)

Fig. 24.4 Chrysopid larva 1 (Photo: Roger Key)

Fig. 24.2 Pale species of *Chrysoperla carnea* complex (Chrysopidae) (Photo: Roger Key)

Fig. 24.5 Chrysopid larva 2 (Photo: Roger Key)

314

important predators on aphids in gardens and on some crop plants. Some larvae cover their dorsal surface with the empty skins of their prey and other debris, and use this covering as a kind of camouflage while they stalk their prey. The eggs of green lacewings are laid at the end of long stalks, either singly or in clusters; these stalks are formed from a hardened secretion of the female and act to deter predators.

One species has been added to the British list since Plant's (1997) key, *Peyerimhoffina gracilis*.

British subfamilies and genera:

Chrysopinae: *Chrysopa, Chrysoperla, Chrysotropia* (or *Chrysopidia*), *Cunctochrysa, Dichochrysa, Nineta, Peyerimhoffina*

Nothochrysinae: *Nothochrysa*

Family Coniopterygidae (6 genera, 13 species)
These tiny lacewings are often known as wax-flies from the powdery coating of wax on the body and wings, which makes them resemble Aleyrodidae (Hemiptera); they are all less than 5 mm long. Traditionally some species are associated with particular plant species, often trees, but the lack of information on this family means that it is still not clear how many of these predatory insects are genuinely confined to prey on a narrow range of host plants.

British subfamilies and genera:

Aleuropteryginae: *Aleuropteryx, Conwentzia, Helicoconis, Semidalis*

Coniopteryginae: *Coniopteryx, Parasemidalis*

Family Hemerobiidae (7 genera, 31 species)
The brown lacewings are smaller and less conspicuous than the chrysopids, but are no less important as predators on aphids and other small invertebrates. The adults are similarly predatory, but there are also records of feeding on honeydew. As with the Coniopterygidae there seem to be associations with prey on a narrow range of host plants, but more work is needed on details of the lifecycles. There are several common species in this family (Figs. 24.6 & 24.7). One species has been added to the British list since Plant's (1997) key, *Hemerobius handschini*.

One species is on the UKBAP list: *Megalomus hirtus* (Bordered brown lacewing).

British subfamilies and genera:

Drepanepteryginae: *Drepanepteryx*
Hemerobiinae: *Hemerobius, Wesmaelius*
Megalominae: *Megalomus*
Microminae: *Micromus*

Fig. 24.6 *Hemerobius marginatus* (Hemerobiidae) (Photo: Roger Key)

Fig. 24.7 *Micromus paganus* (Hemerobiidae) (Photo: Roger Key)

Notiobiellinae: *Psectra*
Sympherobiinae: *Sympherobius*

Family Osmylidae (1 genus, 1 species)
The single British species, *Osmylus fulvicephalus*, is sometimes known as the Giant lacewing; it has a wingspan of about 40 mm and has many dark markings on the wings, which distinguish it from the superficially similar species of *Nothochrysa* (Chrysopidae), which also have an orange head (Fig. 24.8). This species is often found in wooded areas, most commonly in southern Britain, where the semi-aquatic larvae live in mosses by the side of small streams, feeding on small invertebrate prey.

British genus: *Osmylus*

Family Sisyridae (1 genus, 3 species)
The larvae of this group are entirely aquatic; the adults resemble small Hemerobiidae and are known

Fig. 24.8 *Osmylus fulvicephalus* (Osmylidae) (Photo: Roger Key)

as spongeflies because the larvae feed exclusively on freshwater sponges. Records of this group are erratic because the species can appear in huge swarms in one year and then in only low numbers for several subsequent seasons. *Sisyra fuscata* seems to be by far the most common and widespread species in Britain; Elliott (2009) gives detailed information on the ecology of the family.

British genus: *Sisyra*

SUBORDER MYRMELEONTIFORMIA

Family Myrmeleontidae (1 genus, 1 species)
Some very early texts on British insects included the ant-lions, and many later authors assumed this was an error, though there were a handful of supposed sightings of this group. However, the discovery of established populations of *Euroleon nostras* in East Anglia during the 1990s confirmed its presence, and there are more records from the south coast; it has long been known in the Channel Islands. The larva is well-known for its habit of digging cone-shaped pits in sandy soils and waiting buried at the bottom with just its huge mandibles showing, to trap ants

and other small invertebrate prey. Adults have a superficial resemblance to dragonflies, but are easily distinguished by the clubbed antennae.

British genus: *Euroleon*

REFERENCES

Aspöck, H., Aspöck, U. & Hölzel, H. 1980. *Die Neuropteren Europas*. 2 vols. Goecke & Evers, Krefeld.

Elliott, J.M. 2009. Freshwater Megaloptera and Neuroptera of Britain and Ireland. *Scientific Publications of the Freshwater Biological Association* 65: 71 pp.

Fraser, F.C. 1959. Mecoptera, Megaloptera and Neuroptera. *Handbooks for the Identification of British Insects* 1(12, 13): 40 pp.

Grimaldi, D. & Engel, M.S. 2005. *Evolution of the insects*. Cambridge University Press, Cambridge.

Killington, F.J. 1936–7. *A monograph of the British Neuroptera*. 2 vols. Ray Society, London.

Plant, C.W. 1994. *Provisional atlas of the lacewings and allied insects (Neuroptera, Megaloptera, Raphidioptera and Mecoptera) of Britain and Ireland*. Biological Records Centre, Huntingdon.

Plant, C.W. 1997. *A key to the adults of British lacewings and their allies (Neuroptera, Megaloptera, Raphidioptera and Mecoptera)*. Field Studies Council (AIDGAP) no. 245. 90 pp.

Tauber, C.A., Tauber, M.J. & Albuquerque, G.S. 2009. Neuroptera (lacewings, antlions). In: Resh, V.H. & Cardé, R.T. (eds.) *Encyclopedia of insects* (2nd edn.). Academic Press/Elsevier, San Diego & London, pp. 695–707.

WEBSITES

http://nbi.cucaera.co.uk/
A useful overview of the British lacewing fauna, with an up-to-date checklist and distribution records.

http://www.brc.ac.uk/recording_schemes.asp
The home page of the Biological Records Centre contains details of the Neuropterida Recording Scheme.

http://insects.tamu.edu/research/neuropterida/neuroweb.html
Information on world Neuropterida; its content is gradually being migrated to the following site.

http://lacewing.tamu.edu/

25 Order Raphidioptera: the snakeflies

4 species in 1 family

Of the three orders in the Neuropterida, the Raphidioptera may be the most basal, leaving the Neuroptera and Megaloptera as sister-groups; the relevant evidence is discussed by Grimaldi and Engel (2005). They get the common name of snakeflies from their elongate pronotum, which is mobile and can be elevated such that the insect resembles a snake about to strike at its prey (Fig. 25.1); the German name of Kamelhalsfliegen, or camel-neck flies, is equally appropriate. Both the adults and larvae are predatory, though adults are also reported as feeding on pollen. Snakefly larvae (Fig. 25.2) are often found under tree bark where they prey on other insects; currently there are no keys to separate the British larvae, and much more work is needed to elucidate the lifecycles. Adults are not often seen as they spend much of their time high in the tree canopy; newly emerged specimens can be found at lower levels, and the females descend to oviposit, but otherwise any snakeflies seen by the casual observer have probably been dislodged from the canopy by strong winds.

There are 75 species in 2 families in Europe, and around 225 species in 2 families worldwide. A brief world overview of the Raphidioptera is given by Aspöck and Aspöck (2009), and the biology of the group was summarized by Aspöck (2002). The world species were monographed by Aspöck et al. (1991).

HIGHER CLASSIFICATION OF BRITISH RAPHIDIOPTERA

All the British species are in the family Raphidiidae.

SPECIES OF CONSERVATION CONCERN

No species of Raphidioptera have any legal protection and none is on the UKBAP list.

Family Raphidiidae (4 genera, 4 species)
The four British species were formerly all placed in the genus *Raphidia*, but intensive work by European workers has divided this into many small subgroups. Not all authors would accept that these 'genera' have much biological significance, and they should perhaps be treated as subgenera; the consequence is that each of the British species is currently placed in a separate genus. The British species can be separated by details of the wing venation; Fraser's (1959) key is now superseded by that of Plant (1997), and distributional details are given by Plant (1994). Note that *Raphidia* is occasionally mis-spelled as '*Rhaphidia*'.

British genera: *Atlantoraphidia, Phaeostigma, Subilla, Xanthostigma*

The Royal Entomological Society Book of British Insects, First Edition. Peter C. Barnard.
© 2011 Royal Entomological Society. Published 2011 by Blackwell Publishing Ltd.

Fig. 25.1 Adult raphidiid (Photo: Roger Key)

Fig. 25.2 Larva of Raphidiidae (Photo: Peter Barnard)

REFERENCES

Aspöck, H. 2002. The biology of Raphidioptera: a review of present knowledge. *Acta Zoologica Academiae Scientiarum Hungaricae* 48(Suppl. 2): 35–50 [downloadable from: http://actazool.nhmus.hu/48Suppl2/aspockraphi.pdf].

Aspöck, U. & Aspöck, H. 2009. Raphidioptera. In: Resh, V.H. & Cardé, R.T. (eds.) *Encyclopedia of insects* (2nd edn.). Academic Press/Elsevier, San Diego & London, pp. 864–6.

Aspöck, H., Aspöck, U. & Rausch, H. 1991. *Die Raphidiopteren der Erde*. 2 vols. Goecke & Evers, Krefeld.

Fraser, F.C. 1959. Mecoptera, Megaloptera and Neuroptera. *Handbooks for the Identification of British Insects* 1(12, 13): 40 pp.

Grimaldi, D. & Engel, M.S. 2005. *Evolution of the insects.* Cambridge University Press, Cambridge.

Plant, C.W. 1994. *Provisional atlas of the lacewings and allied insects (Neuroptera, Megaloptera, Raphidioptera and Mecoptera) of Britain and Ireland.* Biological Records Centre, Huntingdon.

Plant, C.W. 1997. *A key to the adults of British lacewings and their allies (Neuroptera, Megaloptera, Raphidioptera and Mecoptera).* Field Studies Council (AIDGAP) no. 245. 90 pp.

WEBSITES

http://nbi.cucaera.co.uk/
A useful overview with distribution records.

http://insects.tamu.edu/research/neuropterida/neuroweb.html
This site has information on world Neuropterida; its content is gradually being migrated to the following site.

http://lacewing.tamu.edu/

26 Order Siphonaptera: the fleas

62 species in 7 families

Being obligate ectoparasites on mammals and birds, adult fleas have become highly specialized in their morphology and life histories; as a consequence their affinities with other insect groups have long been debated. Traditionally they have been linked with the Diptera or Mecoptera; recent morphological and molecular studies have suggested a close relationship with the family Boreidae within the Mecoptera, and the consequences of this are discussed by Grimaldi and Engel (2005).

Adult fleas, apart from being entirely wingless, have numerous morphological adaptations, including being strongly flattened laterally to creep between the hairs or feathers of their hosts; they have a very strong and dark cuticle, which protects them from the grooming activities of the host (Fig. 26.1); their long hind legs enable them to jump several centimetres from one host to another. Although generally small, from 1–6 mm long, fleas can cause great irritation when present in large numbers and they can also transmit several diseases, such as typhus and plague, as well as tapeworms. On a world scale around 95% of flea species are found on mammals, with the remainder on birds, but in Britain around 25% of the species are bird parasites. Flea mouthparts are highly adapted to piercing their host skin and sucking blood, although they can live for a long time without feeding, a blood meal is usually essential to stimulate oogenesis and spermatogenesis. Eggs are laid in the host's nest or sleeping quarters; the larvae hatch after a few days and rapidly feed on organic debris. Flea larvae are rather featureless, with no legs or eyes, though they have well-developed biting mandibles (see Fig. 26.3); there are currently no keys for identifying them, even to family level. When fully fed the larvae pupate within cocoons spun from silk, and the length of the pupal stage depends partly on temperature but also crucially on the detection of host activity. Flea pupae can remain dormant for months, but any vibrations in the substrate can trigger a rapid emergence of the adults so that they are immediately ready to infest new hosts.

Many species of fleas are highly specific on a narrow range of hosts and, like many other parasitic groups, they have been the subject of many studies on co-evolution. Because they rely on the nesting activity of their hosts, whether birds or mammals, they are not usually found on roaming mammals such as ungulates. On domestic pets such as dogs and cats they can be a persistent nuisance, the commonest offender being the cosmopolitan cat flea, *Ctenocephalides felis felis* (see Fig. 26.4), which often bites people, sometimes causing allergic reactions and secondary infections. Populations of fleas can be locally huge; over 5000 specimens have been recorded from a bird's nest, and over 7000 from a single hedgehog.

Identification of fleas generally depends on microscopic characters, which means that permanent slide preparations are usually necessary. Living specimens (see Fig. 26.1) or those preserved in alcohol have thick opaque integuments, which have to be softened and made transparent so that internal details can be seen (Fig. 26.2). Some of the more common species can be separated using features such as head shapes and details of the combs of spines (Figs. 26.4 & 26.5). Smit's (1957) handbook

The Royal Entomological Society Book of British Insects, First Edition. Peter C. Barnard.
© 2011 Royal Entomological Society. Published 2011 by Blackwell Publishing Ltd.

Fig. 26.1 Live specimen of *Pulex irritans*, Human flea (Pulicidae) (Photo: Peter Barnard)

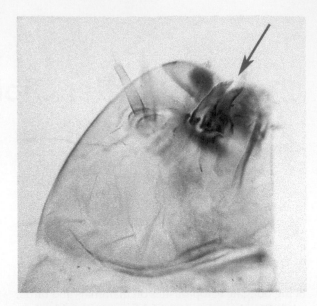

Fig. 26.3 Microscope preparation of flea larva showing mandibles (Photo: Peter Barnard)

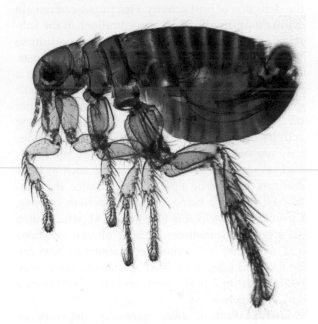

Fig. 26.2 Prepared specimen of *Pulex irritans*, Human flea (Pulicidae) (Photo: Peter Barnard)

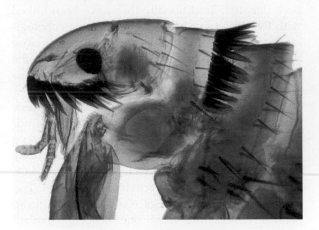

Fig. 26.4 Head of *Ctenocephalides felis felis*, Cat flea (Pulicidae) (Photo: Peter Barnard)

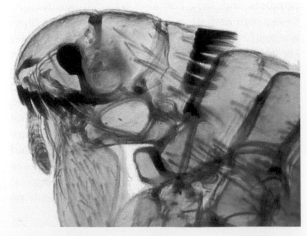

Fig. 26.5 Head of *Ctenocephalides canis*, Dog flea (Pulicidae) (Photo: Peter Barnard)

was completely updated by Whitaker (2007), and the distribution of the British species is covered by George (2008).

There are over 260 species (with many subspecies) in 8 families in Europe, with around 2600 species in 15 families worldwide. A brief world overview of the Siphonaptera is given by Hastriter and Whiting (2009).

SPECIES OF CONSERVATION CONCERN

No species of Siphonaptera have any legal protection and none is on the UKBAP list. However, any that are host-specific would automatically become at risk if their vertebrate host were threatened.

The Families of British Siphonaptera

SUPERFAMILY CERATOPHYLLOIDEA

Family Ceratophyllidae (9 genera, 27 species)
This is the largest family in Britain and it contains both mammal and bird fleas. Some species of *Ceratophyllus*, such as *C. gallinae*, are particularly common on a wide range of bird hosts. *Dasypsyllus gallinulae gallinulae* is similarly widespread, found mainly on passerine birds. Among the mammal fleas, *Nosopsyllus fasciatus*, is commonly found on rats and a few other mammals, as well as occasionally on birds.

This family contains what may be the most restricted flea in Britain: *Ceratophyllus fionnus* is known only from Manx shearwater nests on a single mountain on the Island of Rhum, Inner Hebrides.

British genera: *Amalareus, Callopsylla, Ceratophyllus, Dasypsyllus, Megabothris, Nosopsyllus, Orchopeas, Paraceras, Tarsopsylla*

Family Ischnopsyllidae (2 genera, 8 species)
These are all bat fleas, and most are restricted to just one or two host species.

British genera: *Ischnopsyllus, Nycteridopsylla*

Family Leptopsyllidae (3 genera, 3 species)
The members of this family are found mainly on small mammals like mice, voles and squirrels.

Frontopsylla laeta is unusual in living on house martins, though there are few British records.

British subfamilies and genera:
Amphipsyllinae: *Frontopsylla*
Leptopsyllinae: *Leptopsylla, Peromyscopsylla*

SUPERFAMILY HYSTRICHOPSYLLOIDEA

Family Ctenophthalmidae (4 genera, 12 species)
The members of this family are parasites of small mammals, such as voles and mice. A few species such as *Paleopsylla soricis soricis* are common and widespread, but several others are apparently quite restricted even though their principal host may be very common.

British subfamilies and genera:
Ctenophthalminae: *Ctenophthalmus, Palaeopsylla*
Doratopsyllinae: *Doratopsylla*
Rhadinopsyllinae: *Rhadinopsylla*

Family Hystrichopsyllidae (2 genera, 2 species)
These are found mainly on a variety of small mammals such as mice, voles, shrews and moles. This family includes the largest British flea, *Hystrichopsylla talpae talpae*, whose female can reach a length of about 6 mm; it lives mainly on moles and is a common and widespread species.

British genera: *Hystrichopsylla, Typhloceras*

SUPERFAMILY PULICOIDEA

Family Pulicidae (7 genera, 9 species)
Although this is a relatively small family it includes some of the most common and well-known species. The Human flea, *Pulex irritans* (see Figs. 26.1 & 26.2), is less common on people than in historical times, and its main hosts are badgers, foxes, dogs and cats. The Cat flea, *Ctenocephalides felis felis* (see Fig. 26.4), and the Dog flea, *Ct. canis* (see Fig. 26.5), are closely related but have different head shapes. The Hedgehog flea, *Archaeopsylla erinacei erinacei* (Fig. 26.6), is found on a few mammal species, but the hedgehog is its main host. *Ornithopsylla laetitiae* is a bird flea, and is one of the more uncommon species, found on puffins and storm petrels from just a few islands around the west coast of England, Wales and Ireland. The most notorious member of this family is *Xenopsylla cheopis*, the Plague flea, which is the cosmopolitan vector of bubonic plague; carried mainly by black rats it cannot live for long outdoors in Britain.

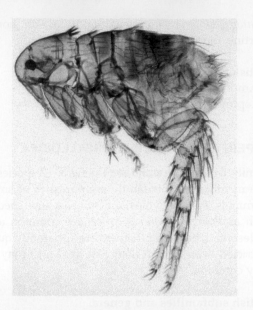

Fig. 26.6 Prepared specimen of *Archaeopsylla erinacei erinacei*, Hedgehog flea (Pulicidae) (Photo: Peter Barnard)

British subfamilies and genera:
 Archaeopsyllinae: *Archaeopsylla, Ctenocephalides*
 Pulicinae: *Echidnophaga, Pulex*
 Spilopsyllinae: *Ornithopsylla, Spilopsyllus*
 Xenopsyllinae: *Xenopsylla*

SUPERFAMILY VERMIPSYLLOIDEA

Family Vermipsyllidae (1 genus, 1 species)
This family is regarded as British on the basis of a single specimen of *Chaetopsylla trichosa* found in Scotland in the 1960s; its host is the badger, and it is quite widespread in Europe.

British genus: *Chaetopsylla*

REFERENCES

GEORGE, R.S. 2008. *Atlas of the fleas (Siphonaptera) of Britain and Ireland*. Biological Records Centre, Huntingdon.

GRIMALDI, D. & ENGEL, M.S. 2005. *Evolution of the insects*. Cambridge University Press, Cambridge.

HASTRITER, M.W. & WHITING, M.F. 2009. Siphonaptera. In: Resh, V.H. & Cardé, R.T. (eds.) *Encyclopedia of insects* (2nd edn.). Academic Press/Elsevier, San Diego & London, pp. 924–8.

SMIT, F.G.A.M. 1957. Siphonaptera. *Handbooks for the identification of British insects* 1(16): 95 pp.

WHITAKER, A.P. 2007. Fleas (Siphonaptera). *Handbooks for the identification of British insects* (2nd edn.) 1(16): 178 pp.

WEBSITES

http://www.zin.ru/Animalia/Siphonaptera/index.htm
 This Russian site is particularly useful for its taxonomic overview of the world Siphonaptera.

27 Order Strepsiptera: the stylops

10 species in 4 families

The Strepsiptera are one of the strangest insect groups, little known to most people, and hardly seen even by entomologists unless they make a special effort to study them. All are obligate endoparasites in other insects and, as so often in parasitic groups, there are many unique morphological and biological adaptations. Their relationship to other orders is still not clear; they have been variously placed within the Coleoptera (as the Stylopoidea), as sister-group to the Coleoptera, as sister-group to the Diptera, or even outside the endopterygotes altogether. These scenarios are discussed by Grimaldi and Engel (2005).

Male Strepsiptera are perhaps the least modified stage of this group, but even they have many peculiar features (Fig. 27.1). Like the Diptera they have only one pair of functional wings, but these are the hind wings, and it is the fore wings that are modified into long halteres, the reverse of the Dipteran situation. In museum specimens these halteres become contorted, which originally gave rise to the alternative common name of twisted-wing insects (though some recent texts attribute this name to a supposed twisting of the hind wings during flight). Males are rarely seen, as they do not feed nor live for more than a few hours but their image is familiar to many entomologists; early in its history the Royal Entomological Society adopted a figure of a male *Stylops kirbii* (currently a synonym of *S. mellittae*) as the emblem for its official seal, now also its logo.

In Britain the hosts of Strepsiptera are either aculeate Hymenoptera (particularly *Andrena* in the Apidae) or homopteran bugs. These 'stylopized'

insects are often recognizable by unusual behaviour, such as becoming increasingly inactive as the parasites develop inside; a closer look will often reveal parasite heads protruding from the host abdomen (Fig. 27.2). Female Strepsiptera are endoparasitic, spending their entire life inside the host insect; they are neotenic with no legs or wings, and only rudimentary eyes, antennae and mouthparts. Her sclerotized head and thorax protrude from the host's abdomen and she emits pheromones to attract males. Mating occurs on the body of the host, the male inseminating the female through the brood canal, which opens between her head and prothorax. The eggs hatch inside the female's body into minute triungulin larvae; these emerge from the brood canal and rapidly disperse using their simple legs; they can sometimes be found on flower-heads. Once they have located a new host they burrow through its cuticle, moult into an apodous stage and then feed on its haemolymph. Pupation occurs inside the host's body, with larval skins forming a puparium, as in some Diptera; the pupal head is visible outside the host. Males emerge and fly off in search of a mate, while females remain in the host, which is still alive even though much of its body may be occupied by the parasite.

The Strepsiptera were traditionally included in some Coleoptera books such as Fowler (1891, as 'Abnormal Coleoptera') and Linssen (1959). Unwin (1984) gave a key to the (then) three British families excluding the Xenidae, though there are no recent accounts of the British species. The last checklist was Kloet and Hincks (1977), which included fifteen species in three families; see Hammond and Hine (1999) for other references. The list used here is

The Royal Entomological Society Book of British Insects, First Edition. Peter C. Barnard.
© 2011 Royal Entomological Society. Published 2011 by Blackwell Publishing Ltd.

Fig. 27.1 Two male adults of Strepsiptera (from Curtis, 1823)

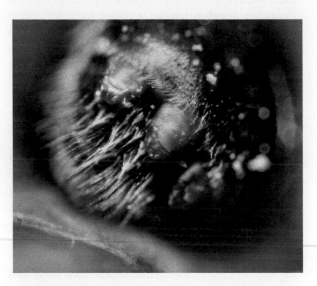

Fig. 27.2 Strepsipteran parasites protruding from abdomen of *Andrena similis* (Hymenoptera: Apidae) (Photo: Robin Williams)

THE HIGHER CLASSIFICATION OF BRITISH STREPSIPTERA

Family Elenchidae (1 genus, 1 species)
Family Halictophagidae (1 genus, 2 species)
Family Stylopidae (3 genera, 5 species)
Family Xenidae (2 genera, 2 species)

SPECIES OF CONSERVATION CONCERN

No species of Strepsiptera have any legal protection and none is on the UKBAP list. However, as obligate parasites they could be at risk if their hosts were threatened.

The Families of British Strepsiptera

Family Elenchidae (1 genus, 1 species)
The only British species is *Elenchus tenuicornis*, a parasite of Delphacidae (Hemiptera).

British genus: *Elenchus*

Family Halictophagidae (1 genus, 2 species)
The members of this family are parasitic on various families of homopteran bugs (Hemiptera).

British genus: *Halictophagus*

Family Stylopidae (3 genera, 5 species)
All the members of this family are parasites on aculeate Hymenoptera, principally the Andreninae, Colletinae and Halictinae in the Apidae; *Andrena* is one of the most common hosts (see Fig. 27.2). All the western Palaearctic species of *Stylops* were synonymized to *S. melittae* by Kinzelbach (1978).

based on Fauna Europaea (http://www.faunaeur. org), but clearly a review of the British species is needed urgently, partly to clarify the taxonomy of some species but also to confirm genuine records from the British Isles. The main differences from the 1977 list are that the family Xenidae has been added, and all the former species of *Stylops* from the western Palaearctic are now synonymized in *S. melittae*, hence the apparent reduction in the number of British species (Kinzelbach, 1978). The European fauna was reviewed by Kinzelbach (1969).

There are around 30 species in 7 families in Europe, with about 600 species in 10 families worldwide; a brief world overview of the Strepsiptera is given by Whiting (2009).

British genera: *Halictoxenos, Hylecthrus, Stylops*

Family Xenidae (2 genera, 2 species)

The species in this family parasitize Sphecidae and Vespidae (Hymenoptera), but British records need confirmation.

British subfamilies and genera:

Paraxeninae: *Paraxenos*
Xeninae: *Pseudoxenos*

REFERENCES

CURTIS, J. 1823. *British entomology*. Vol. 3. Privately published, London.

FOWLER, W.W. 1891. *The Coleoptera of the British Islands*. Vol. 5. Reeve, London.

GRIMALDI, D. & ENGEL, M.S. 2005. *Evolution of the insects*. Cambridge University Press, Cambridge.

HAMMOND, P.M. & HINE, S.J. 1999. Strepsiptera: the stylops. In: Barnard, P.C. (ed.) *Identifying British insects and arachnids: an annotated bibliography of key works*. Cambridge University Press, Cambridge, p. 139.

KINZELBACH, R. K. 1969. Stylopidae, Fächerflügler (= Ordnung: Strepsiptera). *Die Käfer Mitteleuropas* 8: 139–59.

KINZELBACH, R. K. 1978. Strepsiptera. *Tierwelt Deutschlands* 65: 116 pp.

KLOET, G.S. & HINCKS, W.D. 1977. A check list of British insects (2nd edn., revised). Part 3: Coleoptera and Strepsiptera. *Handbooks for the identification of British insects* 11(3): 105 pp.

LINSSEN, E.F. 1959. *Beetles of the British Isles*. 2 vols. Frederick Warne, London, The Wayside and Woodland Series.

UNWIN, D.M. 1984. *A key to the families of British beetles (and Strepsiptera)*. Field Studies Council (AIDGAP) no. 166: 48 pp. [reprinted with minor alterations, 1988].

WHITING, M.F. 2009. Strepsiptera. In: Resh, V.H. & Cardé, R.T. (eds.) *Encyclopedia of insects* (2nd edn.). Academic Press/Elsevier, San Diego & London, pp. 971–2.

WEBSITES

http://www.duhem.com/galerie/streps.en.php
Strepsiptera website, with links to other useful sites.

28 Order Trichoptera: the caddisflies or sedge flies

198 species in 19 families

Trichoptera are closely related to the Lepidoptera, forming the group Amphiesmenoptera, and the two orders are often cited as being one of the best examples of a sister-group pair at the ordinal level. The first fossils of this group appear in the Permian but it seems likely that divergence of life-styles occurred at an early stage; today Lepidoptera are primarily a terrestrial group whereas virtually all the juvenile stages of Trichoptera live in freshwater environments. The case-building habits of the larvae are well known, though more species build nets to filter feed, or else are free-living predators, than actually build tubular cases. However, all caddis larvae construct shelters in which they pupate, and the case-building habit may have evolved through an extension of this pupal retreat. The diversity of caddis larvae and their relative abundance in a wide range of freshwater habitats has led to many species having rather narrow ecological requirements. This means that subtle differences in species diversity can form a useful indicator of freshwater quality, and can in turn detect even low levels of pollution. Some species are tolerant of semi-saline habitats such as salt-marshes, and in the Australasian region a small group lives entirely in sea-water.

As nocturnal or crepuscular insects adult caddisflies have the reputation of being undistinguished in appearance, appearing less diverse and colourful than day-flying Lepidoptera, and therefore difficult to identify (Fig. 28.1). In fact, many of the more common species can be recognized by their wing-patterns and other easily observed characters; these have been given common names by flyfishermen, and all are listed by Barnard and Ross (2008). In contrast, the common names given to family groups are usually based on larval features (Wallace, 2006).

Trichoptera are a cosmopolitan group (though absent from Antarctica), and appear to have evolved in temperate parts of the world, rather than the tropics, with the highest species diversity in the Oriental region. Many caddis larvae require relatively cool, well-oxygenated water, though inevitably there are several exceptions in that some groups can tolerate low oxygen levels; a few species have deserted fresh water completely and live in damp terrestrial habitats.

The first book to cover the identification of adult Trichoptera was by Mosely (1939) and this work still has its uses, at least to the specialist. Macan (1973) provided a more recent handbook but this is now superseded by Barnard and Ross (in press). On a European scale Malicky (2004) is invaluable to the specialist. Following the ground-breaking work on larvae by Hickin (1967), the case-building larvae are now well covered by Wallace et al. (1990), and the free-living larvae by Edington and Hildrew (1995). Wallace (2006) provides a simplified key to all the main larval groups. Because of the interest in this group by anglers, principally flyfishermen, there are several popular books that cover the Trichoptera in lesser detail, including Goddard (1988, 1991) and Price (1989).

Information about the UK Trichoptera Recording Scheme can be found on the website of The Riverfly Partnership (http://www.riverflies.org/index/riverflies/trichoptera.html).

The Royal Entomological Society Book of British Insects, First Edition. Peter C. Barnard.
© 2011 Royal Entomological Society. Published 2011 by Blackwell Publishing Ltd.

Fig. 28.1 *Limnephilus affinis* (Limnephilidae) (Photo: Peter Barnard)

There are around 1100 species in 24 families in Europe, and around 13,000 species in 46 families worldwide. A frequently up-dated World Checklist of Trichoptera on-line at http://entweb.clemson.edu/database/trichopt/ and a good global overview of the group is given by Morse (2009).

HIGHER CLASSIFICATION OF BRITISH TRICHOPTERA

Suborder Annulipalpia
   Superfamily Hydropsychoidea
      Family Ecnomidae (1 genus, 1 species)
      Family Hydropsychidae (3 genera, 11 species)
      Family Polycentropodidae (5 genera, 13 species)
      Family Psychomyiidae (3 genera, 12 species)
   Superfamily Philopotamoidea
      Family Philopotamidae (3 genera, 5 species)
Suborder Integripalpia
   Superfamily Leptoceroidea
      Family Leptoceridae (10 genera, 31 species)
      Family Molannidae (1 genus, 2 species)
      Family Odontoceridae (1 genus, 1 species)
   Superfamily Limnephiloidea
      Family Apataniidae (1 genus, 3 species)
      Family Brachycentridae (1 genus, 1 species)
      Family Goeridae (2 genera, 3 species)
      Family Lepidostomatidae (2 genera, 3 species)
      Family Limnephilidae (19 genera, 55 species)
   Superfamily Phryganeoidea
      Family Phryganeidae (5 genera, 10 species)
   Superfamily Sericostomatoidea
      Family Beraeidae (3 genera, 4 species)
      Family Sericostomatidae (2 genera, 2 species)
Suborder Spicipalpia
   Superfamily Glossosomatoidea
      Family Glossosomatidae (2 genera, 6 species)
   Superfamily Hydroptiloidea
      Family Hydroptilidae (7 genera, 31 species)
   Superfamily Rhyacophiloidea
      Family Rhyacophilidae (1 genus, 4 species)

SPECIES OF CONSERVATION CONCERN

No species of Trichoptera have any legal protection but four are on the UKBAP list: *Glossosoma intermedium* (Glossosomatidae), *Hydropsyche bulgaromanorum* (Hydropsychidae), *Ironoquia dubia* (Limnephilidae) and *Hagenella clathrata* (Phryganeidae). Wallace's (1991) review is very useful, though now slightly out of date.

The Families of British Trichoptera

SUBORDER ANNULIPALPIA

This is the main group of caseless caddis; the larvae are net-spinners or retreat-makers. Essentially they build fixed structures from silk with which to trap food particles or living prey from the water current, and many are therefore found in running water. Most, though not all, can therefore be regarded as external filter-feeders.

SUPERFAMILY HYDROPSYCHOIDEA

Family Ecnomidae (1 genus, 1 species)
Members of this family are sometimes known as the small snare-making caddis. *Ecnomus tenellus* is the only British species, though there are several more in Europe. It is a small but strongly marked species. The larva is unusual in preferring still or slow water, even though it has no external gills; it has all three thoracic segments sclerotized. Ecnomid larvae build a silk snare in which to catch prey.

British genus: *Ecnomus*

Family Hydropsychidae (3 genera, 11 species)
These are the net-making caddis. Nine of the eleven British species are in the genus *Hydropsyche*, the adults being known to anglers as the Grey flags, or Marbled sedges, though *Diplectrona felix* (Fig. 28.2) has a rather different appearance. *Hydropsyche* is a large genus in Europe, with around 60 species. The larvae are often common in running water, where they build nets to trap food items; these nets usually have a well-organized structure with a regular and fairly coarse mesh size. Hydropsychid larvae have all three thoracic segments sclerotized, numerous branched gills along the ventral side of the abdomen and prominent brushes of bristles on the terminal anal prolegs.

Hydropsyche bulgaromanorum (Scarce grey flag) is on the UKBAP list.

Fig. 28.2 *Diplectrona felix* (Hydropsychidae) (Photo: Stuart Crofts)

Fig. 28.4 *Philopotamus montanus* (Philopotamidae) (Photo: Stuart Crofts)

Fig. 28.3 *Polycentropus flavomaculatus* (Polycentropodidae) (Photo: Stuart Crofts)

reddish brown *T. waeneri* is the most common and widespread, known to anglers as the Small red sedge, but most species are small and dark. The larvae build fixed tubular galleries on rocks or submerged branches, made from silk but often with other material incorporated, from which they graze algae from the surrounding substrate. As the food supply becomes exhausted the larva extends the front of its gallery into a new territory and destroys the back end, which is similar to the way in which case-building species extend their tubular cases as they grow larger. In general appearance psychomyiid larvae resemble those of the Polycentropodidae.

British genera: *Lype, Psychomyia, Tinodes*

British genera: *Cheumatopsyche, Diplectrona, Hydropsyche*

Family Polycentropodidae (5 genera, 13 species)
These are the snare-making caddis; the family name is often mis-spelled Polycentropidae. A small but quite diverse family; all species are small but sometimes heavily marked, and at least three species are known to anglers as Dark-spotted (or Yellow-spotted) sedges (Fig. 28.3). The larvae are found in a variety of freshwater types, though not usually fast-flowing, but all construct rather elongate silken tubes in which they rest. The tubes are coarse-meshed and act as snares to catch swimming, drifting or walking live prey. Polycentropodid larvae have only the prothorax sclerotized and have no abdominal gills, resembling Psychomyiidae.

British genera: *Cyrnus, Holocentropus, Neureclipsis, Plectrocnemia, Polycentropus*

Family Psychomyiidae (3 genera, 12 species)
These are the gallery-making caddis. Of the 12 British species eight are in the genus *Tinodes*; the

SUPERFAMILY PHILOPOTAMOIDEA

Family Philopotamidae (3 genera, 5 species)
These are known as the bag-making caddis. *Philopotamus montanus* is the best known species, especially in upland areas. Anglers in these regions know it as the Yellow-spotted sedge (Fig. 28.4), which can cause confusion with the polycentropodids given the same name. The larvae construct long tubular bag-like nets with a very fine mesh, with which they filter diatoms and small pieces of detritus from the fast-flowing streams in which they live. Like the Polycentropodidae and Psychomyiidae, philopotamid larvae have only the prothorax sclerotized and no abdominal gills (Fig. 28.5).

British genera: *Chimarra, Philopotamus, Wormaldia*

SUBORDER INTEGRIPALPIA

These are the true cased caddis, whose larvae build the familiar tubular cases. The group is apparently

Fig. 28.5 *Philopotamus montanus* larva (Philopotamidae) (Photo: Stuart Crofts)

Fig. 28.6 *Athripsodes albifrons* (Leptoceridae) (Photo: Stuart Crofts)

monophyletic, and they probably represent the most recently evolved families. Although the case may have a defensive or camouflage value, its main function is probably to allow the larva to generate a permanent water current through the case by undulating its abdomen. This brings a constant supply of oxygen, which allows the larvae to live in still or even stagnant water; thus the evolution of the case has enabled a whole group of Trichoptera to invade entirely new freshwater habitats. Most larvae in this suborder are either grazers on algae, fungi or bacteria on dead leaves and wood fragments, but many are omnivorous.

SUPERFAMILY LEPTOCEROIDEA

Family Leptoceridae (10 genera, 31 species)
These are known as the long-horned caddis because of the extremely long adult antennae, often three times the length of the fore wing. From their habit of swarming in large numbers around trees or bushes in the late afternoon they can be mistaken for day-flying moths in the family Adelidae, which have similarly long antennae. Many are quite large and conspicuous species with distinct markings (Figs. 28.6 & 28.7) and several have been given common names by anglers. Curiously, most leptocerid larvae also have relatively long antennae, but since these appendages are normally very short in Trichoptera, they are still somewhat inconspicuous. The larvae build a variety of tubular cases, some quite short and strongly curved, covered with small stones; others are longer and straighter with a fine sand covering, and some are composed of vegetable material often arranged in a spiral pattern. Species that build lighter cases have the ability to swim quite rapidly with their case, and this process is helped by the presence of long fringes of hairs on their elongated hind legs. Most species seem to have omnivorous larvae, but members of the genus

Fig. 28.7 *Athripsodes cinereus* (Leptoceridae) (Photo: Stuart Crofts)

Ceraclea are specialized feeders on freshwater sponges.

British genera: *Adicella, Athripsodes, Ceraclea, Erotesis, Leptocerus, Mystacides, Oecetis, Setodes, Triaenodes, Ylodes*

Family Molannidae (1 genus, 2 species)
The two species in the genus *Molanna* have no common name; they are moderate-sized pale species whose wings wrap around the body at rest. The larvae live in still or slow-flowing water, and are unusual amongst caddis larvae in being able to live on fine sandy substrates. Their shield-shaped cases have broad anterior and lateral extensions, which seem to protect the larvae from predators in such an exposed environment.

British genus: *Molanna*

Family Odontoceridae (1 genus, 1 species)
The single British species *Odontocerum albicorne* is known to anglers as the Silver or Grey sedge (Fig. 28.8). It is a fairly large species with conspicuous white and grey markings, and the antennae have a toothed appearance when examined closely. The larva has an anchor-shaped marking on its pale head; it lives in stony streams and constructs a

Fig. 28.8 *Odontocerum albicorne* (Odontoceridae)
(Photo: Stuart Crofts)

Fig. 28.9 *Brachycentrus subnubilus* (Brachycentridae)
(Photo: Stuart Crofts)

Fig. 28.10 Egg-masses of *Brachycentrus subnubilus*
(Brachycentridae) (Photo: Stuart Crofts)

Fig. 28.11 Larva of *Brachycentrus subnubilus*
(Brachycentridae) (Photo: Stuart Crofts)

slightly curved, sand-grain case, which has its rear
end partially blocked by a single large sand grain.

British genus: *Odontocerum*

SUPERFAMILY LIMNEPHILOIDEA

Family Apataniidae (1 genus, 3 species)
The three species of *Apatania* were previously placed
as a subfamily within the Limnephilidae, but recent
opinion favours the elevation of this group to family
level. The adults are small, with prominent dark
venation and a large dark pterostigma. The larvae
live in cold lakes or spring-fed streams and construct short conical cases of large sand grains.

British genus: *Apatania*

Family Brachycentridae (1 genus, 1 species)
The Grannom, *Brachycentrus subnubilus*, is the only
British species in this family, and is well-known to
flyfishermen as being one of the first freshwater
insects to appear on the wing, in April or even
earlier. It has broad, darkly marked wings (Fig. 28.9)
and swarms in large numbers on some rivers. As in

most Trichoptera the eggs are laid in gelatinous
clumps under the water surface (Fig. 28.10). The
young larva builds a square section case, which
becomes circular in section in later life and may be
attached to living plants to anchor it in a strong
current. The larva is an active feeder, seizing passing
particles with its grasping legs (Fig. 28.11).

British genus: *Brachycentrus*

Family Goeridae (2 genera, 3 species)
Despite being uniformly coloured the members of
this family are well-known to anglers, *Goera pilosa*

Fig. 28.12 *Lepidostoma hirtum* (Lepidostomatidae) (Photo: Peter Barnard)

Fig. 28.13 *Limnephilus lunatus* (Limnephilidae) (Photo: Stuart Crofts)

being the Medium sedge and the two species of *Silo* the Black sedges. The larvae build very distinctive cases that have large stones along the sides, which presumably act as ballast in fast water currents. They are typically algal grazers on the surface of large stones.

British genera: *Goera, Silo*

Family Lepidostomatidae (2 genera, 3 species)
The three British species are placed in two genera (*Lasiocephala* was recently synonymized with *Lepidostoma*), with two species known to the angler as the Silver sedge. The adults are fairly small and nondescript (Fig. 28.12). Lepidostomatid larvae are found in cool running water, and occasionally lake shores and have a bewildering array of case types. The young larvae build slender circular cases from sand grains, and this pattern sometimes persists throughout the larval life. Sometimes they change over to using plant material, arranged in a square section, giving the case a unique appearance with the back half circular, and the front half square. Some of these larvae change to a completely square case made solely from plant material.

British genera: *Crunoecia, Lepidostoma*

Family Limnephilidae (19 genera, 55 species)
This is the largest family in Britain and it is the dominant family throughout the higher latitudes of the northern hemisphere. Limnephilids exhibit a wide variety of biological and morphological adaptations in all stages of their lifecycle. Adults range in size from small to very large; typically they have rather angular wings with characteristic wing patterns (Figs. 28.13 & 28.14). Unusually for Trichoptera,

Fig. 28.14 *Halesus radiatus* (Limnephilidae) (Photo: Stuart Crofts)

some species have an extended adult life, sometimes lasting several months, and they spend much of this time in a state of aestivation or summer diapause, which seems to be a response to the drying up of temporary water bodies in the warmer part of the year. The length of this diapause, or even its existence, is dependent on latitude, even within the British Isles. Limnephilid larvae build a variety of cases, sometimes recognizable at least at generic level, but often depending on what building materials are available. In faster water the cases are usually constructed from sand grains and small stones (Fig. 28.15), whereas in still, eutrophic water plant material is often used (Fig. 28.16). Some species include snail shells in their case construction (Fig. 28.17), and some add very long sticks along the sides of the case, possibly to deter fish predators; others place numerous small twigs or stems arranged tangentially around the case in overlapping rows, a style called appropriately 'log-cabin' cases (Fig. 28.18).

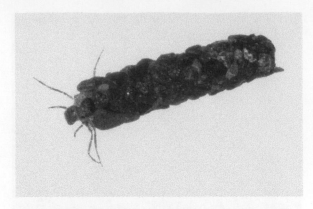

Fig. 28.15 Limnephilid case built from small stones (Photo: Stuart Crofts)

Fig. 28.18 'Log-cabin' case of a limnephilid (Photo: Stuart Crofts)

Fig. 28.16 Limnephilid case using some plant material (Photo: Peter Barnard)

Fig. 28.17 Limnephilid case incorporating snail shells (Photo: Roger Key)

Fig. 28.19 Empty limnephilid cases (Photo: Roger Key)

Being strongly made, the empty cases can persist for several seasons after the insects have emerged (Fig. 28.19). *Enoicyla pusilla* is the only terrestrial caddisfly in Britain, restricted to the West Midlands where its larva lives in damp leaf-litter (Fig. 28.20);

Fig. 28.20 Terrestrial larvae of *Enoicyla pusilla* (Limnephilidae) (Photo: Roger Key)

the female of this species is the only wingless caddis in Britain.

Ironoquia dubia (Scarce brown sedge) is on the UKBAP list.

British genera: *Allogamus, Anabolia, Chaetopteryx, Drusus, Ecclisopteryx, Enoicyla, Glyphotaelius, Grammotaulius, Halesus, Hydatophylax, Ironoquia, Limnephilus, Melampophylax, Mesophylax, Micropterna, Nemotaulius, Potamophylax, Rhadicoleptus, Stenophylax*

SUPERFAMILY PHRYGANEOIDEA

Family Phryganeidae (5 genera, 10 species)
Although a relatively small family the Phryganeidae include the largest species in Britain, with a wingspan of up to 60 mm. Most species have distinctive wing-patterns and many have anglers' names (Fig. 28.21). Like the Limnephilidae they are predominantly a northern hemisphere group, especially in the higher latitudes. The larvae are mostly omnivorous, and their large cases are usually built from plant material, sometimes arranged spirally as in some Leptoceridae (Fig. 28.22). One species,

Fig. 28.21 *Agrypnia varia* (Phryganeidae) (Photo: Roger Key)

Fig. 28.22 Phryganeid larva and case (Photo: Robin Williams)

Agrypnia pagetana, occasionally uses a single hollow plant stem of an appropriate diameter.

Hagenella clathrata (Window winged sedge) is on the UKBAP list.

British genera: *Agrypnia, Hagenella, Oligotricha, Phryganea, Trichostegia*

SUPERFAMILY SERICOSTOMATOIDEA

Family Beraeidae (3 genera, 4 species)
A small family of small-sized and uniformly black species, but the fact that they are placed in three genera is an indication of the considerable morphological differences in both adults and larvae. The adults are not conspicuous enough to have common names. The larvae live in small streams or marshes, and build small curved conical cases of fine sand grains, often superficially resembling some Leptoceridae.

British genera: *Beraea, Beraeodes, Ernodes*

Family Sericostomatidae (2 genera, 2 species)
A small family with just two species in two genera, although *Sericostoma personatum* is by far the most common; it has the delightful anglers' name of the Welshman's button. The male maxillary palps are enormously enlarged, forming a kind of mask over the front of the head. Sericostomatid larvae are usually found in running water, though sometimes also on lake margins, where they feed mainly on plant detritus, and build sand-grain cases that are short, curved and conical. These can resemble the cases of Odontoceridae, but the latter has the posterior end partially closed by a single large sand grain, whereas those of sericostomatids have a membrane with a central round hole.

British genera: *Notidobia, Sericostoma*

SUBORDER SPICIPALPIA

The larvae in this group are either free-living, or else build a variety of primitive shelters that are probably precursors of pupal cases rather than true larval cases. All build a thick-walled enclosed cocoon from silk in which to pupate, giving the group the name of closed-cocoon makers, though the suborder is almost certainly not monophyletic; it includes the families usually considered the most primitive in the Trichoptera.

SUPERFAMILY GLOSSOSOMATOIDEA

Family Glossosomatidae (2 genera, 6 species)
Known as the saddle-case caddis, the adults in this family are mostly small greyish species that are often extremely abundant in running water, and are generally known as the Tiny grey sedges by anglers (Fig. 28.23). The larvae are grazers of algae and diatoms on the upper surfaces of rocks. They build rather loose cases of small pebbles, which are called saddle-cases but rather resemble tortoise shells. The front and rear openings are interchangeable in that the larva often reverses within the case, and they will abandon their case if disturbed. As in most Trichoptera, the pupae actively swim to the surface for the adult to emerge and the wing colours can be seen clearly at this stage (Fig. 28.24).

Fig. 28.23 *Agapetus fuscipes* (Glossosomatidae) (Photo: Stuart Crofts)

Fig. 28.24 Mature pupa of *Agapetus ochripes* (Glossosomatidae) (Photo: Stuart Crofts)

Glossosoma intermedium (Small grey sedge) is on the UKBAP list.

British genera: *Agapetus, Glossosoma*

SUPERFAMILY HYDROPTILOIDEA

Family Hydroptilidae (7 genera, 31 species)
Known as the micro-caddis these are the smallest of the caddisflies, some species having a wing-span of only 5 mm. Despite their minuteness this is one of the most diverse families, more speciose than any other family worldwide, with nearly 2000 species described. The adults often have very long fringes of hairs on the wings (Fig. 28.25) and some are reluctant to fly, instead being seen to run around rather like ants. They are easily confused with the small 'microlepidoptera' moths. The first four larval instars are free-living, though rarely seen because of their small size, but in the final fifth instar the larva changes its appearance, developing an enlarged abdomen, and constructing a portable case from silk. These cases vary greatly in form, some being bottle-shaped or bean-shaped, others like domes or purses. Some genera are adapted to feed on filamentous algae and they also incorporate algal strands into their cases. In addition to the general references cited above, Marshall (1978) has more detail on adult identification in this family.

British genera: *Agraylea, Allotrichia, Hydroptila, Ithytrichia, Orthotrichia, Oxyethira, Tricholeiochiton*

SUPERFAMILY RHYACOPHILOIDEA

Family Rhyacophilidae (1 genus, 4 species)
These are the free-living caddis. Although there are just four British species in the genus *Rhyacophila*,

Fig. 28.25 Pinned specimen of *Agraylea sexmaculata* (Hydroptilidae) (Photo: Peter Barnard)

Fig. 28.26 Ovipositing female of *Rhyacophila dorsalis* (Rhyacophilidae) (Photo: Stuart Crofts)

Fig. 28.27 Larva of *Rhyacophila dorsalis* (Rhyacophilidae) (Photo: Stuart Crofts)

there are around 90 species in Europe. The moderately large adults are various shades of grey and yellow, and the most common species, *Rh. dorsalis*, is known to anglers as the Sandfly sedge (Fig. 28.26). The larvae live in cool flowing water where they are active predators; they are distinguished by their bright green colour with pinkish bunches of gills on the sides of the abdomen (Fig. 28.27). Though they are free-living, they build a loose shelter of small stones in which to pupate.

British Genus: *Rhyacophila*

REFERENCES

BARNARD, P.C. & ROSS, E. 2008. *Guide to the adult caddisflies or sedge flies (Trichoptera).* Field Studies Council (AIDGAP) OP129.

BARNARD, P.C. & ROSS, E. (in press). Adult caddisflies of the British Isles. *Handbooks for the identification of British insects.*

EDINGTON, J.M. & HILDREW, A.G. 1995. Caseless caddis larvae of the British Isles. *Scientific Publications of the Freshwater Biological Association* 53: 134 pp.

GODDARD, J. 1988. *John Goddard's waterside guide.* Unwin Hyman, London.

GODDARD, J. 1991. *Trout flies of Britain and Europe.* A & C Black, London.

HICKIN, N.E. 1967. *Caddis larvae: larvae of the British Trichoptera.* Hutchinson, London.

MACAN, T.T. 1973. A key to the adults of the British Trichoptera. *Scientific Publications of the Freshwater Biological Association* 28: 151 pp.

MALICKY, H. 2004. *Atlas of European Trichoptera* (2nd edn.). Springer, Dordrecht.

MARSHALL, J.E. 1978. Trichoptera: Hydroptilidae. *Handbooks for the identification of British insects* 1(14a): 31 pp.

MORSE, J.C. 2009. Trichoptera (caddisflies). In: Resh, V.H. & Cardé, R.T. (eds.) *Encyclopedia of insects* (2nd edn.). Academic Press/Elsevier, San Diego & London, pp. 1015–20.

MOSELY, M.E. 1939. *The British caddis flies (Trichoptera).* Routledge, London.

PRICE, T. 1989. *The angler's sedge: tying and fishing the caddis.* Blandford Press, London.

WALLACE, I.D. 1991. A review of the Trichoptera of Great Britain. *Research and Survey in Nature Conservation* 32: 59 pp. Nature Conservancy Council, Peterborough.

WALLACE, I.D. 2006. *Simple key to caddis larvae.* Field Studies Council (AIDGAP) OP105.

WALLACE, I.D., WALLACE, B. & PHILIPSON, G.N. 1990. A key to the case-bearing caddis larvae of Britain and Ireland. *Scientific Publications of the Freshwater Biological Association* 51: 237 pp.

WEBSITES

http://www.riverflies.org/index/riverflies/trichoptera.html
The UK Trichoptera Recording Scheme.

http://entweb.clemson.edu/database/trichopt/
A well-maintained world checklist of Trichoptera.

Index to all names of insect genera, families and higher taxa in the UK

Orders and higher groups are in bold (e.g. **Diptera**); family group names and suborders are in plain text (e.g. Nymphalidae); genera are in *italics* (e.g. *Carabus*); common names of family and higher groups are in lower case (e.g. dragonfly). References to pp. 10–15 are the outline classification in Chapter 1.

Aaroniella, 115
Abax, 130
Abdera, 152
Abia, 262
Abiinae, 262
Ablabesmyia, 215
Ablaxia, 245
Ablerus, 241
Abraeinae, 169
Abraeus, 169
Abraxas, 281, 295
Abrostola, 288
Absyrtus, 256
Acaenitinae, 253, 255
Acaenitus, 255
Acalles, 149
Acalypta, 79
Acalyptratae, 13, 178, 188
Acalyptus, 149
Acampsis, 252
Acamptocladius, 215
Acanopsilus, 258
Acanosema, 258
Acanthiophilus, 200
Acanthocinus, 135
Acanthocnema, 204
Acanthococcus, 100
Acanthodelphux, 93
Acantholyda, 260
Acanthopsilus, 258
Acanthopsyche, 298
Acanthoscelides, 136, 137
Acanthosoma, 87
Acanthosomatidae, 11, 74, 87
Acanthoxyla, 65

Acartophthalmidae, 13, 178, 193
Acartophthalmus, 194
Acasis, 281
Acaudella, 97
Acaudinum, 97
Acentria, 296
Acentropinae, 296
Acentrotypus, 147
Aceratoneuromyia, 242
Acerella, 28
Acerentomata, 10, 28
Acerentomidae, 10, 28
Acerentomon, 28
Acerentulus, 28
Acericerus, 91
Acerotella, 257
Acetropis, 78
Achaius, 256
Achalcinae, 182
Achalcus, 182
Achanthiptera, 203
Achanthipterinae, 203
Achenium, 173
Acherontia, 273
Acheta, 60
Achlya, 274
Achroia, 296, 297
Achrysocharoides, 242
Achyrolimonia, 221
Acidia, 201
Acidoproctus, 110
Acidota, 173
Acigona, 296
Acilius, 130, 131

Acinia, 200
Aclastus, 255
Aclerda, 99
Aclerdidae, 11, 75, 99
Acleris, 299, 300
Acletoxenus, 192
Aclista, 258
Aclypea, 172
Acnemia, 213
Acodiplosis, 212
Acolobus, 256
Acompocoris, 76
Acompsia, 278
Acompus, 86
Aconias, 255
Acontia, 288
Acontiinae, 288
Acosmetia, 269, 271, 287, 288
Acricotopus, 215
Acrididae, 11, 58, 59
Acridoidea, 11, 58
Acrisis, 252
Acritus, 169
Acrobasis, 297
Acrocera, 184
Acrocercops, 281
Acroceridae, 13, 178, 184
Acroclita, 300
Acrocormus, 245
Acrodactyla, 256
Acrolepia, 300
Acrolepiidae, 15, 270, 300
Acrolepiopsis, 300
Acrolocha, 173
Acrolyta, 255

The Royal Entomological Society Book of British Insects, First Edition. Peter C. Barnard.
© 2011 Royal Entomological Society. Published 2011 by Blackwell Publishing Ltd.